TRINIDAD AND TOBAGO

MATHS CONNECT 3

Irmin Edwards-Kennedy

Gwendolyn Pope

HODDER EDUCATION
AN HACHETTE UK COMPANY

Although every effort has been made to ensure that website addresses are correct at time of going to press, Hodder Education cannot be held responsible for the content of any website mentioned in this book. It is sometimes possible to find a relocated web page by typing in the address of the home page for a website in the URL window of your browser.

Hachette UK's policy is to use papers that are natural, renewable and recyclable products and made from wood grown in well-managed forests and other controlled sources. The logging and manufacturing processes are expected to conform to the environmental regulations of the country of origin.

Orders: please contact Bookpoint Ltd, 130 Park Drive, Milton Park, Abingdon, Oxon OX14 4SE. Telephone: +44 (0)1235 827827. Fax: +44 (0)1235 4004501. Email education@bookpoint.co.uk Lines are open from 9 a.m. to 5 p.m., Monday to Saturday, with a 24-hour message answering service. You can also order through our website: www.hoddereducation.com

First published by Pearson Education Limited
Published from 2015 by Hodder Education,
An Hachette UK Company
Carmelite House
50 Victoria Embankment
London EC4Y 0DZ
www.hoddereducation.com

20
IMP 10 9 8 7 6 5 4 3 2

British Library Cataloguing in Publication Data is available
from the British Library on request.

ISBN: 978 0 435 895 97 6

Typeset by Tech-Set Ltd, Gateshead, Tyne and Wear
Original illustrations © Pearson Education Limited, 2011
Illustrated by Adrian Barclay and Tech-Set Ltd.
Cover design by mccdesign ltd.
Printed in Dubai

Cover photo: © Getty images/Brand X Pictures

Acknowledgements
Every effort has been made to contact copyright holders of material reproduced in this book. Any omissions will be rectified in subsequent printings if notice is given to the publishers.

Maths Connect for Trinidad and Tobago has been designed to support the Secondary Education Modernization Programme (SEMP) for the lower secondary school maths curriculum of Trinidad and Tobago. It has built on the research and consultation with maths teachers, educators and curriculum officers across the Caribbean region that was conducted for *Caribbean Maths Connect* and in the UK for *Maths Connect*. The publishers would like to acknowledge the contribution made by the authors of *Caribbean Maths Connect*: Benita Byer-Bowen, Rosalyn Kelly and Delia D. Samuel.

The publishers would like to thank Saskia Besier, Gwen Burns and Harry Smith for contributing additional material.

The publishers would like to thank the following for permission to use photographs:
pp. 16 and 18 © Corbis; pp. 55, 62, 97 lower and 126 © Mike van der Wolk; p. 56 © Harcourt Education Ltd/Tudor Photography; p. 93 © Corbis/Neil Rabinowitz; p. 94 © Getty Images/Altrendo; p. 97 top © iStockPhoto.com/John Arder; p. 145 © Getty Images/PhotoDisc

Contents

To the teacher

Trinidad and Tobago Maths Connect is a unique course organised to fit the **General Curriculum Framework for Trinidad and Tobago (Form 3)**. It covers key areas of knowledge, understanding and skills which provide a firm foundation to raise competence and confidence in mathematics. The lively approach of the text is designed to make the study of mathematics both interesting and enjoyable.

The table indicates how the material in this course fits with the General Curriculum Framework for Trinidad and Tobago (Form 3).

Trinidad and Tobago Maths Connect, with its wealth of stimulus material, will help students to understand the relevance of mathematics to everyday problems. There are investigations and suggestions for project work at the end of each chapter. The book also contains illustrated case studies to show how mathematics is used by real people in different jobs.

The modules are designed to be used sequentially. Each 'Looking back' relates to the previous section, and 'Revise and consolidate' applies to all previous material. The 'Test your knowledge' tests at the end relate to topics throughout the book.

	Topic and chapters	Pages
Term 1	**Exploring real numbers**	
	Chapter 1 – Real numbers	1–25
	Chapter 2 – Representation of real numbers	26–41
	Applying measurement in 2-D and 3-D	
	Chapter 3 – Measurement: circles and prisms	42–62
	Taking basic algebra further	
	Chapter 4 – Algebraic expressions and identities	63–71
Term 2	**Using equations to model mathematics**	
	Chapter 5 – Modelling real-life situations using equations	74–83
	Consumer arithmetic: Business/financial institutions	
	Chapter 6 – Interest, depreciation and investment	84–97
	Making connections in basic geometry	
	Chapter 7 – Pythagoras' theorem	98–112
	Chapter 8 – Polygons	113–126
	Chapter 9 – Rotation	127–133
	Chapter 10 – Enlargement and similar triangles	134–148
Term 3	**Functions and inequalities**	
	Chapter 11 – Set theory	153–157
	Chapter 12 – Relations and functions	158–166
	Chapter 13 – Inequalities	167–179
	Statistical analysis of data: discrete and continuous	
	Chapter 14 – Frequency distribution	180–195

How to use this book

This book is divided up into 14 colour-coded chapters. Number chapters are orange; Algebra chapters are green; Geometry chapters are blue; Measurement chapters are purple; and Data Handling chapters are red. Each chapter is divided up into units.

Unit 2.2 Roots

In this unit you will:

- find the square roots of positive integers
- find other roots of positive integers
- use a calculator to find roots.

Key words
Square root
Cube root
Root

Directly under the unit heading are the **lesson targets**. These tell you what you will learn in the lesson. To help you remember important vocabulary, there is a **Key words** box here.

> You already know that 16 is a square number and that $16 = 4 \times 4$. We say that 4 is the **square root** of 16. Finding the square root of a square number is like working out the length of the side of a square when you know its area.

In the **explanation box**, you can see a summary of the key ideas that are covered in the lesson. The **key words** are printed in **bold type**.

Worked examples

$18^2 = 324$. Use this to find **a.** 19×18 **b.** 17×18.

a. $19 \times 18 = (18 \quad 1) \times 18 = (18 \times 18) \quad (1 \times 18)$

$= 324 \quad 18 = 342$

Remember the associative law.

b. $17 \times 18 = (18 \quad 1) \times 18) = (18 \times 18) \quad (1 \times 18)$

$= 324 \quad 18 = 306$

The **worked examples** show you how to answer the exercise questions. On the blue paper you can see the kind of working you should be writing in your exercise book. The green **hint boxes** help explain how to calculate the answers.

The exercises for each lesson are made up of three kinds of question:

- **practice** questions, which allow you to practise the basic skills
- **problem** questions, which ask you to apply the skills you have learned
- **investigation** questions, which give you practice at solving open-ended problems.

At the end of each chapter are 'Looking back' exercises to help you review what you have learned. There are also three 'Revise and consolidate' exercises spread throughout the book for more review and practice work, and finally, ten 'Tests of your knowledge' at the end.

Key words
Natural
Integers
Rational
Factor
Multiple
Prime
BOMDAS

Chapter 1

Real numbers

You have already developed a wide range of skills in working with numbers. In this chapter we will briefly review some of the number concepts covered in Year 1 and 2 and consolidate and combine the skills you have developed so that you can apply these in new contexts and to real-life situations. The aim is to become a competent mathematician who can solve number-related problems quickly and efficiently.

Unit 1.1 Basic numeracy

In this unit you will:

- revise and apply the basic number concepts you will use as you work through this book
- use the four operations (+, −, × and ÷) to perform simple computations.

The set of **natural** numbers consists of the counting numbers N = {1, 2, 3 …}.
The set of positive **integers** includes 0: {0, 1, 2, 3 …}.
The set of integers includes numbers less than 0, the negative numbers Z = {… −3, −2, −1, 0, 1, 2, 3 …}.

Rational numbers are numbers that can be written in the form of a fraction where the numerator and denominator are both whole numbers. For example: $\frac{2}{4}$, $\frac{6}{7}$ and 1.5 are all rational numbers.
All whole numbers are rational because they can be written in the form $\frac{n}{1}$.

Even numbers can be divided exactly by 2. For example, 4, 8, 100, 1 000 008.
Odd numbers cannot be divided exactly by 2. For example, 3, 9, 11, 121, 1 000 121.

A **factor** is a number which divides exactly into a given number. For example, 5 is a factor of 20.

A **multiple** is formed by multiplying a given number by any integer. For example, 10 is a multiple of 1, 2 and 5 (1 × 10, 2 × 5 and 5 × 2).

Prime numbers are numbers which have only two factors, the number itself and 1. The first five prime numbers are 2, 3, 5, 7 and 11. **Note that 1 is not a prime number**.

When we perform operations on numbers we work in a very specific order.
Brackets are always solved and removed first. Remember the fraction line acts like a bracket!
Multiplication and division are performed next.
Addition and subtraction are performed from left to right.

This order is sometimes called **BOMDAS** – this is a mnemonic to help you remember the order of operations, not a rule in itself.

The answer to an addition problem can be called the **sum**.
The answer to a subtraction problem can be called the **difference**.
The answer to a multiplication problem is the **product**.
The answer to a division problem is the **quotient**.

Worked example

12 17 26 −21 49 −100 100

From the numbers above, write down

a. a multiple of 3

b. a prime number

c. all the odd numbers

d. a negative integer

e. the difference between the first and the last number

f. the sum of all the numbers.

a. 12

b. 17

c. 17, −21, 49

d. −21 or −100

e. 88

f. 83

Exercise

1. From the numbers 6, 17, 32, 41 and 49, write down
   a) a multiple of 4
   b) two prime numbers
   c) a factor of 36
   d) a number which is the sum of two other numbers in the set
   e) the product of the largest and smallest numbers.

2. What is the sum of the first six odd numbers?

3. Find the product of 100 and 4 023. Write your answer in figures and words.

4. Write down all the prime numbers smaller than 60.

5. Find the difference between 1 000 and 878.

6. Write down all the factors of 19. What does this tell you about this number?

7. Mrs James has \$300.35 in her bank account. She writes a cheque for \$388.75 and the bank accepts the payment. What would be the new balance in her account?

8. Apply the correct order of operations to calculate
   a) $4(12 - 6) + (6) - (3)$
   b) $3(16 - 12) - (11)(22 - 16)$
   c) $\frac{1}{2}$ of $36 - 21 + 72$
   d) $12 - (7 - 3) - 2 \times 8 + 2$

Key words
Fraction
Denominator
Numerator
Proper fraction
Improper fraction
Mixed number

Unit 1.2 Fractions

In this unit you will:

- learn how to use fractions to describe part of something
- write an improper fraction as a mixed number
- write a mixed number as an improper fraction.

One way to think of a **fraction** is as a number of parts of a whole.

The bottom number in a fraction is the number of parts we divide the whole into. These parts are the same size. The bottom number is called the **denominator**.

The top number in a fraction is the number of these parts we are counting. The top number is called the **numerator**.

These are **proper fractions**: the denominator is greater than the numerator. So the fraction is made of some of the parts of the whole.

$\frac{3}{4}$

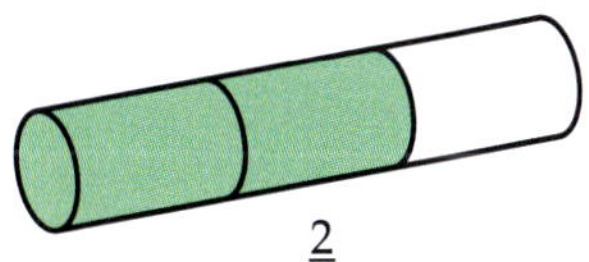

$\frac{2}{3}$

These are **improper fractions**: the denominator is less than the numerator. An improper fraction can be converted to a **mixed number** – part whole number and part fraction.
$\frac{5}{2}$ as a mixed number is $2\frac{1}{2}$.

Whole numbers have a denominator of 1. For example, $5 = \frac{5}{1}$.

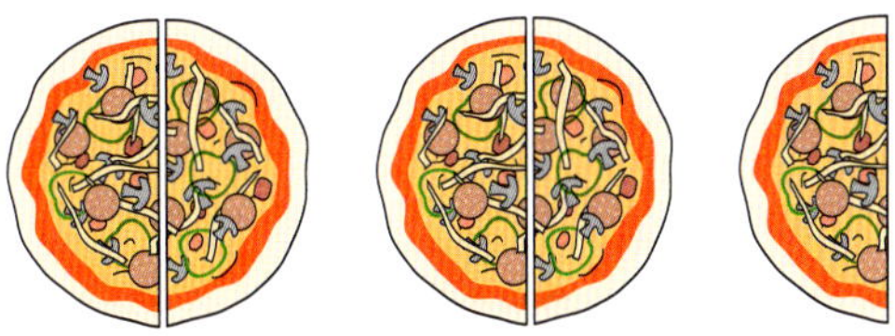

$\frac{5}{2}$ or $2\frac{1}{2}$

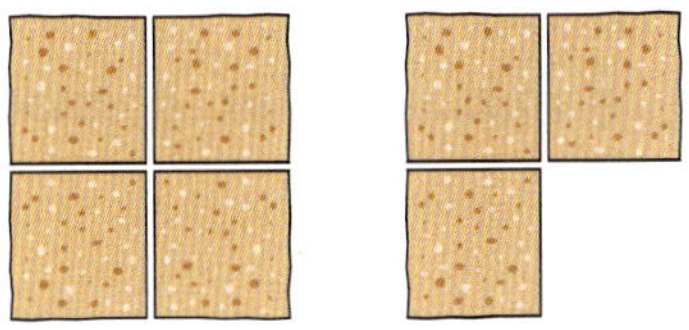

$\frac{7}{4}$ or $1\frac{3}{4}$

Worked examples

1. Convert $3\frac{5}{8}$ into an improper fraction.
2. Convert $\frac{32}{5}$ into a mixed number.

1. $3 \times \frac{8}{8} = \frac{24}{8}$.
   $\frac{24}{8} + \frac{5}{8} = \mathbf{\frac{29}{8}}$

   Start by finding the number of eighths in 3 wholes by multiplying 8 by 3.

2. $32 \div 5 = 6$ remainder 2
   therefore $\frac{32}{5} = \mathbf{6\frac{2}{5}}$

   Divide 32 by 5 to find how many wholes can be made from $\frac{32}{5}$.

Exercise

1 **a)** Look at each shape. Write the fraction that is shaded.

i) 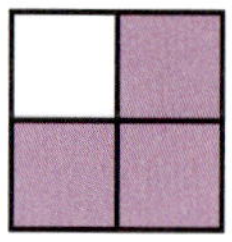

ii) 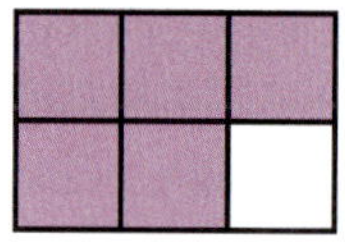

iii) 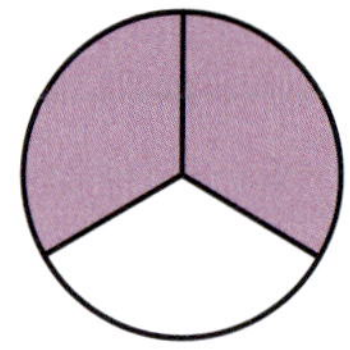

iv) 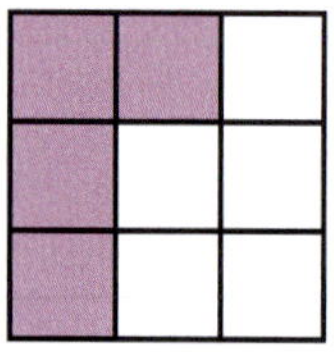

v) 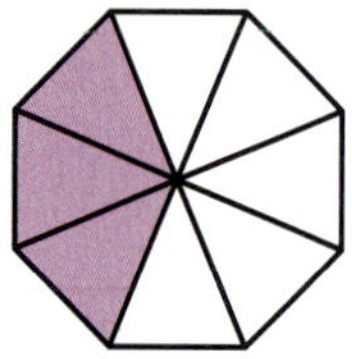

vi) 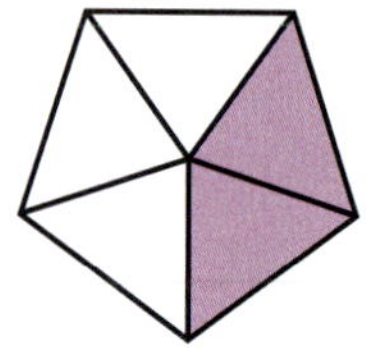

b) Write the fraction of the unshaded part of each shape.

2 Convert these mixed numbers into improper fractions:

a) $2\frac{3}{4}$ **b)** $3\frac{1}{3}$ **c)** $5\frac{2}{5}$ **d)** $4\frac{1}{6}$ **e)** $1\frac{5}{8}$

f) $1\frac{4}{7}$ **g)** $4\frac{1}{2}$ **h)** $2\frac{3}{10}$ **i)** $3\frac{11}{15}$ **j)** $6\frac{2}{3}$

3 Convert these improper fractions into mixed numbers:

a) $\frac{11}{5}$ **b)** $\frac{22}{3}$ **c)** $\frac{7}{4}$ **d)** $\frac{18}{7}$ **e)** $\frac{9}{2}$

f) $\frac{11}{6}$ **g)** $\frac{32}{9}$ **h)** $\frac{43}{11}$ **i)** $\frac{27}{8}$ **j)** $\frac{40}{12}$

4 Write your answers to the following as mixed numbers:

a) How many hours equal 90 minutes? **b)** How many weeks equal 10 days?

c) How many minutes in 130 seconds? **d)** How many metres in 130 cm?

5 Copy the pairs of numbers, writing $<$, $>$ or $=$ between each pair:

a) $\frac{11}{2}$ and $\frac{25}{7}$ **b)** $2\frac{1}{5}$ and $3\frac{1}{5}$ **c)** $6\frac{3}{8}$ and $\frac{53}{8}$ **d)** $\frac{17}{4}$ and $4\frac{3}{4}$

e) $5\frac{4}{5}$ and $\frac{27}{5}$ **f)** $\frac{31}{12}$ and $2\frac{7}{12}$ **g)** $6\frac{2}{3}$ and $\frac{22}{3}$ **h)** $\frac{46}{10}$ and $4\frac{3}{5}$

i) $8\frac{1}{2}$ and $\frac{33}{4}$ **j)** $\frac{23}{11}$ and $2\frac{3}{11}$

> Try converting both numbers to a mixed number to compare them.
> $<$ means 'less than' and $>$ means 'greater than'.

6 Choose any odd number between 20 and 40, for example 23.
Now halve it. Write it as a number of halves: $\frac{23}{2}$, and then as a mixed number: $11\frac{1}{2}$.
Next write it as a number of thirds: $\frac{23}{3} = 7\frac{2}{3}$.
Continue like this, up to tenths.
Repeat for a different number.

7 Oranges are cut into quarters. At half-time in a football match, eleven players and four substitutes each have a quarter of orange. How many oranges do they have between them? How many are left from a bag of ten oranges?

Key words
Numerator
Denominator
Equivalent fraction
Cancel
Lowest terms

Unit 1.3 Equivalent fractions

In this unit you will:

- recognise sets of equivalent fractions
- simplify a fraction by cancelling
- reduce a fraction to its lowest terms.

The fractions $\frac{1}{3}$, $\frac{2}{6}$ and $\frac{4}{12}$ are **equivalent**. The same fraction of each grid is shaded.

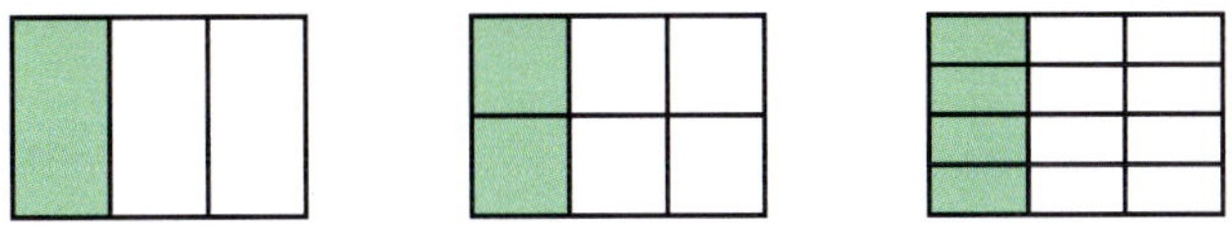

You can use fraction walls to find sets of **equivalent fractions**.

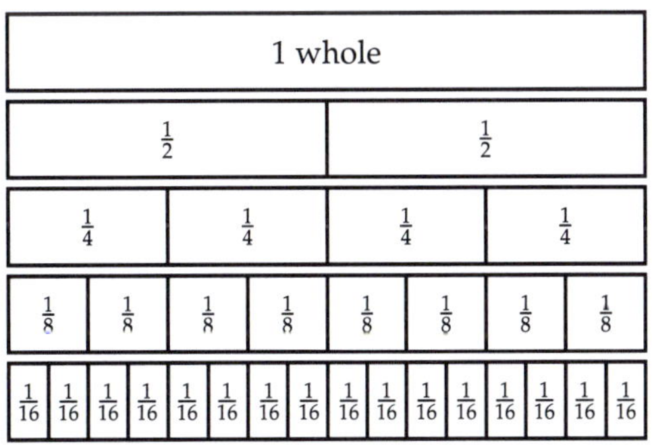

This fraction wall illustrates the following sets of equivalent fractions:

$\frac{1}{2} = \frac{2}{4} = \frac{4}{8} = \frac{8}{16}$, $\frac{1}{4} = \frac{2}{8} = \frac{4}{16}$, $\frac{3}{4} = \frac{6}{8} = \frac{12}{16}$

You can also use a multiplication square to find sets of equivalent fractions.

1	2	3	4	5	6	7	8	9	10
2	4	6	8	10	12	14	16	18	20
3	6	9	12	15	18	21	24	27	30
4	8	12	16	20	24	28	32	36	40
5	10	15	20	25	30	35	40	45	50
6	12	18	24	30	36	42	48	54	60

The two highlighted rows show the following set of equivalent fractions: $(\frac{2}{5}, \frac{4}{10}, \frac{6}{15}, \frac{8}{20}, \ldots)$
Some of these fractions can be simplified by dividing both numerator and denominator by the same number – this is called **cancelling**. When a fraction can't be simplified by cancelling (e.g. $\frac{2}{5}$), we say it is in its **lowest terms**.

Worked examples

1. Find the missing number: $\frac{3}{4} = \frac{\square}{20}$
2. Write $\frac{16}{40}$ in its lowest terms.

1. The denominator has been multiplied by 5. ($4 \times 5 = 20$)
   The numerator, 3, must also be multiplied by 5 ($5 \times 3 = 15$), giving $\frac{3}{4} = \frac{15}{20}$
2. $\frac{16}{40} = \frac{4}{10}$
   $\frac{4}{10} = \frac{2}{5}$

Cancel by dividing both the numerator and denominator by 4.
Divide again by 2. Now the fraction is in its lowest terms.

Exercise

1 Copy each pair of equivalent fractions and find the missing numbers:

a) $\frac{2}{3} = \frac{\square}{6}$ **b)** $\frac{3}{5} = \frac{6}{\square}$ **c)** $\frac{\square}{12} = \frac{2}{6}$ **d)** $\frac{5}{6} = \frac{\square}{24}$ **e)** $\frac{3}{4} = \frac{\square}{12}$

f) $\frac{5}{8} = \frac{10}{\square}$ **g)** $\frac{\square}{10} = \frac{21}{30}$ **h)** $\frac{6}{\square} = \frac{18}{21}$ **i)** $\frac{5}{\square} = \frac{15}{27}$ **j)** $\frac{\square}{20} = \frac{35}{100}$

2 Write the following, in their lowest terms:

a) $\frac{6}{10}$ **b)** $\frac{12}{16}$ **c)** $\frac{16}{24}$ **d)** $\frac{20}{25}$ **e)** $\frac{15}{18}$ **f)** $\frac{18}{60}$ **g)** $\frac{28}{49}$ **h)** $\frac{11}{33}$ **i)** $\frac{35}{45}$ **j)** $\frac{20}{32}$

3 In each set four fractions are equivalent. Write down the fraction which is not.

a) $\frac{6}{8}, \frac{3}{4}, \frac{30}{40}, \frac{15}{24}, \frac{9}{12}$ **b)** $\frac{6}{9}, \frac{12}{18}, \frac{4}{6}, \frac{14}{27}, \frac{16}{24}$ **c)** $\frac{12}{15}, \frac{4}{5}, \frac{8}{10}, \frac{24}{35}, \frac{40}{50}$ **d)** $\frac{5}{6}, \frac{35}{42}, \frac{20}{24}, \frac{10}{12}, \frac{15}{20}$

e) $\frac{42}{60}, \frac{56}{70}, \frac{21}{30}, \frac{14}{20}, \frac{7}{10}$ **f)** $\frac{9}{21}, \frac{18}{42}, \frac{15}{28}, \frac{3}{7}, \frac{6}{14}$ **g)** $\frac{4}{18}, \frac{8}{36}, \frac{14}{63}, \frac{16}{54}, \frac{2}{9}$ **h)** $\frac{6}{16}, \frac{15}{40}, \frac{24}{72}, \frac{12}{32}, \frac{3}{8}$

4 Use a set of 1 – 30 number cards. Shuffle them and deal out ten pairs of cards. For each pair, place the smaller number as the numerator and the larger as the denominator of a fraction. Reduce each of the ten fractions to its lowest terms.

$$\frac{\boxed{1\ 0}}{\boxed{2\ 4}}$$

5 Who am I?

a) I am a fraction equivalent to $\frac{3}{4}$. My denominator is 12.
b) I am a fraction equivalent to $\frac{4}{5}$. My numerator is 20.
c) I am a fraction equivalent to $\frac{2}{3}$. My numerator and denominator have a total of 15.
d) I am a fraction. You'll need four digits to write me down. My digits are all even. My numerator and denominator have a total of 22. I reduce to $\frac{7}{12}$ in my lowest terms.

Investigation

6 You need a set of 1 – 9 digit cards.
Draw a large version of this 'equivalent fractions board'.
Investigate different ways of placing the cards to create pairs of equivalent fractions.

$$\frac{\square}{\square\square} = \frac{\square}{\square}$$

Here is one example:

$$\frac{\boxed{9}}{\boxed{1}\boxed{2}} = \frac{\boxed{3}}{\boxed{4}}$$

Work systematically. For example: keep the denominator of the right hand fraction fixed, and explore different ways of arranging the cards to make a fraction which is equivalent.

Key words

Equivalent fraction
Lowest terms
Improper fractions
Mixed number
Denominator

Unit 1.4 Adding and subtracting fractions 1

In this unit you will:

- add two fractions
- subtract two fractions
- break up a calculation into steps.

If fractions have the same denominator, drawing grids or number lines can help when adding and subtracting them.

$\frac{3}{8} + \frac{1}{8} = \frac{4}{8}$

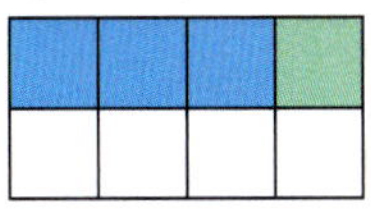

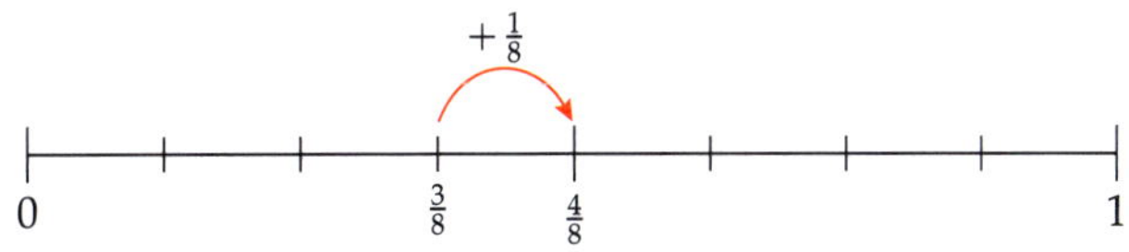

If the fractions have different denominators, convert them to **equivalent fractions** that have the same denominator:

Rewrite $\frac{3}{4} + \frac{1}{2}$ as $\frac{3}{4} + \frac{2}{4}$

$\frac{3}{4} + \frac{2}{4} = \frac{5}{4}$

$\frac{5}{4} = 1\frac{1}{4}$

So $\frac{3}{4} + \frac{1}{2} = 1\frac{1}{4}$.

Worked examples

1. Work out $\frac{7}{8} + \frac{5}{8}$
2. Work out $\frac{3}{4} - \frac{1}{8}$

1. $\frac{7}{8} + \frac{5}{8} = \frac{12}{8}$

   $\frac{12}{8} = 1\frac{4}{8}$

   $1\frac{4}{8} = 1\frac{1}{2}$

   Once you have the answer $\frac{12}{8}$, change the improper fraction to a mixed number ($1\frac{4}{8}$). Reduce the fraction to its lowest terms ($1\frac{1}{2}$).

2. $\frac{3}{4} - \frac{1}{8} = \frac{6}{8} - \frac{1}{8}$

   $\frac{6}{8} - \frac{1}{8} = \frac{5}{8}$

   so $\frac{3}{4} - \frac{1}{8} = \mathbf{\frac{5}{8}}$

   Find an equivalent fraction to $\frac{3}{4}$ so that both fractions in the subtraction have the same denominator.

Exercise

1 Work these out, and express the answer in its lowest terms:

a) $\frac{3}{5} + \frac{1}{5}$ **b)** $\frac{7}{8} - \frac{1}{8}$ **c)** $\frac{8}{10} + \frac{4}{10} - \frac{3}{10}$ **d)** $\frac{5}{6} + \frac{1}{6} + \frac{4}{6}$

e) $1\frac{1}{8} + \frac{3}{8}$ **f)** $2\frac{7}{10} - \frac{3}{10}$ **g)** $3\frac{5}{6} + \frac{5}{6}$ **h)** $4\frac{1}{4} - \frac{3}{4}$

i) $3\frac{1}{5} - 2\frac{2}{5}$ **j)** $1\frac{3}{4} + 2\frac{3}{4}$

2 Find $\square$ in each of the following:

a) $\frac{3}{4} + \square = 1$

b) $\frac{7}{10} - \square = \frac{6}{10} = \frac{3}{5}$

c) $\frac{3}{8} + \square + \frac{1}{8} = 1\frac{1}{8}$

d) $\frac{7}{12} - \frac{1}{12} - \square = \frac{1}{12}$

e) $1\frac{1}{5} + \square = 2\frac{3}{5}$

f) $3\frac{11}{12} - \square = 2\frac{1}{2}$

g) $\square + \frac{3}{4} = 4$

h) $\square - \frac{2}{3} = 3\frac{2}{3}$

What is $2\frac{1}{2}$ as an improper fraction with 12 as the denominator?

3 Work out:

a) $\frac{3}{4} + \frac{1}{2}$

b) $\frac{2}{3} + \frac{1}{6}$

c) $\frac{5}{8} - \frac{1}{4}$

d) $\frac{7}{10} - \frac{1}{5}$

e) $\frac{3}{5} + \frac{1}{5} + \frac{1}{10}$

f) $\frac{1}{2} + \frac{1}{4} + 1\frac{1}{4}$

g) $1\frac{1}{2} + \frac{3}{4}$

h) $\frac{1}{8} + 1\frac{3}{4}$

i) $\frac{1}{3} + \frac{2}{3} + \frac{1}{6}$

j) $\frac{3}{5} + \frac{1}{4} + \frac{2}{5}$

Remember to convert to equivalent fractions first. For example in part **a**, $\frac{1}{2} = \frac{2}{4}$.

4 In Mary's flowerbed there are red, yellow and purple flowers. $\frac{3}{10}$ of the flowerbed is filled with red flowers and $\frac{2}{10}$ of the flowerbed is filled with yellow flowers. The remainder of the flowerbed is filled with purple flowers. What fraction of the flowerbed is filled with purple flowers?

5 We ate half of my birthday cake on Saturday, and one third of it on Sunday. What fraction is left?

6 Janice went on a two-week holiday. She spent $\frac{3}{5}$ of her holiday money in the first week, and only $\frac{1}{10}$ in the second week. What fraction of her holiday money did she bring home?

7 Who am I?

a) I am a fraction. I am $\frac{1}{2}$ more than $\frac{1}{3}$.

b) I am a fraction. I am $\frac{1}{4}$ less than $\frac{5}{8}$.

c) I am an odd number of fifths. When $\frac{1}{2}$ is subtracted from me the result is equivalent to $\frac{2}{20}$.

Investigations

6 Start with a number, for example $1\frac{3}{4}$.
Explore different additions which have this number as an answer.
For example: two fractions with the same denominator: $1\frac{1}{4} + \frac{2}{4}$ or $\frac{7}{8} + \frac{7}{8}$
three fractions with the same denominator: $\frac{3}{4} + \frac{3}{4} + \frac{1}{4}$ or $\frac{7}{8} + \frac{3}{8} + \frac{4}{8}$
two fractions with different denominators: $\frac{1}{2} + 1\frac{1}{4}$

9 Here are some pairs of fractions which have a difference of $\frac{1}{2}$:

$\frac{3}{4}$ and $\frac{1}{4}$ $\frac{7}{8}$ and $\frac{3}{8}$ $1\frac{1}{6}$ and $\frac{2}{3}$

Investigate pairs of fractions which have a difference of $\frac{1}{3}$.

Repeat for $\frac{3}{4}$.

Start with thirds and write down about five pairs. Then look at sixths, ninths, twelfths, and so on.

Key words
- Decimal
- Fraction
- Simplify
- Cancel
- Equivalent fraction
- Recurring decimal

Unit 1.5 Fractions and decimals

In this unit you will:

- convert fractions to decimals with and without a calculator
- convert decimals to fractions
- recognise recurring decimals
- understand the relationship between fractions and decimals.

To convert a **decimal** to a **fraction**:

Write it as a number of tenths, hundredths and thousandths, then **simplify** it as much as possible by **cancelling**.

$0.28 = \frac{28}{100} = \frac{14}{50} = \frac{7}{25}$ $\qquad$ $0.675 = \frac{675}{1000} = \frac{135}{200} = \frac{27}{40}$

To convert a **fraction** to a **decimal** there are three methods.

- **Method 1:**
  Find an **equivalent fraction** with a denominator of 10, 100, 1000, …
- **Method 2:**
  Divide the numerator by the denominator using written division.
- **Method 3:**
  Divide the numerator by the denominator using a calculator.

Sometimes the fraction is equivalent to a **recurring decimal**.

$\frac{1}{3} = 0.3333333\ldots$, written as $0.\dot{3}$

$3\overline{)1.0000\ldots}$ = $0.3333\ldots$

The 3's continue indefinitely and we show this by writing a dot above the 3.
We read the decimal as 'zero point three recurring'.
Some decimals have a set of recurring digits:

$\frac{5}{11} = 0.4545454\ldots$, written as $0.\dot{4}\dot{5}$

We write a dot over the first and last digit in the recurring set.

Worked examples

1. Write the fraction $\frac{71}{125}$ as a decimal.
2. Convert $\frac{7}{8}$ to a decimal using written division.

1. $\frac{71}{125} = \frac{142}{250} = \frac{284}{500} = \frac{568}{1000} = 0.568$

Multiply numerator and denominator by 8, i.e. by 2, then 2, then 2.

2.
```
   0.875
8)7.0
  6.4     0.8 x 8
  0.60
  0.56    0.07 x 8
  0.04
  0.04    0.005 x 8
  0.00    0.875 x 8
```

$\frac{7}{8} = 0.875$

Exercise

1. Write each decimal as a fraction in its lowest terms:

a) 0.8	**b)** 0.64	**c)** 0.76	**d)** 0.485	**e)** 0.516
f) 0.06	**g)** 0.375	**h)** 0.82	**i)** 0.608	**j)** 0.084

2. Write each fraction as a decimal by converting to an equivalent fraction with a denominator of 10, 100 or 1000:

a) $\frac{4}{5}$	**b)** $\frac{3}{4}$	**c)** $\frac{15}{25}$	**d)** $\frac{17}{20}$	**e)** $\frac{31}{50}$
f) $\frac{9}{25}$	**g)** $\frac{111}{250}$	**h)** $\frac{79}{200}$	**i)** $\frac{85}{500}$	**j)** $\frac{125}{20}$

3. Convert these fractions to decimals, using written division.

a) $\frac{8}{5}$ **b)** $\frac{3}{8}$ **c)** $\frac{11}{4}$ **d)** $\frac{13}{8}$

Check each answer using a calculator.

4. Convert these fractions to recurring decimals using written division.

a) $\frac{2}{3}$	**b)** $\frac{2}{9}$	**c)** $\frac{8}{9}$	**d)** $\frac{1}{6}$
e) $\frac{7}{9}$	**f)** $\frac{5}{6}$	**g)** $\frac{4}{9}$	**h)** $\frac{1}{3}$

5. Copy and complete this chart, showing the decimal equivalent of fractions with denominators up to 10.
Use the dot notation for recurring decimals.
If the decimal does not recur, round it to two decimal places.

$\frac{2}{3} = 0.6\dot{6}$

Numerator (columns) / **Denominator** (rows)

	1	2	3	4	5	6	8	9	10
1	1	2	3	4	5	6			
2		1							
3		$0.6\dot{6}$				2			
4									
5			0.60		1			1.80	
6									
8									
9		$0.2\dot{2}$							
10									

6. $\frac{3}{11} = 0.2727272\ldots$ Find a fraction that is equivalent to $0.0272727\ldots$

7. Copy this number line.
Draw a pointer to show the position of $0.\dot{1}$.

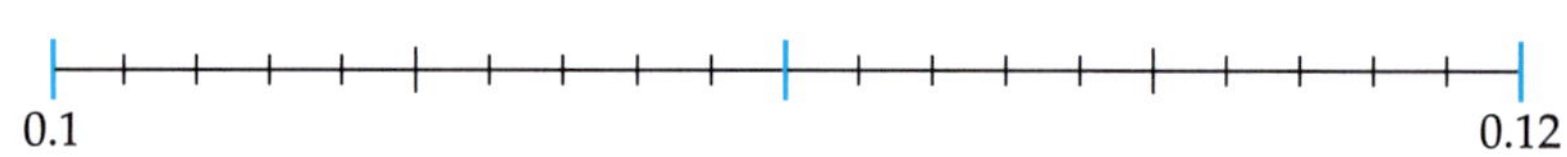

Investigations

8. Use a calculator to convert each of these fractions into decimals:
$\frac{1}{9}, \frac{2}{9}, \frac{3}{9}, \frac{4}{9}, \frac{5}{9}, \frac{6}{9}, \frac{7}{9}, \frac{8}{9}, \frac{9}{9}, \frac{10}{9}, \frac{11}{9}$
What patterns do you notice? Explore similar patterns in elevenths.
Find another denominator that produces patterns like this.

9. Use a calculator to check that $\frac{1}{3} = 0.3333333$ or $0.\dot{3}$
The fraction is equivalent to a recurring decimal with one recurring digit.
Check with a calculator that $\frac{1}{11} = 0.090909$ or $0.\dot{0}\dot{9}$
The fraction is equivalent to a recurring decimal with two recurring digits.
Investigate fractions that are equivalent to recurring decimals with:
a) one recurring digit **b)** two recurring digits.

Key words
Terminating decimal
Non-terminating decimal
Recurring decimal

Unit 1.6 Terminating and non-terminating decimals

In this unit you will:

- look at the difference between terminating and non-terminating decimals
- use recurring decimal notation
- convert fractions to terminating and non-terminating decimals
- convert terminating and recurring decimals to fractions.

When a fraction of the type $\frac{a}{b}$ is converted into a decimal, the answer will either be a terminating decimal or a non-terminating decimal.

A **terminating decimal**, such as 0.5, 0.123, 0.67, 0.747474, ends after a definite number of digits have been given.

A **non-terminating decimal**, such as 0.33333…, 0.45454545…, goes on repeating itself without end.

A non-terminating decimal such as the examples above, is also known as a **recurring decimal**. What do you think the word recurring means?

> A recurring decimal is written in abbreviated form by using the **recurring decimal notation**.
>
> 0.33333333... is normally written as $0.\dot{3}$
> 0.45454545... is normally written as $0.\dot{4}\dot{5}$
>
> The dot or dots above the number show what is to be repeated.

REVISION (1): Converting a terminating or non-terminating fraction to a decimal.

Find an equivalent fraction with denominator of 10, 100, 1000, etc.

Examples:

$\frac{3}{5} = \frac{6}{10} = 0.6$

$\frac{3}{20} = \frac{15}{100} = 0.15$

You can also convert a fraction to a decimal by dividing the numerator by the denominator. You can do this using a written method or by using a calculator.

Examples:

$\frac{1}{5} = 1 \div 5 = 0.2$

$\frac{5}{12} = 5 \div 12 = 0.416666\ldots = 0.41\dot{6}$

$\frac{1}{6} = 1 \div 6 = 0.166666\ldots = 0.1\dot{6}$

REVISION (2): Converting a terminating decimal to a fraction.

You convert a terminating decimal to a fraction by using powers of 10 – 10, 100, 1000, etc. – as the denominator, and then simplify as much as possible.

Examples:

$0.6 = \frac{6}{10} = \frac{3}{5}$

$0.12 = \frac{12}{100} = \frac{6}{50} = \frac{3}{25}$

$0.345 = \frac{345}{100} = \frac{69}{200}$

Worked example

We are going to look at a method to **convert a non-terminating decimal to a fraction.**

1. Convert the recurring decimal $0.\dot{3}$ to a fraction.
2. Convert the recurring decimal $0.\dot{3}\dot{4}$ to a fraction.

1. $0.\dot{3} = 0.333333\ldots$

   Let $a = 0.333333\ldots$

   $a = 0.333333\ldots$ (1), if we multiply both sides by 10 we get

   $10a = 3.333333\ldots$ (2), if we subtract (1) from (2) we get

   $9a = 3$, if we divide both sides by 9 we get

   $a = \frac{3}{9} = \frac{1}{3}$

2. $0.\dot{3}\dot{1} - 0.313131\ldots$

   Let $a = 0.343434\ldots$

   $a = 0.343434\ldots$ (1), if we multiply both sides by 100 we get

   $100a = 34.343434\ldots$ (2), if we subtract (1) from (2) we get

   $99a = 34$, if we divide both sides by 99 we get

   $a = \frac{34}{99}$

Why did we multiply by 10 in problem 1 but by 100 in problem 2?

Exercise

1 Write these recurring decimals using the recurring dot notation:

a) 0.111111… **b)** 3.62222… **c)** 19.323232…

d) 7.6191919… **e)** 0.181818… **f)** 0.14285714285714…

2 Convert these terminating decimals to fractions in their lowest terms:

a) 0.4 **b)** 0.32 **c)** 0.27 **d)** 0.682 **e)** 0.124

f) 0.75 **g)** 0.28 **h)** 0.06 **i)** 0.025

3 Convert these fractions to decimals. Write your answers as recurring decimals.

a) $\frac{2}{9}$ **b)** $\frac{5}{6}$ **c)** $\frac{2}{6}$ **d)** $\frac{8}{9}$ **e)** $\frac{3}{14}$

4 Convert these recurring decimals to fractions:

a) $0.\dot{6}$ **b)** $0.\dot{1}$ **c)** $0.\dot{5}$ **d)** $0.\dot{7}$

e) $0.\dot{1}\dot{8}$ **f)** $0.\dot{9}\dot{1}$ **g)** $0.\dot{5}\dot{1}\dot{3}$ **h)** $0.\dot{1}\dot{2}\dot{5}$

Investigations

5 Find the fraction that is equivalent to the recurring decimal $0.\dot{9}$.
Comment on your answer.

6 Write out the recurring decimals which are equivalent to the fractions:

$\frac{1}{7}$ $\frac{2}{7}$ $\frac{3}{7}$ $\frac{4}{7}$ $\frac{5}{7}$ $\frac{6}{7}$

Explain the relationship between the pattern of the recurring decimals.

Key words
Denominator
Numerator
Equivalent fraction
Common denominator
Lowest common multiple
Improper fraction
Mixed number

Unit 1.7 Adding and subtracting fractions 2

In this unit you will:

- add and subtract fractions by converting them to equivalent fractions with a common denominator
- add and subtract mixed numbers.

To add or subtract fractions that have the same **denominator**, simply add or subtract the **numerators**. For example:

$\frac{4}{11} + \frac{5}{11} = \frac{9}{11}$ $\qquad \frac{11}{15} - \frac{4}{15} = \frac{7}{15}$

To add or subtract fractions that have different denominators, convert them to **equivalent fractions** which all have the same (common) denominator.

Choose the smallest possible **common denominator**, i.e. the **lowest common multiple** (LCM) of the denominators.

To add the fractions $\frac{13}{18}$ and $\frac{17}{24}$:
Their denominators are 18 and 24.
The multiples of 18 are 18, 36, 54, 72, 90, …
The multiples of 24 are 24, 48, 72, 96, …
The lowest common multiple (LCM) of 18 and 24 is 72.
So, find equivalent fractions with denominators of 72:

$\frac{13}{18} = \frac{52}{72}$

$\frac{17}{24} = \frac{51}{72}$

Add: $\frac{52}{72} + \frac{51}{72} = \frac{103}{72} = 1\frac{31}{72}$

This fraction is an **improper fraction**

Mixed numbers are part whole numbers and part fractions.

To add or subtract mixed numbers, either calculate the whole number part and the fraction parts separately, and then combine them, or convert each mixed number to an improper fraction, then add or subtract them as fractions.

Worked examples

1. Work out $1\frac{5}{8} + 2\frac{3}{5}$
2. Work out $\frac{4}{15} + \frac{7}{12} - \frac{3}{10}$

1. **Method A** - Add whole numbers first

$1 + 2 = 3$ $\qquad \frac{5}{8} + \frac{3}{5} = \frac{25}{40} + \frac{24}{40}$

LCM of 8 and 5 = 40.

$= \frac{49}{40}$

$= 1\frac{9}{40}$

$3 + 1\frac{9}{40} = 4\frac{9}{40}$

Add the whole number part.

Method B - Make into **improper fractions** first.

$1\frac{5}{8} = \frac{13}{8}$ $\quad 2\frac{3}{5} = \frac{13}{5}$

$$\frac{13}{8} + \frac{13}{5} = \frac{65}{40} + \frac{104}{40}$$
$$= \frac{169}{40}$$
$$= \frac{49}{40}$$

LCM of 8 and 5 = 40.

2.

$\frac{4}{15} = \frac{16}{60}$ (×4) $\quad \frac{7}{12} = \frac{35}{60}$ (×5) $\quad \frac{3}{10} = \frac{18}{60}$ (×6)

Multiples of 15: 15, 30, 45, 60, 75.
Multiples of 12: 12, 24, 36, 48, 60, 72.
Multiples of 10: 10, 20, 30, 40, 50, 60.
LCM of 15, 12 and 10 = 60.

$\frac{16}{60} + \frac{35}{60} - \frac{18}{60} = \frac{33}{60} = \frac{11}{20}$

Cancel by dividing by 3.

Exercise

1 Complete the following calculations, giving your answer in its lowest terms.

a) $\frac{5}{8} + \frac{7}{8}$ **b)** $\frac{7}{15} - \frac{2}{15}$ **c)** $1\frac{3}{7} + 3\frac{4}{7}$ **d)** $3\frac{1}{12} - 1\frac{7}{12}$ **e)** $2\frac{3}{10} + 4\frac{1}{10} - 3\frac{9}{10}$

2 **i)** Find the LCM (lowest common multiple) for these denominators.

a) $\frac{3}{5} - \frac{3}{8}$ **b)** $\frac{4}{7} + \frac{5}{6}$ **c)** $\frac{11}{15} + \frac{5}{9}$ **d)** $\frac{15}{24} - \frac{8}{16}$ **e)** $\frac{7}{12} + \frac{3}{10} - \frac{8}{30}$

ii) Now complete the calculations.

3 Evaluate:

a) $1\frac{3}{5} + 2\frac{3}{7}$ **b)** $3\frac{2}{5} - 1\frac{4}{9}$ **c)** $2\frac{2}{15} - 1\frac{5}{6}$ **d)** $4\frac{2}{11} + 1\frac{3}{8}$ **e)** $2\frac{7}{18} + 3\frac{5}{24}$

4 If $A = 2\frac{5}{12}$ $B = 1\frac{3}{8}$ $C = 3\frac{11}{15}$ $D = 1\frac{7}{10}$, find the value of:

a) $A + B$ **b)** $C - A$ **c)** $D - B$ **d)** $A + C - D$ **e)** $A - D$

5 A photograph is $6\frac{3}{8}$ inches wide and $8\frac{5}{6}$ inches high. What is its perimeter ?

6 In a box of chocolates, $\frac{1}{3}$ were toffees, $\frac{3}{8}$ were creams, $\frac{1}{6}$ were plain chocolate and the rest were nut chocolates. What fraction were nut chocolates?

7 John has $\frac{7}{8}$ of a length of timber. Liam has $\frac{4}{5}$ of a similar length of timber.

a) Who has the longest piece of timber?

b) What is the difference?

8 Name two fractions with different denominators which have a total of $\frac{7}{12}$.

9 Kerry spent $\frac{1}{3}$ of her pocket money on magazines.
She gave $\frac{3}{7}$ of it to charity and spent $\frac{2}{9}$ at the cinema.
If she has 1.50$ left, how much pocket money did she have?

Find the total fraction she used first.

Key words
Cancel
Numerator
Denominator
Mixed number
Improper fraction

Unit 1.8 Multiplying fractions

In this unit you will:

- multiply a fraction by a fraction, cancelling common factors.

$\frac{3}{5} \times \frac{5}{6}$ is equivalent to $\frac{3}{5}$ of $\frac{5}{6}$

From the grids, $\frac{3}{5}$ of $\frac{5}{6} = \frac{3 \times 5}{5 \times 6} = \frac{15}{30}$ or $\frac{1}{2}$

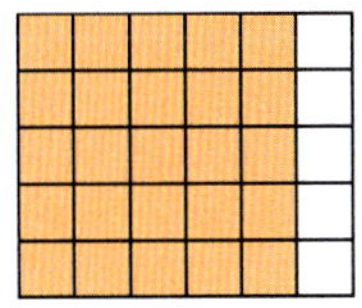

$\frac{5}{6}$

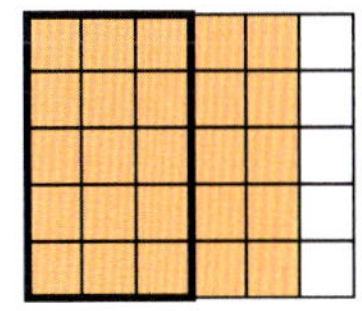

$\frac{3}{5}$ of $\frac{5}{6}$

To multiply fractions:

Cancel by dividing any **numerator** and any **denominator** by the same number.

When no further cancelling is possible, multiply the numerators together, and the denominators together.

For example: $\frac{4}{15} \times \frac{5}{12} = \frac{\overset{1}{\cancel{4}} \times \overset{1}{\cancel{5}}}{\underset{3}{\cancel{15}} \times \underset{3}{\cancel{12}}} = \frac{1}{9}$

To multiply **mixed numbers**:
First change the mixed numbers to improper fractions, then multiply the fractions.

For example: $1\frac{3}{7} \times 2\frac{4}{5} = \frac{10}{7} \times \frac{14}{5} = \frac{\overset{2}{\cancel{10}} \times \overset{2}{\cancel{14}}}{\underset{1}{\cancel{7}} \times \underset{1}{\cancel{5}}} = \frac{4}{1}$ or 4

Worked examples

1. A picture is $6\frac{1}{4}$ inches long and $8\frac{4}{5}$ inches wide. Calculate its area.
2. What is $\frac{2}{3} \times \frac{2}{3} \times \frac{2}{3}$?

1. Area $= 6\frac{1}{4} \times 8\frac{4}{5}$ square inches

Area = length × width.

$= \frac{\overset{5}{\cancel{25}}}{\underset{1}{\cancel{4}}} \times \frac{\overset{11}{\cancel{44}}}{\underset{1}{\cancel{5}}}$ square inches

Mixed numbers must be converted to improper fractions.

$= \frac{55}{1} = 55$ square inches

2. $\frac{2}{3} \times \frac{2}{3} \times \frac{2}{3} = \frac{8}{27}$

Exercise

1 Work out:

a) $\frac{3}{4} \times \frac{8}{9}$ **b)** $\frac{2}{3} \times \frac{4}{5}$ **c)** $\frac{7}{15} \times \frac{20}{21}$

d) $\frac{8}{9} \times \frac{7}{16}$ **e)** $\frac{3}{4} \times \frac{5}{6}$ **f)** $\frac{10}{12} \times \frac{3}{8}$

2 Evaluate:

a) $1\frac{1}{3} \times 2\frac{1}{4}$ **b)** $5\frac{1}{4} \times 2\frac{2}{3}$ **c)** $4\frac{2}{3} \times 3\frac{3}{7}$

d) $1\frac{3}{5} \times 3\frac{1}{8}$ **e)** $1\frac{5}{6} \times 1\frac{7}{11}$ **f)** $3\frac{3}{5} \times 2\frac{4}{9}$

3 Work out:

a) $\frac{5}{14} \times \frac{9}{10} \times \frac{7}{18}$ **b)** $(\frac{4}{5})^2$ **c)** $\frac{16}{21} \times \frac{35}{48} \times \frac{21}{35}$

d) $1\frac{1}{11} \times 3\frac{3}{4} \times 1\frac{1}{5}$ **e)** $(1\frac{1}{4})^3$

4 A room measures $3\frac{3}{4}$ metres by $2\frac{2}{5}$ metres. What is its area?

Look at Example 1.

5 John spends $\frac{5}{8}$ of his pocket money at the cinema. He spends $\frac{2}{3}$ of what is *left* on sweets. What fraction of his total pocket money does he spend on sweets?

6 What fraction am I?

a) I am $\frac{1}{6}$ of $\frac{2}{3}$

b) I am $\frac{1}{2}$ of $\frac{1}{4}$ of $\frac{4}{5}$

c) I am $1\frac{2}{3}$ cubed.

7 This square has sides of 1m.

What are the areas of the 2 shaded squares?

8 To find the area of a circle we use the formula

$A = \pi \times r \times r$, where A is area, r is radius, and we can take an approximate value of π as $\frac{22}{7}$.

If a circle has a radius of $3\frac{1}{2}$ cm, what is its area?

9 A cuboid has a length of $2\frac{1}{3}$ inches, a width of $2\frac{2}{5}$ inches and a height of $1\frac{7}{8}$ inches.

Volume of a cuboid is length × width × height

a) What is the volume of the cuboid?

b) Each cubic inch has a mass of $3\frac{1}{3}$ grams, what is the mass of the cuboid?

Investigations

10 A ball is dropped from a height of $2\frac{1}{3}$ m. It bounces to $\frac{3}{4}$ of the height from which it fell at each bounce.

a) Find the height after one bounce.

b) How many bounces does it take to bounce less than 1 metre?

Key words
Numerator
Denominator
Inverted
Mixed number
Improper fraction
Cancel

Unit 1.9 Dividing fractions

In this unit you will:

- divide a fraction by a fraction, cancelling common factors.

$\frac{3}{4} \div \frac{1}{8}$ can be read as: 'three quarters divided by one eighth'
and: 'how many eighths in three quarters?'

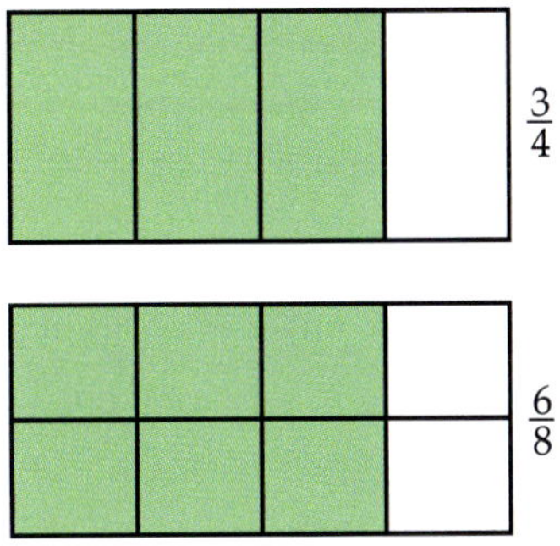

'There are six eighths in three-quarters'

To divide one fraction by another:
Dividing by 4 ($\frac{4}{1}$) is equivalent to multiplying by $\frac{1}{4}$
Dividing by $\frac{1}{4}$ is equivalent to multiplying by 4 ($\frac{4}{1}$)
Dividing by $\frac{4}{5}$ is equivalent to multiplying by $\frac{5}{4}$

Dividing by a fraction is equivalent to multiplying by that fraction with the **numerator** and **denominator inverted** (turned upside down).

For example: $\frac{9}{10} \div \frac{3}{5} = \frac{9}{10} \times \frac{5}{3}$

$$= \frac{\overset{3}{\cancel{9}} \times \overset{1}{\cancel{5}}}{\underset{2}{\cancel{10}} \times \underset{1}{\cancel{3}}} = \frac{3}{2} \text{ or } 1\frac{1}{2}$$

To divide one **mixed number** by another:
First change the mixed numbers to **improper fractions**, then divide the fractions.

For example: $3\frac{3}{4} \div 2\frac{1}{12} = \frac{15}{4} \div \frac{25}{12} = \frac{15}{4} \times \frac{12}{25}$, and then **cancel**.

$$\frac{15}{4} \times \frac{12}{25} = \frac{\overset{3}{\cancel{15}} \times \overset{1}{\cancel{12}}}{\underset{2}{\cancel{4}} \times \underset{1}{\cancel{25}}} = \frac{6}{5} \text{ or } 1\frac{1}{5}$$

Worked examples

1. Work out $\frac{9}{14} \div \frac{3}{4}$

2. The kennels have $5\frac{1}{2}$ kg of dog food. If each dog can eat $\frac{2}{5}$ kg of food, how many dogs can they feed?

1. $\frac{9}{14} \div \frac{3}{4} = \frac{9}{14} \times \frac{4}{3}$

Remember: ÷ by $\frac{3}{4}$ is the same as × by $\frac{4}{3}$.

$= \frac{\cancel{9}^{3}}{\cancel{14}_{7}} \times \frac{\cancel{4}^{2}}{\cancel{3}_{1}} = \frac{6}{7}$

2. $5\frac{1}{5} \div \frac{2}{5} = \frac{26}{5} \div \frac{2}{5}$

Change to an improper fraction.

$= \frac{\cancel{26}^{13}}{\cancel{5}_{1}} \times \frac{\cancel{5}^{1}}{\cancel{2}_{1}} = 13$ dogs.

Exercise

1 What would I multiply by in order to get the same answer as dividing by:

a) $\frac{3}{5}$ **b)** $1\frac{1}{4}$ **c)** $\frac{4}{3}$?

2 Work out:

a) $\frac{3}{5} \div \frac{9}{10}$ **b)** $\frac{3}{4} \div \frac{9}{16}$ **c)** $\frac{5}{8} \div \frac{15}{16}$

d) $\frac{4}{7} \div \frac{2}{3}$ **e)** $\frac{2}{3} \div \frac{5}{6}$

3 Evaluate:

a) $1\frac{4}{5} \div \frac{3}{10}$ **b)** $8\frac{1}{4} \div 1\frac{3}{8}$ **c)** $1\frac{2}{3} \div 1\frac{1}{9}$

d) $5\frac{1}{4} \div 1\frac{1}{6}$ **e)** $4\frac{1}{6} \div 1\frac{2}{3}$

4 I have $3\frac{3}{4}$ metres of rope.
How many lengths of $\frac{5}{8}$ m can I cut the rope into?

5 The area of a rectangle is $7\frac{3}{5}$ cm^2.
If the length is $4\frac{2}{9}$ cm, what is its width?

6 Who am I?

a) I am the result of finding how many $\frac{5}{6}$ there are in $2\frac{1}{2}$

b) I am the multiplication equivalent of dividing by $1\frac{2}{5}$

c) I am half the result of dividing $\frac{3}{10}$ by $\frac{3}{5}$

7 Find the missing values.

a) $\square \div \frac{2}{3} = \frac{3}{8}$ **b)** $\square \div \frac{9}{10} = \frac{2}{3}$

c) $\square \times \frac{3}{10} = \frac{2}{5}$ **d)** $1\frac{1}{8} \times \square = \frac{1}{4}$

8 **a)** At a party, there are $6\frac{1}{4}$ bars of chocolate and each child is given $\frac{3}{8}$ of a bar. How many children can be given chocolate?

b) If there are 25 children, what fraction of a bar should each child be given?

9 To find the diameter of a wheel you divide the circumference by π.

a) If the circumference of a wheel is $4\frac{5}{7}$ cm, what is its diameter?

b) What is the radius?

Use $\frac{22}{7}$ as an approximation for π.

Circumference $= \pi d = 2\pi r$

Key words
Digit
Tenths
Hundredths

Unit 1.10 Multiplying and dividing by numbers between 0 and 1

In this unit you will:

- multiply and divide by numbers between 0 and 1.

For positive numbers, dividing by a number between 0 and 1 makes the number greater; multiplying by a number between 0 and 1 makes the number smaller.

When multiplying by 10, 100 or 1000, the **digits** move 1, 2 or 3 places to the left. When dividing by 10, 100 or 1000, the digits move 1, 2 or 3 places to the right.

Dividing by a fraction is equivalent to multiplying by its inverse.

$$\div \frac{a}{b} \text{ is equivalent to } \times \frac{b}{a}$$

Dividing by 0.1 ($\frac{1}{10}$, one **tenth**), is equivalent to multiplying by 10 ($\frac{10}{1}$)
Dividing by 0.01 ($\frac{1}{100}$, one **hundredth**), is equivalent to multiplying by 100 ($\frac{100}{1}$)

Multiplying by 0.1 ($\frac{1}{10}$), is equivalent to dividing by 10 ($\frac{10}{1}$)
Multiplying by 0.01 ($\frac{1}{100}$), is equivalent to dividing by 100 ($\frac{100}{1}$)

$\times 0.1$	is equivalent to	$\div 10$	$\div 0.1$	is equivalent to	$\times 10$
$\times 0.01$	is equivalent to	$\div 100$	$\div 0.01$	is equivalent to	$\times 100$

To multiply by 0.03, for example, first multiply by 0.01, and then multiply by 3.
To divide by 0.4, for example, first divide by 0.1, and then divide by 4.

Worked examples

1. Work out $5.13 \div 0.3$
2. Work out 23.4×0.07

1. $5.13 \div 0.3 = 5.13 \div 0.1 \div 3$
   $= 51.3 \div 3$
   $= 17.1$

   $5.13 \div 0.1 = 5.13 \times 10$
   $= 51.3$

2. $23.4 \times 0.07 = 23.4 \times 0.01 \times 7$
   $= 0.234 \times 7$
   $= 1.638$

   $23.4 \times 0.01 = 23.4 \div 100$
   $= 0.234$

Exercise

1. Copy and complete these statements.
   a) Multiplying by 0.1 is the same as .... by 10.
   b) Dividing by 0.1 is the same as ....
   c) Multiplying by 0.01 is the same as ....
   d) Dividing by 0.01 is the same as ....
   e) Multiplying by 0.08 is the same as .... by 0.01, then multiplying by ....
   f) Dividing by 0.6 is the same as dividing by .... then .... by 6.

2 Copy and complete the table.

	×1000	÷10	×0.1	×0.01	÷0.1	÷0.01
4.29						
13.6						
4.3						
7						

3 For each of these calculations, say whether the first number will be greater or smaller than the answer.

a) 39.3×4 **b)** $72.3 \div 3$ **c)** 3.2×0.1
d) $13.7 \div 0.1$ **e)** $37.3 \div 0.46$ **f)** 126×0.25

4 Work out:

a) $4.8 \div 0.4$ **b)** 69.3×0.3 **c)** $36 \div 0.6$
d) 3.2×0.7 **e)** $4.21 \div 0.5$ **f)** 15×0.9

5 Evaluate

a) 6.6×0.06 **b)** $7.5 \div 0.05$ **c)** 63.2×0.07
d) $3.28 \div 0.04$ **e)** 65×0.03 **f)** 21×0.08

6 A book has 282 pages, each 0.02 mm thick.
How thick is the book?

Always check that your answer is sensible.

7 A box of paper clips has a mass of 12 g.
If each paper clip has a mass of 0.03 g, how many paper clips are in the box?

8 An eraser is 0.6 cm thick. How many would be needed to make a pile at least 1 m high?

9 Make up some questions for your partner. Construct a grid like this, putting in numbers of your choice.

Number	×	÷	=

Swap grids for your partner to complete, then check each other's answers.
Do you agree?

Investigations

10 p and q are numbers less than 1.
a and b can be any number.

a) Using this information, can you complete the grid?

Number	$\times p$	$\div q$
a	32.4	☐
b	☐	5.8

$a \times p$ must equal 32.4 and $b \div q$ must equal 5.8

b) Is there only one solution?

Key words

Strategy
Multiplier
Power of 10
Divisor

Unit 1.11 Multiplying and dividing decimals

In this unit you will:

- learn strategies for multiplying and dividing with decimals.

There are nearly always several methods (or **strategies**) for doing calculations.

To calculate 0.26×0.3, for example:

Method 1 Change the decimals to whole numbers
$0.26 \times 0.3 = 26 \times 3 \div 1000 = 78 \div 1000 = 0.078$

Method 2 Change the **multiplier** to include a **power of 10**. For example:
$0.26 \times 0.3 = 0.26 \times 0.1 \times 3 = (0.26 \div 10) \times 3 = 0.026 \times 3 = 0.078$

Method 3 Change to fractions and multiply, cancelling where possible.
$0.26 \times 0.3 = \frac{26}{100} \times \frac{3}{10} = \frac{26 \times 3}{100 \times 10} = \frac{78}{1000} = 0.078$

To calculate $0.84 \div 0.4$, for example:

Method 1 Change the decimals to whole numbers
$0.84 \div 0.4 = 84 \div 40 = 84 \div 4 \div 10 = 21 \div 10 = 2.1$

Method 2 Change the **divisor** to include a power of 10. For example:
$0.84 \div 0.4 = 0.84 \div 0.1 \div 4 = 0.84 \times 10 \div 4 = 8.4 \div 4 = 2.1$

Method 3 Change to fractions, and divide by converting it to a multiplication:
$0.84 \div 0.4 = \frac{84}{100} \div \frac{4}{10} = \frac{84}{100} \times \frac{10}{4} = \frac{84 \times 10}{100 \times 4}$ which cancels to $\frac{21}{10} = 2.1$

Worked examples

1. A book is 2.8 cm thick. If each page is 0.04 cm thick, how many pages does the book have?
2. A window is 0.95 m wide by 0.6 m high. What is its area?

1. **Method 1** $2.8 \div 0.04 = 280 \div 4$
$= 70$

To change the divisor into whole numbers, multiply everything by 100.

Method 2 $2.8 \div 0.04 = 2.8 \div 0.01 \div 4$
$= (2.8 \times 100) \div 4$
$= 280 \div 4$
$= 70$

Change the divisor to include a power of 10.
$\div 0.04 = \div 0.01 \div 4$

Method 3 $2.8 \div 0.04 = \frac{28}{10} \div \frac{4}{100}$
$= \frac{\cancel{28}^{7}}{\cancel{10}_{1}} \times \frac{\cancel{100}^{10}}{\cancel{4}_{1}}$
$= 70$

Change to fractions and multiply by the inverse of $\frac{4}{100}$

2. Method 1 $0.95 \times 0.6 = 95 \times 6 \div 1000$

$= 570 \div 1000$

$= 0.57\ \text{m}^2$

Change the decimals into whole numbers: $\times$ 100 and $\times$ 10, then $\div$ 1000.

Method 2 $0.95 \times 0.6 = 0.95 \times 0.1 \times 6$

$= (0.95 \div 10) \times 6$

$= 0.57\ \text{m}^2$

Change the multiplier to include a power of 10: $\times\ 0.6 = \times\ 0.1 \times 6$

Method 3 $0.95 \times 0.6 = \frac{\overset{19}{\cancel{95}}}{100} \times \frac{\overset{3}{\cancel{6}}}{\underset{2}{\cancel{10}}} = \frac{57}{100}$ (2, 1)

$= 0.57\ \text{m}^2$

Change into fractions.

Exercise

1 Complete these calculations using any of the methods shown on page 21.

a) 0.3 × 0.4	**b)** 0.24 × 0.3	**c)** 0.12 × 0.21
d) 0.23 × 0.3	**e)** 0.54 × 0.5	**f)** 0.7 × 0.04
g) 0.27 × 0.3	**h)** 0.6 × 0.64	**i)** 0.25 × 0.06
j) 0.56 ÷ 0.7	**k)** 0.81 ÷ 0.9	**l)** 0.34 ÷ 0.02
m) 0.64 ÷ 0.8	**n)** 0.8 ÷ 0.05	**o)** 0.36 ÷ 0.03
p) 0.49 ÷ 0.7	**q)** 0.72 ÷ 0.12	**r)** 0.11 ÷ 0.44

2 Copy and complete this multiplication pyramid using any method you choose. Each block is found by multiplying the number underneath on the left by the number underneath on the right.

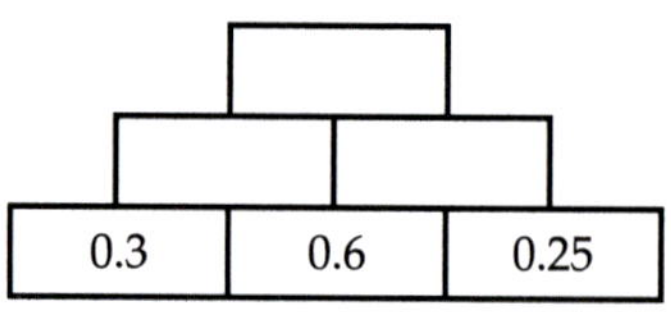

3 Copy and complete this division pyramid. Each block is found by dividing the number underneath on the left by the number underneath on the right.

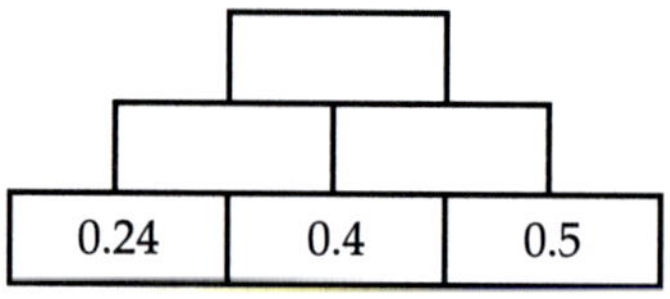

4 Using mental methods:

a) Find the area of a picture 0.3 m long by 0.25 m wide.

b) I have 0.6 litres of cream. If each individual trifle needs 0.05 litres of cream, how many trifles can I make?

c) A coin is 0.3 cm thick. How many coins will there be in a pile 4.8 cm tall?

5 Using 0–9 digit cards, shuffle and deal two numbers, each with 1 or 2 decimal places. e.g. 0.3 and 0.52. Now multiply your decimals by any mental method. Swap with your partner to check each other's answers. Repeat for several calculations.

Key words
Less than <
Greater than >
Ordering
Ascending
Descending

Unit 1.12 Problems involving fractions and decimals

In this unit you will:

- convert decimals to fractions
- convert fractions to decimals
- order decimals and fractions
- solve decimal and fraction problems.

Now that you are able to convert decimals into fractions, and fractions into decimals, you will be able to solve a variety of problems involving both decimals and fractions.

The types of problem that you will be given will require you to consider the relative size of the decimal and fraction so that you are able to place them in a specified order.

Problem 1: Use of the < (**less than**) and > (**greater than**) symbols.

Write the correct symbol < or > between the following:

a) 0.27 0.56 ANSWER: $0.27 < 0.56$

b) $\frac{6}{13}$ $\frac{4}{13}$ ANSWER: $\frac{6}{13} > \frac{4}{13}$

c) 0.23 $\frac{3}{16}$ Can you answer this by looking at the decimal and fraction?

We are able to solve parts a) and b) as we are dealing with either decimals or fractions. However, for part c) we have a decimal and a fraction, which are not so easy to compare. What we need to do is to convert so that we have either decimals or fractions. The easier method is to **convert the fraction to a decimal**. So:

$\frac{3}{16} = 3 \div 16 = 0.1875$

ANSWER: $0.23 > 0.1875$ or $0.23 > \frac{3}{16}$

Problem 2: Ordering a set of values.

An extension of problem 1 involves the **ordering** of a larger set of values. You will be required to arrange a list of mixed decimals and fractions in either **ascending** (smallest to biggest) or **descending** (biggest to smallest) order.

While the question may look more complicated, the method to be used is the same as that explained in problem 1.

Step by step method:

1. Convert all the fractions into decimals.
2. Arrange the decimals in size order.
3. Rewrite the list using the original fraction notation.

Worked example

1. Put the following into ascending order:

   $0.\dot{6}$ $\frac{6}{11}$ 0.71 $\frac{5}{6}$ 0.57

   Convert all fractions into decimals:

   $\frac{6}{11} = 6 \div 11 - 0.\dot{5}\dot{4}$

   $\frac{5}{6} = 5 \div 6 = 0.8\dot{3}$

   $0.\dot{6}$ $0.\dot{5}\dot{4}$ 0.71 $0.8\dot{3}$ 0.57

   Arrange all the decimals in ascending order. Remember: biggest to smallest.

   $0.\dot{5}\dot{4}$ 0.57 $0.\dot{6}$ 0.71 $0.8\dot{3}$

   Finally, rewrite the list using the original fraction notation.

   $\frac{6}{11}$ 0.57 $0.\dot{6}$ 0.71 $\frac{5}{6}$

Exercise

1 Here are some number patterns. Fill in the empty spaces with the correct decimal value.

a) … $\frac{1}{2}$ … … 0.8 …

b) … $\frac{9}{10}$ … … 1.5 …

c) 0.62 … … … … $\frac{18}{25}$

d) 0.49 … … … … $\frac{47}{50}$

e) $\frac{2}{3}$ … … … … 1.5

2 Repeat question **1** but fill the empty spaces with the correct fraction.

3 z is a decimal with one decimal place. Write a list of its possible values, if both these conditions are satisfied:

$\frac{1}{3} < z < \frac{2}{3}$ $\quad$ $\frac{1}{6} < z < \frac{5}{6}$

4 Which of $\frac{5}{6}$ or $\frac{6}{5}$ is nearer to 1? Explain your reasoning.

5 Without use of a calculator, say which of the following fractions have exact decimal equivalents:

a) $\frac{4}{5}$ **b)** $\frac{5}{9}$ **c)** $\frac{7}{25}$ **d)** $\frac{4}{13}$ **e)** $\frac{8}{17}$

Explain the reason for each decision. Use a written method to check if you were correct.

6 Put the following into ascending order:

a) $0.\dot{6}$ $\frac{6}{11}$ 0.71 0.57 $\frac{4}{7}$

b) $\frac{9}{13}$ 0.7 0.58 $\frac{9}{16}$

c) 0.9 $\frac{20}{21}$ 0.89 $\frac{15}{16}$ 0.8

d) $\frac{1}{5}$ 0.19 $\frac{3}{8}$ 0.17 $\frac{3}{20}$

e) 0.4 $\frac{3}{10}$ 0.38 $\frac{3}{8}$ 0.3

Looking back

1. Convert these fractions to decimals. Write your answer as recurring decimals.

   a) $\frac{5}{12}$ b) $\frac{1}{24}$ c) $\frac{7}{15}$ d) $\frac{11}{30}$ e) $\frac{14}{15}$

2. Evaluate

   a) $\frac{2}{5}+\frac{1}{3}$ b) $\frac{5}{14}-\frac{1}{6}$ c) $1\frac{3}{4}+2\frac{5}{9}$

3. A cuboid has sides of length 5.1 cm, 2.9 cm and 15 mm. What is its volume in cm^3?

4. I have a piece of timber of length $2\frac{1}{4}$ m. How many $\frac{3}{8}$ m pieces can I cut it into?

5. Change each of these fractions to a decimal. Write between them the correct symbol < or > or =.

   a) 0.375 $\frac{3}{6}$ b) $\frac{3}{7}$ 0.41 c) $\frac{2}{3}$ 0.74 d) 0.625 $\frac{5}{8}$

   e) 0.1 $\frac{1}{6}$ f) $\frac{11}{12}$ 0.9 g) 0.79 $\frac{7}{9}$

6. Put the following into descending order:

   a) $0.\dot{6}$ $\frac{8}{15}$ 0.59 $\frac{9}{17}$ $\frac{13}{22}$ b) $\frac{1}{13}$ 0.1 0.08 $\frac{1}{8}$ 0.05

   c) 0.7 $\frac{13}{20}$ 0.68 $\frac{7}{9}$ $\frac{14}{19}$ d) 0.86 $0.\dot{8}$ $\frac{7}{8}$ $\frac{9}{11}$ 0.8

   e) $\frac{4}{9}$ 0.4 0.39 $\frac{5}{13}$ 0.45

7. The following give terminating decimals:

   $\frac{1}{2}$ $\frac{1}{4}$ $\frac{1}{5}$ $\frac{1}{8}$ $\frac{1}{10}$ $\frac{1}{16}$ $\frac{1}{20}$

   The following give recurring decimals:

   $\frac{1}{3}$ $\frac{1}{6}$ $\frac{1}{7}$ $\frac{1}{9}$ $\frac{1}{11}$ $\frac{1}{12}$ $\frac{1}{13}$

   By considering the denominator can you find the 'rule' for identifying which fraction will convert into a recurring decimal?

8. Convert the following into decimals, and indicate which of the decimals are terminating and which are recurring.

   a) $\frac{3}{11}$ b) $\frac{23}{100}$ c) $\frac{1}{8}$ d) $\frac{4}{5}$ e) $\frac{3}{7}$

   f) $\frac{6}{25}$ g) $\frac{3}{16}$ h) $\frac{17}{20}$ i) $\frac{34}{125}$

9. Find the recurring decimals equivalent to the fractions:

   $\frac{1}{13}$ $\frac{2}{13}$ $\frac{3}{13}$ … $\frac{10}{13}$ $\frac{11}{13}$ $\frac{12}{13}$

   Separate the recurring decimals into two distinct sets of numbers. Explain the connection between them.

Investigation

10. $1+\frac{1}{2}=1\frac{1}{2}$

    $1+\frac{1}{2}+\frac{1}{4}=1\frac{3}{4}$

    $1+\frac{1}{2}+\frac{1}{4}+\frac{1}{8}=1\frac{7}{8}$

    Continue the pattern for 4 more rows.

    a) What patterns do you notice?

    b) Is there a quick way of working out the answer to the next line?

    c) What would the answer be if you could keep on adding forever in this sum?

Chapter 2

Representation of real numbers

In this chapter you will develop your understanding of index numbers and learn to manipulate roots and indices. You will derive the rules of indices and practise using scientific notation.

Key words
- Base
- Index
- Indices
- Power
- Exponent

Unit 2.1 Writing numbers using indices

In this unit you will:

- write expressions using index form
- evaluate expressions written in index form
- use your calculator to find powers.

You have already learnt about square numbers. Look at this number pattern:

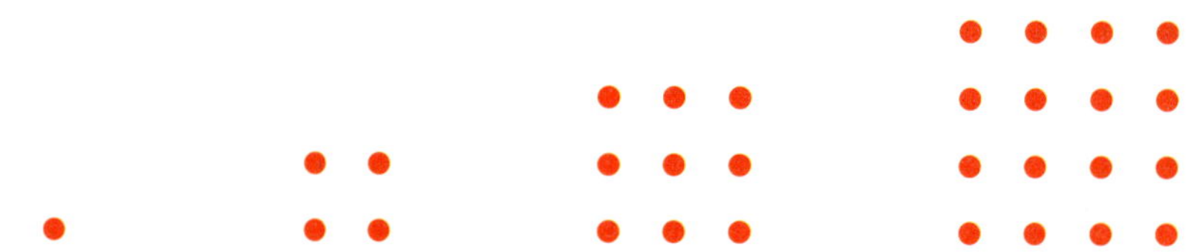

The numbers 1, 4, 9, 16, 25, 36 and so on, are called square numbers. Square numbers are formed by multiplying a whole number by itself.
To check whether a number is a square number, ask whether you could represent it as the area of a square shape. For example:

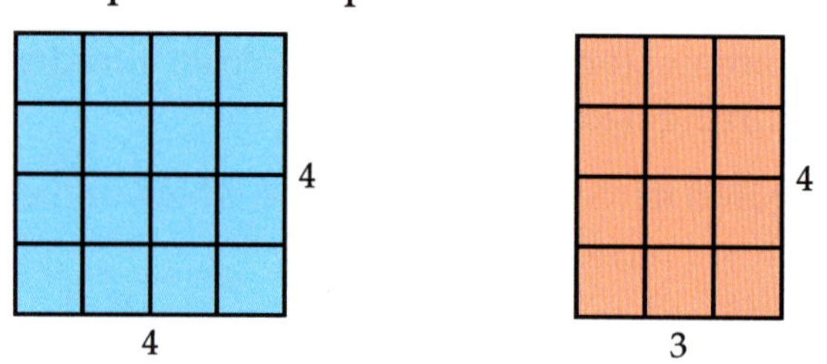

16 is a square number. 12 is not a square number.

12 is a rectangular number.

$16 = 4 \times 4 = 4^2$.

$4^2 = 16$ We read this as 'four squared equals 16'.

4^2 is 16 written in **index** form.

Any number can be written in index form. For example, a^b is 'a to the power of b'.

'a' is called the **base**.
'b' is called the index, **power** or **exponent**. The plural of index is **indices**.

A number written in the index form can be expanded as a product. The index tells us the number of times we multiply the base by itself.

$a^b = a \times a \ldots \times a \times a$ $\qquad$ b times

$X^Y = X \times X \times \ldots \times X$ $\qquad$ Y times

Worked examples

1. Write the following numbers out in full as products.
   a. 3^4 **b.** 2^5
2. Rewrite the following in index form.
   a. 4×4 **b.** $6 \times 6 \times 6$
3. Evaluate
   a. 2^3 **b.** 5^4 **c.** $2^2 \times 3^2$.

1. **a.** $3^4 = 3 \times 3 \times 3 \times 3$ **b.** $2^5 = 2 \times 2 \times 2 \times 2 \times 2$
2. **a.** $4 \times 4 = 4^2$ **b.** $6 \times 6 \times 6 = 6^3$
3. **a.** $2^3 = 2 \times 2 \times 2 = 8$ **b.** $5^4 = 5 \times 5 \times 5 \times 5 = 625$
   c. $2^2 \times 3^2 = 2 \times 2 \times 3 \times 3 = 4 \times 9 = 36$

On your calculator, the key x^y is used to evaluate expressions in index form.
To evaluate 6^4:

Press 6 (the base)

Press x^y

Press 4 (the power)

Press =

Exercise

1 Identify the base and index in each of the following expressions.

a) 11^5 **b)** 2^4 **c)** 3^1 **d)** 7^0 **e)** 5^{-3}
f) 4^8 **g)** 1^9 **h)** $(\frac{3}{4})^6$ **i)** $(-4)^2$ **j)** $(\frac{1}{3})^{-5}$

2 Rewrite the following in full as products.

a) 5^3 **b)** 4^4 **c)** 2^6 **d)** 3^7 **e)** 8^2
f) 1^5 **g)** 2^8 **h)** 10^3 **i)** 11^4 **j)** 6^2

3 Rewrite in index form:

a) 6×6 **b)** $3 \times 3 \times 3 \times 3$
c) $7 \times 7 \times 7 \times 7 \times 7$ **d)** $2 \times 2 \times 2$
e) $5 \times 5 \times 5 \times 5$ **f)** $4 \times 4 \times 4 \times 4 \times 4$
g) 8 **h)** $2 \times 2 \times 2 \times 2 \times 2 \times 2 \times 2$
i) $9 \times 9 \times 9 \times 9 \times 9 \times 9$ **j)** $4 \times 4 \times 4$

4 Evaluate the following and use your calculator to check your answers.

a) 2^4 **b)** 3^4 **c)** 4^3 **d)** 12^2
e) 2^5 **f)** 5^3 **g)** 1^7 **h)** $(-4)^2 \times 2^3$
i) $5^2 \times 3^1$ **j)** $(-3)^2 \times (-2)^3$

Unit 2.2 The basic rules of indices

Key words
Index/indices
Powers
Base

In this unit you will:

- determine some basic laws of indices
- apply the laws of indices in simplifying expressions.

You already know how to use **powers** or **indices** to write numbers in mathematics. For example 6^3 is read as '6 cubed' or '6 to the third power', and v^m is read as 'v to the m power'.

3 and m are indices. — Singular – index, plural – indices

When we write expressions out in full, we are able to find quick methods of working with indices and this has allowed mathematicians to develop rules for operating with indices.

$4^5 = 4 \times 4 \times 4 \times 4 \times 4$ and $a^3 = a \times a \times a$

$5^2 \times 5^4 = 5^6$ — $(5 \times 5) \times (5 \times 5 \times 5 \times 5) = 5 \times 5 \times 5 \times 5 \times 5 \times 5 = 5^6$

This leads us to the first rule.

Rule 1 $p^a \times p^b = p^{a+b}$ — With the same **base**, add indices when multiplying. p is the base.

$7^{10} \div 7^4 = 7^6$ — $\dfrac{7 \times 7 \times 7 \times 7 \times 7 \times 7 \times 7 \times 7 \times 7 \times 7}{7 \times 7 \times 7 \times 7} = 7 \times 7 \times 7 \times 7 \times 7 \times 7$

This leads us to the second rule.

Rule 2 $m^a \div m^b = m^{a-b}$ — With the same base, subtract indices when dividing. m is the base.

$(4^2)^3 = 4^6$ — $(4 \times 4) \times (4 \times 4) \times (4 \times 4) = 4^6$

This leads us to the third rule.

Rule 3 $(g^a)^b = g^{ab}$ — Multiply indices when one index is raised to another power.

Worked example

Simplify

a. $2n^6 \times m^3n^2$ **b.** $(h^8g^4)^2$ **c.** $x^4 \div x^3y^2$

a. $2n^6 \times m^3n^2 = 2 \times n^6 \times n^2 \times m^3 = 2n^8m^3$ — Multiplication is commutative.

b. $(h^8g^4)^2 = (h^8g^4)(h^8g^4) = (h^8)^2 \times (g^4)^2 = h^{16}g^8$

c. $x^4 \div x^3y^2 = x^{4-3}y^2 = xy^2$

Exercise

1. Simplify the following algebraic expressions.

a) $x^2 \times 3x^2$ b) $2m^3 \times m^4$ c) $4z^3 \times 5z^3$

d) $5n^3 \times 2n \times n$ e) $q \times q \times 8q$ f) $2a^2 \times 2a^2 \times 2a^2$

2. Simplify the following algebraic expressions.

a) $\frac{x^3}{x^2}$ b) $\frac{4a^5}{a}$ c) $\frac{6x^4}{w^3}$

d) $\frac{4n^3}{2n^2}$ e) $\frac{14y^2}{7y^5}$ f) $\frac{21z^3}{3z^3}$

3. Simplify the following.

a) $t^3 \times t^4 \times t^5$ b) $k^4 \times k^6 \div k^7$ c) $\frac{p^5g^2}{p^3}$

d) $\frac{(y^5)^2z^3}{y^7z}$ e) $4a^x \times 2a^y$ f) $\frac{16q^9r^6}{(8q^3r)^2}$

4. If $a = 3$, $b = 2$ and $c = 4$, find the value of the following.

a) $a^b\mathrm{c}$ b) $(c^b)^{\mathrm{a}}$ c) $(ac)^b \div bc^a$

d) $a^b \times a^c$ e) $\frac{(ab)^c}{a^b}$ f) $4a^b \times 6a^c \div 12a^c$

5. Write the answers to the following in index form.

a) $5^4 \div 5^4$ b) $42^{10} \div 42^{10}$

c) $a^6 \div a^6$ d) $y^m \div y^m$

Check by expanding.

6. Give a numerical value for each problem in question 5.

7. Tell a friend what you observe about anything 'to the zero power' e.g. $10^0 = \square$?

8. Simplify the following.

a) $9^{\frac{1}{2}} \times 9^{\frac{1}{2}}$ b) $8^{\frac{1}{3}} \times 8^{\frac{1}{3}} \times 8^{\frac{1}{3}}$ c) $h^{\frac{1}{4}} \times h^{\frac{1}{4}} \times h^{\frac{1}{4}} \times h^{\frac{1}{4}}$

9. What do you observe about the questions in 8 and the results?

Investigation

Challenge a classmate to state the squares of {1, 2, 3, 4, … 20}.

Challenge a classmate to state the cubes of {1, 2, 3, 4, … 10}.

Explain how to use your calculator to find the value of numbers raised to any power.

Unit 2.3 Roots

Key words
Square root
Cube root
Root

In this unit you will:

- find the square roots of positive integers
- find other roots of positive integers
- use a calculator to find roots.

You already know that 16 is a square number and that $16 = 4 \times 4$. We say that 4 is the **square root** of 16. Finding the square root of a square number is like working out the length of the side of a square when you know its area.

For example:

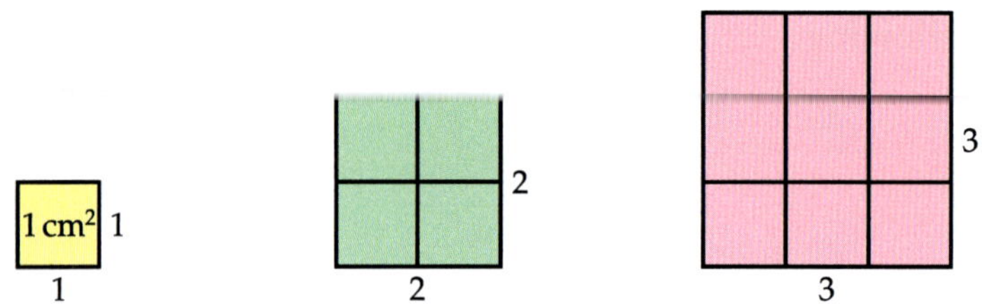

area = $1\,\text{cm}^2$	area = $4\,\text{cm}^2$	area = $9\,\text{cm}^2$
side = 1 cm	side = 2 cm	side = 3 cm
$\sqrt{1} = 1$	$\sqrt{4} = 2$	$\sqrt{9} = 3$

We use a special sign to show square root: $\sqrt{\ }$

For example $\sqrt{81} = 9$.

We read this as 'the square root of 81 is 9'.

Worked examples

1. Find the square roots of **a.** 16 **b.** 49.

2. Evaluate the following. **a.** $\sqrt{(4 \times 9)}$ **b.** $\sqrt{(144 \times 81)}$ **c.** $\sqrt{\left(\frac{36}{49}\right)}$

3. $18^2 = 324$. Use this to find **a.** 19×18 **b.** 17×18.

1. **a.** $\sqrt{16} = 4$ **b.** $\sqrt{49} = 7$

2. **a.** $\sqrt{(4 \times 9)} = \sqrt{4} \times \sqrt{9}$
$= 2 \times 3$
$= 6$

b. $\sqrt{(144 \times 81)} = \sqrt{144} \times \sqrt{81}$
$= 12 \times 9$
$= 108$

c. $\sqrt{\left(\frac{36}{49}\right)} = \frac{\sqrt{36}}{\sqrt{49}} = \frac{6}{7}$

3. **a.** $19 \times 18 = (18 + 1) \times 18 = (18 \times 18) + (1 \times 18)$
$= 324 + 18 = 342$

Remember the associative law.

b. $17 \times 18 = (18 - 1) \times 18) = (18 \times 18) - (1 \times 18)$
$= 324 - 18 = 306$

Exercise

1 Find the square roots of the following positive integers.

a) 4 **b)** 121 **c)** 36 **d)** 49 **e)** 81
f) 144 **g)** 9 **h)** 16 **i)** 64 **j)** 10 000

2 Write the square roots of the products of these numbers.

a) 16×25 **b)** 4×81 **c)** 100×36 **d)** 9×49
e) 400×9 **f)** 64×100 **g)** 121×4 **h)** 81×400
i) 49×900 **j)** $36 \times 10\,000$

3 Find the value of the following by factorising into well-known square numbers.

a) $\sqrt{900}$ **b)** $\sqrt{400}$ **c)** $\sqrt{2\,500}$
d) $\sqrt{196}$ **e)** $\sqrt{4\,900}$ **f)** $\sqrt{360\,000}$

$\sqrt{2\,500} = \sqrt{25} \times \sqrt{100}$

4 Evaluate

a) $\sqrt{\left(\frac{81}{64}\right)}$ **b)** $\sqrt{\left(\frac{4}{25}\right)}$ **c)** $\sqrt{\left(\frac{1}{9}\right)}$

d) $\sqrt{\left(\frac{25}{16}\right)}$ **e)** $\sqrt{\left(\frac{100}{121}\right)}$

5 A square tile has an area of 49 cm^2. What is the length of one side?

6 Find the perimeter of a square field which has an area of 625 m^2.

7 Use a calculator to determine whether the following are square numbers or not.

a) 225 **b)** 7 921 **c)** 16 348
d) 55 696 **e)** 0.9 **f)** 0.3249

8 $12^2 = 144$, $16^2 = 256$ and $40^2 = 1\,600$. Use these to write the value of

a) 17×16 **b)** 41×40 **c)** 13×12
d) 15×16 **e)** 39×40 **f)** 11×12.

Investigation

9 The rules for finding the length of the missing side of this type of triangle are

- find the square of both of the given sides
- add the two answers together
- find the square root of this answer to get the length of the missing side.

Follow these steps to find the lengths of the missing sides of these triangles.

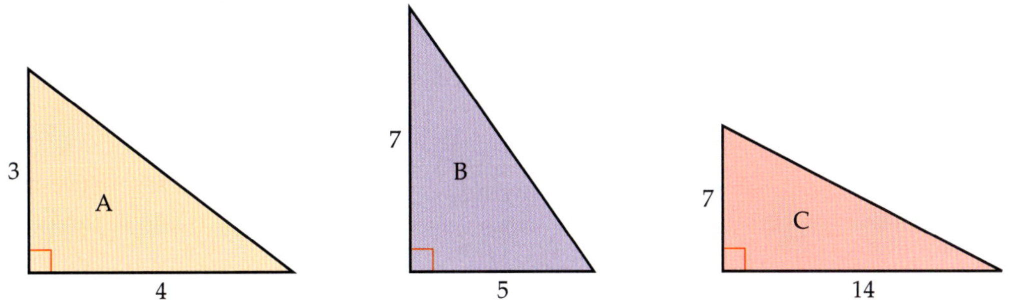

You will learn more about this in Chapter 7 when you study Pythagoras' theorem.

Other roots

It is possible to find roots of numbers other than square roots.
For example, $2 \times 2 \times 2 = 8$ can be written as $2^3 = 8$. We say '2 cubed equals 8'. The **cube root** of 8 is 2. Again, you can think of this as finding the length of the side of a cube if you know the volume of the cube.
To show that we are working with a cube root, we write a small 3 in front of the root sign.

So $\sqrt[3]{8} = 2$

This is read as 'the third root of 8 is 2' or 'the cube root of 8 is 2'.

Because we can write numbers to any index, it follows that mathematically, we must be able to find any root:
so $5^4 = 5 \times 5 \times 5 \times 5 = 625$ and $\sqrt[4]{625} = 5$.

This is read as "the fourth root of 625 is 5".

On your calculator, the key $\sqrt[x]{y}$ is used to find roots other than squared or cubed roots.

The method you will use on your calculator will depend on the make and model you are using. These are the two most common methods but you should refer to your calculator manual to make sure you know how to do this.

To evaluate $\sqrt[4]{625}$:

Method 1:

Press 4 (the root)

Press $\sqrt[x]{y}$

Press 6 2 5
(the number you are finding the root of)

Press =

Method 2:

Press 6 2 5

Press $\sqrt[x]{y}$

Press 4

Press =

Exercise

1 In the expressions below, there are missing numbers. Complete the expressions.

a) $2^4 = 16$, so $\sqrt[4]{16} = \square$

b) $5^3 = \square$, so $\sqrt[3]{125} = 5$

c) $10^{\square} = 100\,000$, so $\sqrt[5]{100\,000} = 10$

d) $3^5 = 243$, so $\sqrt[\square]{243} = 3$

e) $4^3 = 64$, so $\sqrt[3]{\square} = 4$

2 Use your calculator to evaluate the following:

a) $\sqrt[3]{216}$ b) $\sqrt[4]{2\,401}$ c) $\sqrt[8]{256}$ d) $\sqrt[5]{3\,125}$ e) $\sqrt[6]{729}$

Key words
Scientific notation
Standard form

Unit 2.4 Scientific notation

In this unit you will:

- write and work with numbers in scientific notation (standard form).

The range of numbers we work with in mathematics is very large. For example, the population of the world in 2005 was estimated to be 6.4 billion, whilst the diameter of atoms that scientists are working with can be less than one-millionth of a centimetre. When we have to write or type these numbers out using numerals, they take up space, it takes a lot of time and it is easy to make mistakes – plus, calculators can only display a certain number of digits – for bigger or smaller numbers they have to express them differently.

For example, if we enter 6 000 000 $\times$ 12 000 on a calculator, the display shows 7.2^{10}.
This means the answer is 7.2×10^{10} or 72 000 000 000.
7.2×10^{10} is a method of writing numbers called **scientific notation**, or **standard form**.

When a number is written in scientific notation, two parts are multiplied together:

- a number between 1 and 10 (including 1, but not including 10)
- a power of 10.

So, we could write the numbers 7 000 and 8 600 in scientific notation like this:

7 000
7×10^3
number from 1–10 ↗ ↖ power

8 600
8.6×10^3
number from 1–10 ↗ ↖ power

This means that 7 000 is 7 groups of 10^3 or 7 groups of 1 000 and 8 600 is 8.6 groups of 10^3. The power (index) tells you how many times you must multiply the number by ten.

"The population of Trinidad and Tobago is 1.3×10^6"

The population of Trinidad and Tobago is written here in standard form or scientific notation.

$1.3 \times 10^6 = 1\,300\,000$

Therefore, we can also say that the population of Trinidad and Tobago is 1 300 000.

Very small numbers can also be written in scientific notation using negative powers. For example, it takes a laser printer 2.5×10^{-5} seconds to print a single letter on a page. In this case, the negative power (index) tells you by what you should divide to find the value of the number (work from the decimal point). In this case,
$2.5 \div 10\,000 = 0.000\,025$ seconds.

Worked examples

1. Express the following numbers in standard form.

   a. 1600   b. 0.56   c. 73 400   d. 0.0089

2. Express these numbers in ordinary form.

   a. 7.2×10^4 b. 4.91×10^{-2}

1. a. 1600 — Shift the digits 3 places to the right to obtain 1.6.

   $1600 = 1.6 \times 10^3$

   b. 0.56 — Shift the digits 1 place to the left to obtain 5.6.

   $0.56 = 5.6 \times 10^{-1}$

   c. 73 400 — Shift the digits 4 places to the right to obtain 7.34.

   $73\,400 = 7.34 \times 10^4$

   d. 0.0089 — Shift the digits 3 places to left to obtain 8.9.

   $0.0089 = 8.9 \times 10^{-3}$

2. a. $7.2 \times 10^4 = 7.2 \times 10\,000 = 72\,000$ b. 0.0491

Exercise

1. The number 3 600 000 can be written as 3.6 million. Write these numbers in millions.

   a) 60 000 000   b) 9 900 000   c) 12 890 000
   d) 1 342 000   e) 500 000   f) 234 000

2. Write as powers of 10:

   a) 1 million   b) 1 hundred   c) 10 thousand
   d) 100 000   e) ten.

3. Write in scientific notation:

   a) 49   b) 500   c) 1 800
   d) 14.2   e) 165 000   f) 0.7
   g) 0.0095   h) 0.04   i) 0.000 832
   j) 0.000 076   k) six billion   l) three million.

4. Express the following numbers in ordinary form.

   a) 1.2×10^3 b) 3×10^2 c) 8×10^5
   d) 2.59×10^6 e) 9.1×10^4 f) 5×10^{-1}
   g) 4.73×10^{-4} h) 3.6×10^{-2} i) 1.8×10^{-3}
   j) 2.07×10^{-5}

5 The table shows the estimated population of some countries in 2003. Write each population in standard form.

Country	Afghanistan	The Bahamas	Barbados	Canada	Cuba
Population	29 000 000	297 000	277 000	32 000 000	11 000 000
Country	Dominica	Fiji	Iceland	Japan	New Zealand
Population	69 600	868 500	281 000	127 000 000	3 950 000

6 The mass of the Earth is approximately 5.98×10^{24} kg and the mass of the Sun is approximately 1.97×10^{30} kg.

a) Write each mass in full.

b) Write your own mass in scientific notation.

c) About how many times heavier is the Earth than you?

7 Give three examples of jobs in which people might use scientific notation regularly. For each one, say how you think they might use it.

Investigation

8 In astronomy, scientists have to deal with vast distances between planets and other bodies in the universe. To talk about these distances, they use a unit called a light year.

Find out what a light year is and how it is calculated.

Draw a diagram of the Solar System.

Name the planets and give their approximate distance from the Sun in light years.

Unit 2.5 Estimation

In this unit you will:

- estimate by rounding to a number
- estimate by rounding to a number of decimal places
- estimate by rounding to a number of significant figures
- apply estimation techniques to real-life situations.

Key words
Estimate
Decimal places
Significant figures
Rounding
Approximate

In many instances, an estimation or approximate value is used instead of the actual value. We **estimate** the population of a country, the attendance at a concert.

The attendance at a carnival fete was 15 327, which represents the exact number of tickets sold. However, the headline in the daily paper reads:

15 000 attend Machel Montano fete

15 327 has been **rounded** to the nearest thousand. We can round to the nearest ten, hundred, thousand, an so on ….

15 327 rounded to the nearest ten is 15 330.

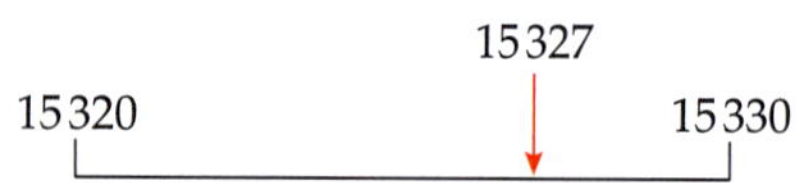

15 327 lies between 15 320 and 15 330 but is closer to 15 330. It is rounded up.

15 327 rounded to the nearest hundred is 15 300.

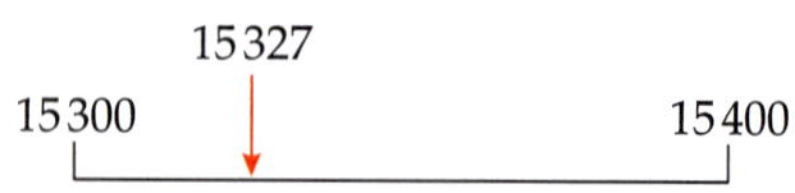

15 327 lies between 15 300 and 15 400 but is closer to 15300. It is rounded down.

15 327 rounded to the nearest thousand is 15 000.

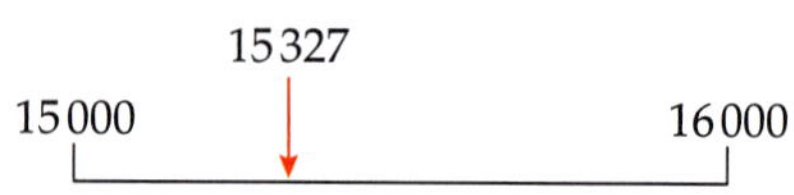

15 327 lies between 15 000 and 16 000 but is closer to 15 000. It is rounded down.

Decimal numbers can also be rounded by specifying the number of **decimal places**. In the number 23.568, the digits 5, 6 and 8 are all less than 1 and are called decimal values.

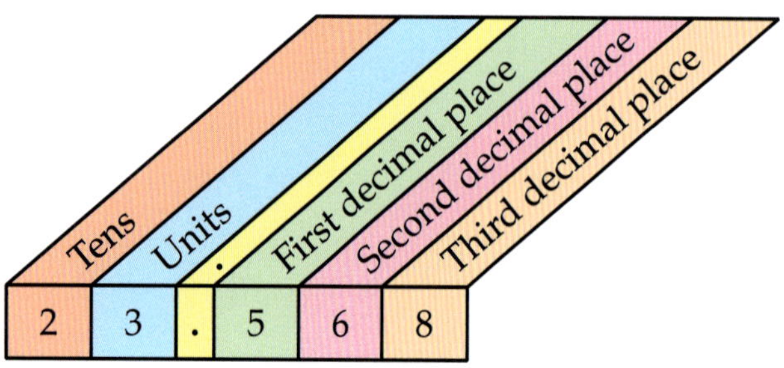

23.568 rounded to 2 decimal places is 23.57
23.568 rounded to 1 decimal place is 23.6
23.568 rounded to a whole number is 24

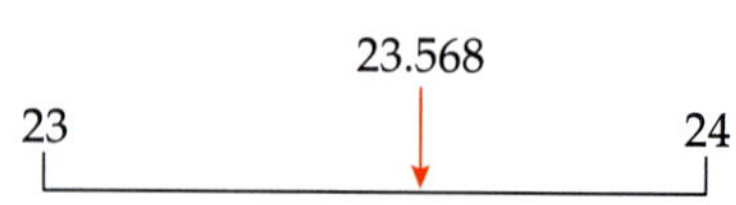

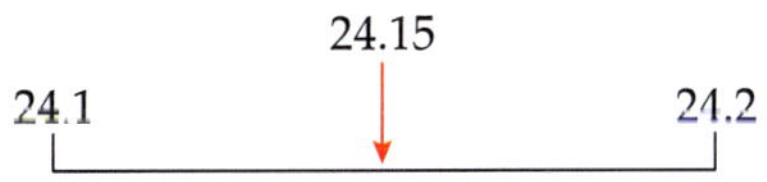

24.15 lies exactly midway between 24.1 and 24.2. When the place value is 5 or more we always round up, so 24.15 to 1 decimal place is 24.2.

You may be asked to round numbers to a prescribed number of **significant figures**. Significant means important. When estimating the number of people at a cricket match, you do not need to say there are exactly 12 343 people there. You can give your answer to two significant figures (in this case, 12 000 people).

Use these rules for dealing with significant figures:

- All non-zero digits are significant.
- Zeros between two non-zero digits are significant. For example, 103.24 has **five** significant figures (1, 0, 3, 2 and 4).

  To round to n significant figures:

  1) Start with the leftmost non-zero.
  2) Keep n digits and replace the rest with zeros.
  3) Round up if necessary.

  Rounding to two significant figures (2 s.f.):

43 200	becomes	43 000
19.1	becomes	19
19.9	becomes	20
0.0253	becomes	0.025
0.00154	becomes	0.0015

Leading zeros are *not* significant, so start counting at the 2 in 0.0253.

Worked examples

1. Round the amount $35 467 to the nearest

a. $10 **b.** $100 **c.** $1000

2. Round the numbers 4.4315, 7.3962 to

a. 3 decimal places **b.** 2 decimal places

3. Correct 642.803 to

a. 1 s.f. **b.** 2 s.f. **c.** 3 s.f. **d.** 4 s.f. **e.** 5 s.f.

1. a.	$35 470	$35 467 lies between $35 460 and $35 470 but is closer to $35 470
b.	$35 500	$35 467 lies between $35 400 and $35 500 but is closer to $35 500
c.	$35 000	$35 467 lies between $35 000 and $36 000 but is closer to $35 000
2. a.	4.4315	
	4.432	4315 lies exactly midway between 4.431 and 4.432 and is therefore rounded up
	4.43	4.4315 lies between 4.43 and 4.44 but is closer to 4.43
b.	7.3962	
	7.396	7.3962 lies between 7.396 and 7.397 but is closer to 7.396
	7.40	7.3962 lies between 7.39 and 7.40 but is closer to 7.40

3. a. 600 — 0 replaces the other significant digits to maintain the place value

b. 640

c. 643

d. 642.8

e. 642.80 — 0 is only significant when it lies between 2 other non-zero significant digits

Exercise

1. Identify the number of decimal places (d.p.) given in each number

   a) 26.3 b) 35.41 c) 0.00862 d) 7.00063 e) 5976

2. Identify the number of significant figures (s.f.) given in each number.

   a) 196 b) 98.70 c) 276.11 d) 0.134 e) 0.0076 f) 0.017009

Trailing zeros in numbers with a decimal point are significant.

3. Round the number 4.5734526 to

   a) 2 d.p. b) 5 d.p. c) 4 d.p. d) 6 d.p. e) 1 d.p.

4. Round the number 0.043827 to

   a) 2 d.p. b) 6 d.p. c) 4 d.p. d) 3 d.p. e) 1 d.p.

5. Round the following numbers to 2 decimal places.

   a) 126.34526 b) 31.4253 c) 0.3217 d) 0.04651

6. Round the following numbers to 3 decimal places.

   a) 6.51445 b) 0.394137 c) 231.14553 d) 0.0028

7. Correct the number 12623.372 to

   a) 1 s.f. b) 2 s.f. c) 3 s.f. d) 6 s.f. e) 7 s.f.

8. Correct the number 0.043827 to

   a) 6 s.f. b) 5 s.f. c) 4 s.f. d) 3 s.f. e) 2 s.f. f) 1 s.f.

9. Represent the result to 1 decimal place

   a) 1.96×7 b) 21.4×6 c) 0.45×0.12

10. Represent the result to 1 decimal place

   a) 0.45×0.12 b) 2.57×1.9 c) $\frac{0.971}{3.2}$

Project

Use the following representations: $25 = \frac{100}{4}$, $50 = \frac{100}{2}$, $99 = 100 - 1$; evaluate the following

900×25 91×50 53×99

How can you express 125 and 97 in a similar manner to evaluate 88×125 and 88×97?

Key words
Roots
Square root
Fractional indices
Negatve indices

Unit 2.6 Roots, fractional and negative indices

In this unit you will:

- differentiate between powers and roots
- use a calculator to find powers and roots of numbers
- perform calculations that include fractional and negative indices.

Remember that $\sqrt{25}$ is read as 'the **square root** of 25'. The square root is that number which when multiplied by itself gives 25.
Therefore, $\sqrt{25} = 5$.

$\sqrt[5]{32} = 2$ is read as 'the fifth **root** of 32 = 2'. — Because $2^5 = 32$.

$\sqrt[n]{m}$ is read 'the nth root of m'.

In Question 8 of the previous Unit, you have seen that:

a) $9^{\frac{1}{2}} \times 9^{\frac{1}{2}} = 9^1 = 9$ b) $8^{\frac{1}{3}} \times 8^{\frac{1}{3}} \times 8^{\frac{1}{3}} = 8$ c) $h^{\frac{1}{4}} \times h^{\frac{1}{4}} \times h^{\frac{1}{4}} \times h^{\frac{1}{4}} = h$

We add indices when multiplying same base.

a) $\sqrt{9^{\frac{1}{2}} \times 9^{\frac{1}{2}}} = \sqrt{9} = 9^{\frac{1}{2}}$ — For square roots, it is not necessary to put the 2 in $\sqrt[2]{\ }$.

b) $\sqrt[3]{8^{\frac{1}{3}} \times 8^{\frac{1}{3}} \times 8^{\frac{1}{3}}} = \sqrt[3]{8} = 8^{\frac{1}{3}}$

c) $\sqrt[4]{h^{\frac{1}{4}} \times h^{\frac{1}{4}} \times h^{\frac{1}{4}} \times h^{\frac{1}{4}}} = \sqrt[4]{h} = h^{\frac{1}{4}}$

So $m^{\frac{1}{n}} = \sqrt[n]{m}$ — The denominator of the fraction tells us the root.

We also know:

1. $7^5 \div 7^7 = \dfrac{7 \times 7 \times 7 \times 7 \times 7}{7 \times 7 \times 7 \times 7 \times 7 \times 7 \times 7} = \dfrac{1}{7^2}$ — Expand and simplify.

   $= 7^{5-7} = 7^{-2}$ — Subtract indices when dividing.

   Therefore $m^{-n} = \dfrac{1}{m^n}$ — Divide and use a positive index

2. $6^2 \div 6^2 = 6^{2-2} = 6^0$ but $6^2 \div 6^2$ is also $\dfrac{6^2}{6^2} = 1$, therefore $6^0 = 1$.

So $p^0 = 1$ — Anything to the zero power = 1 except 0^0, which is 0.

Worked examples

1. What are the two possible values of $\sqrt{64}$?
2. Find the value of $\sqrt[3]{-27}$.
3. Use your calculator to find
   a. $\sqrt{36}$ b. $\sqrt{3.60}$

1. 8 and -8

$8 \times 8 = 64$ and $(-8) \times (-8) = 64$

2. $\sqrt[3]{-27} = -3$

$(-3) \times (-3) \times (-3) = -27$

$-3 \times -3 = +9$ and $+9 \times -3 = -27$

Check by using the calculator to do $(-3)^3$.

3. a. $\sqrt{36} = \pm 6$
   b. $\sqrt{3.60} = \pm 1.897$

Remember -6×-6 is also 36.
The calculator may not give the negative result.

Before you go to the calculator, take a guess. $1^2 = 1$ and $2^2 = 4$, $1 < 3.60 < 4$ so the result is between 1 and 2. Learn to use your calculator to find the powers and roots of numbers.

The rules we learnt before are also used with **fractional** and **negative indices**.

For example:

- $5^{\frac{1}{2}} \times 5^{\frac{1}{4}} = 5^{\frac{3}{4}}$ — Using a positive index.
- $(a^{-\frac{2}{3}})^6 = a^{-\frac{12}{3}} a^{-4} = \dfrac{1}{a^4}$ — Invert to get a positive index.
- $\dfrac{h^{-8}k^{-2}}{h^{-5}} = \dfrac{h^5}{h^8k^2} = \dfrac{h^{5-8}}{k^2} = \dfrac{h^{-3}}{k^2} = \dfrac{1}{h^3k^2}$

Write as positive indices.

Subtract indices when dividing.

Exercise

1 Find the value of the following.

a) $\sqrt{16}$ b) $\sqrt{1600}$ c) $\sqrt{4}$ d) $\sqrt{400}$ e) $\sqrt{169}$

Remember there are two answers for each of these.

2 Find the value of the following.

a) $\sqrt[3]{8}$ b) $\sqrt[3]{125}$ c) $\sqrt[3]{64}$ d) $\sqrt[3]{216}$ e) $\sqrt[3]{512}$

3 Use your calculator to find the value of

a) $\sqrt{40}$ b) $\sqrt{160}$ c) $\sqrt{6.4}$ d) $\sqrt{1.69}$ e) $\sqrt{16.9}$

4 Rewrite the following using fractional indices.

a) $\sqrt[5]{243}$ b) $\sqrt[10]{b}$ c) $\sqrt{g}$ d) $\sqrt[3]{h}$ e) $\sqrt[6]{100}$

5 Rewrite the following using positive indices.

a) a^{-5} b) $p^{-3} \times q^3 \div p^{-5}$ c) $(c^{-6})^{\frac{1}{2}}$ d) $49^{-\frac{1}{2}}$ e) $\left(\dfrac{b^{-3}}{c^{-4}}\right)$

Remember $4 \div \frac{1}{2} = 4 \times \dfrac{2}{1}$

6 Simplify the following, giving your answers with positive indices.

a) $(d^6)^{\frac{1}{2}} \times d^2$ b) $f^{-7} \times f^{-6} \div f^9$ c) $(y^{\frac{2}{3}})^{-12}$

d) $\dfrac{p^3q^{-3}}{p^3q^{-6}}$ e) $\dfrac{r^8s^{-4}}{r^3s^{-5}}$ f) $\dfrac{8t^{-9}u^4}{4^{-2}t^{-9}u^{-3}}$

Investigation

a) Find two ways of simplifying the following.

i) $\sqrt{144 \times 81}$ ii) $\sqrt{\frac{36}{4}}$ iii) $\sqrt{25 \times 64}$ iv) $\sqrt{\frac{64}{16}}$

b) Write a general statement about what you observed.

c) Try other examples to test if your statement works all the time.

Looking back

1 Simplify:

a) $x^4 \times x^7$ b) $y^2 \times y^3$ c) $4x^2 \times x^3$

d) $a^3 \times 2a^3$ e) $2d^4 \times 3d^2 \times 5$ f) $3s^2 \times 3s^4 \times 5s$

g) $3x^2 \times x^3y^4$ h) $3a^2b^3 \times 3a^2 \times 4b^2$ i) $xy \times xyz^2$

j) $2x^7y \times 3x^2z^3 \times 4z^2$ k) $3m^2 \times 7n^2$ l) $(x^2)^4$

m) $(3a^2)^3$ n) $(4x^4)^3$ o) $2(m^3)^3$.

2 Simplify:

a) $\dfrac{x^6}{x^4}$ b) $\dfrac{8a^4}{4a^3}$ c) $\dfrac{15x^2y^3}{3xy}$

d) $\dfrac{a^8}{a}$ e) $\dfrac{12x^7}{3x^9}$

3 Find the value of m.

a) $a^m = 1$ b) $x^m = \sqrt{x}$ c) $x^m = \dfrac{1}{x^3}$

d) $3^m = 27$

4 Simplify:

a) $(4x^0)^2$ b) $x^{-2} \div \dfrac{1}{x^4}$ c) $(\sqrt{mn})^3 \times m$

d) $(9x^2)^{\frac{1}{2}}$ e) $(8b)^{\frac{1}{3}}$.

Chapter 3

Measurement: circles and prisms

In this chapter you will review what you have already learnt about perimeter and area of shapes, including the circle and composite shapes. You will apply your skills to solve problems related to perimeter and area. In the second half of the chapter, you will work with 3-D shapes, calculating volume, capacity and surface area.

Key words
Perimeter
Circumference
Radius
Diameter

Unit 3.1 Perimeter and circumference

In this unit you will:

- determine the perimeter of different polygons
- find the circumference of circles.

Do you remember that the **perimeter** of a shape is the total distance around the shape? If the shape is a regular polygon, such as an equilateral triangle, a square or a pentagon, the perimeter is calculated by multiplying the length of one side, l, by the number of sides.

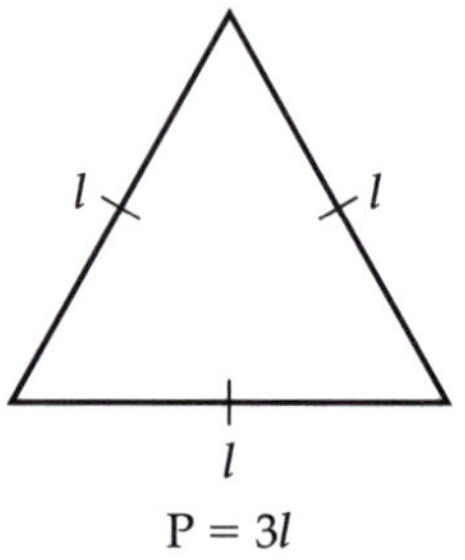

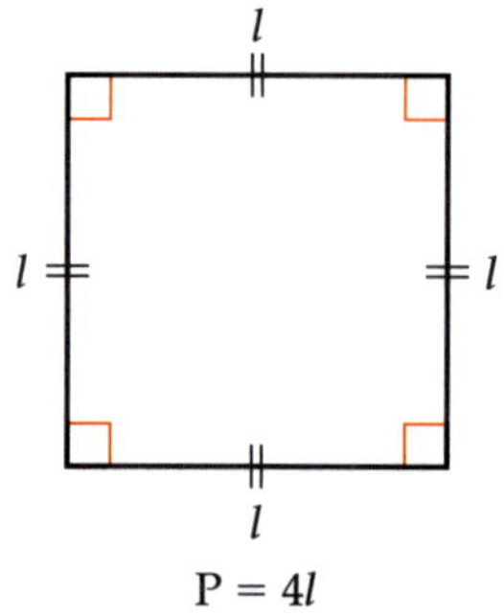

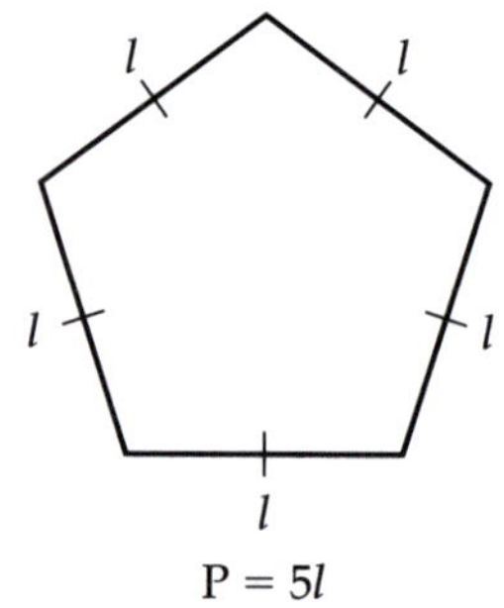

The distance around the edge of a circle is called the **circumference**. We use the irrational number *pi* when we are working with circles. *Pi* or π is a constant ratio of $\frac{22}{7}$. The circumference of a circle C is 2π times the **radius** which is written as $C = 2\pi r$. The **diameter** is twice the length of any radius, so the formula may also be written as $C = \pi d$. π has an approximate value of 3.14.

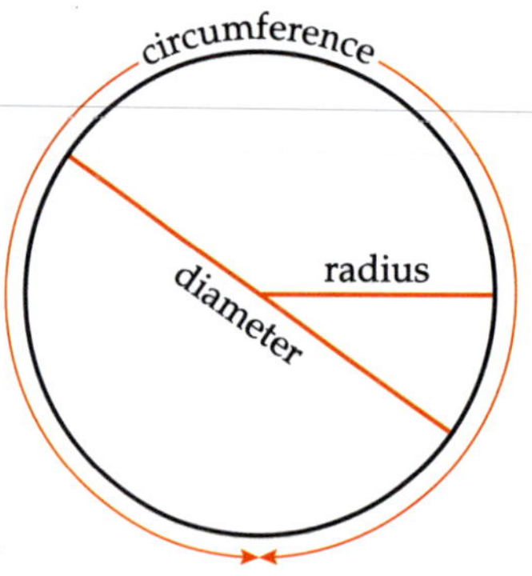

Worked example

Calculate the perimeter of the shapes below.

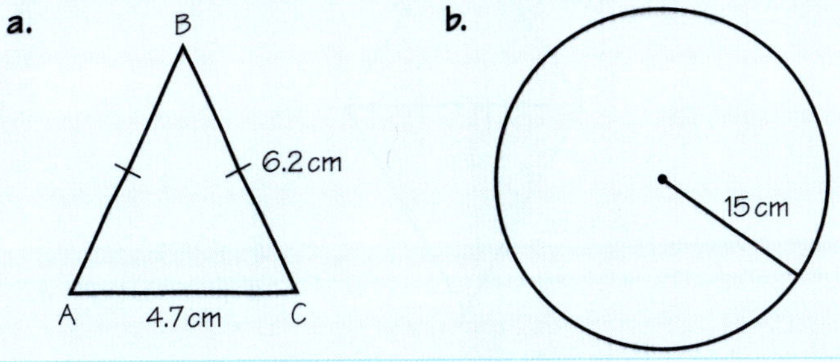

a. The triangle is isosceles, so AB = BC = 6.2 cm and AC = 4.7 cm.

The perimeter is: $6.2 + 6.2 + 4.7 = 17.1$ cm.

b. The circumference is: $2\pi r = 2 \times 3.14 \times 15 = 94.2$ cm.

Exercise

1 Calculate the circumference of the following circles, using $\pi = 3.14$. Give your answers correct to one decimal place.

a)

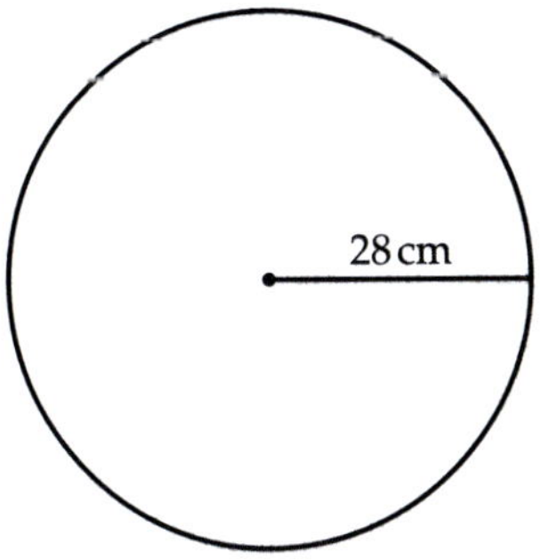

b)

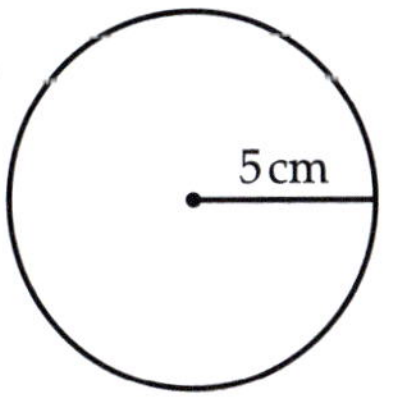

c)

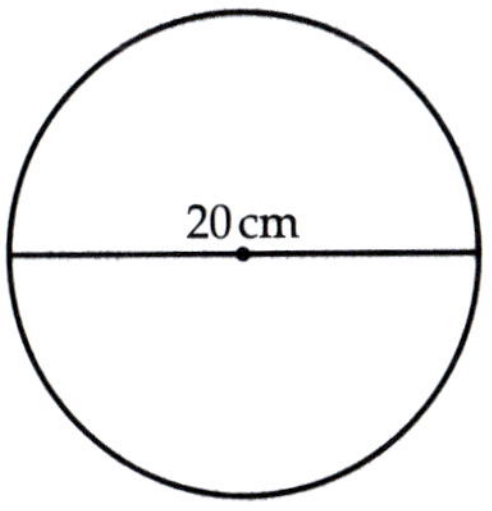

d)

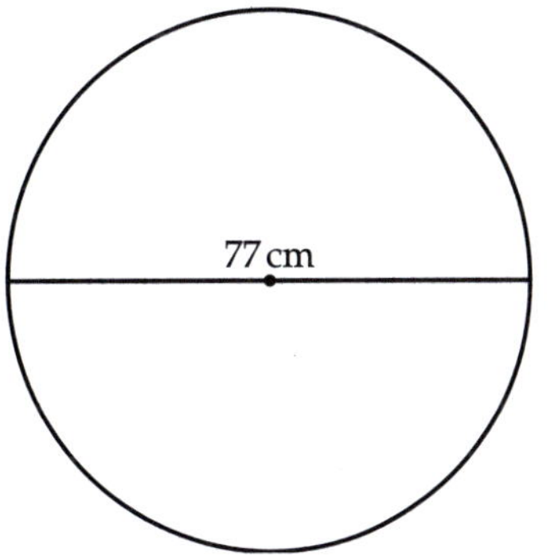

2 Work out the perimeter of each shape below.

a)

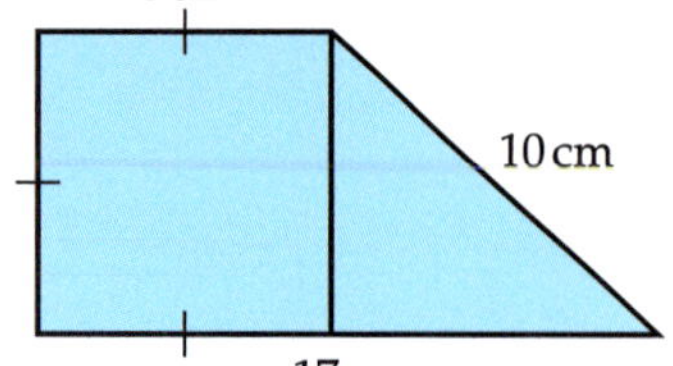

b)

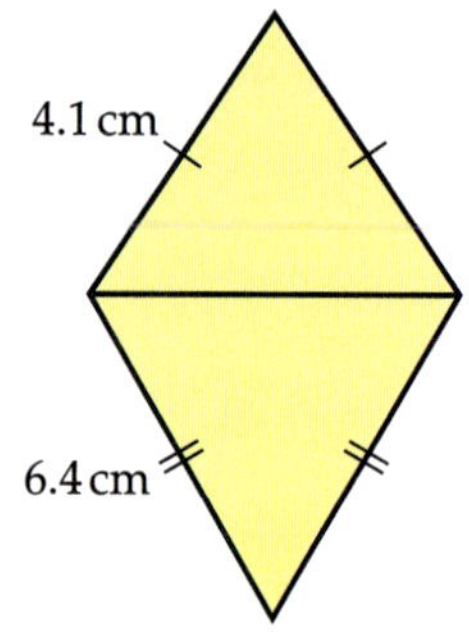

c)

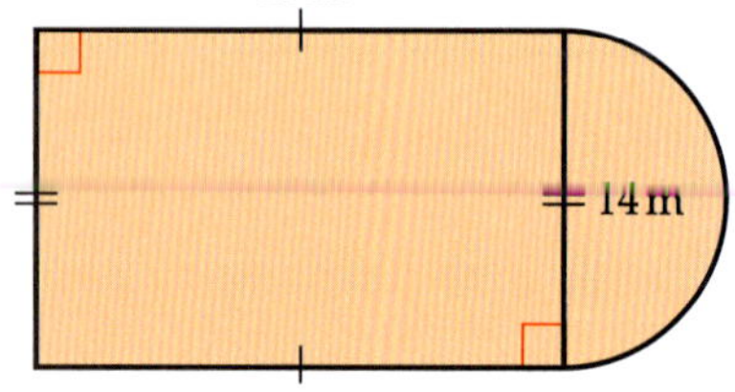

The perimeter of a semi-circle is half the circumference of a circle with the same radius.

d)

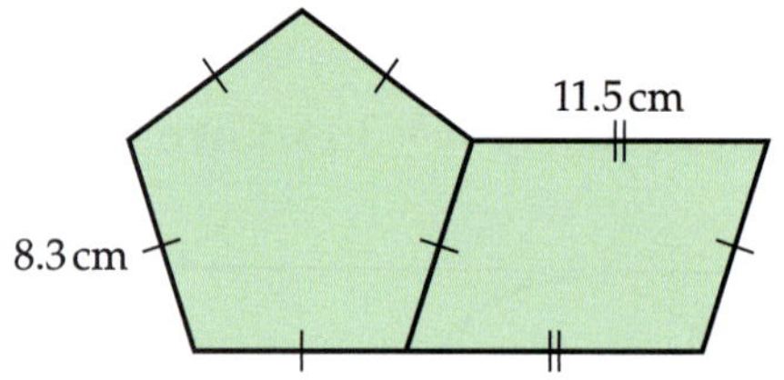

e)

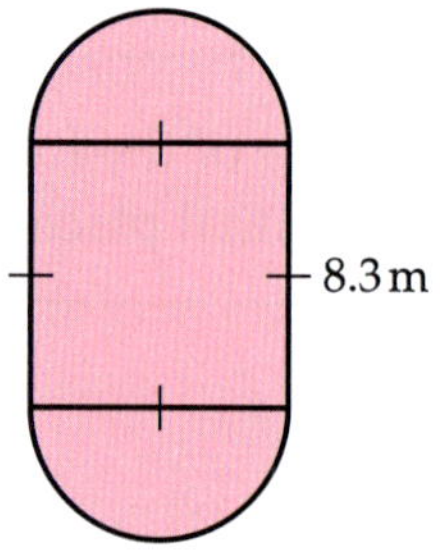

f)

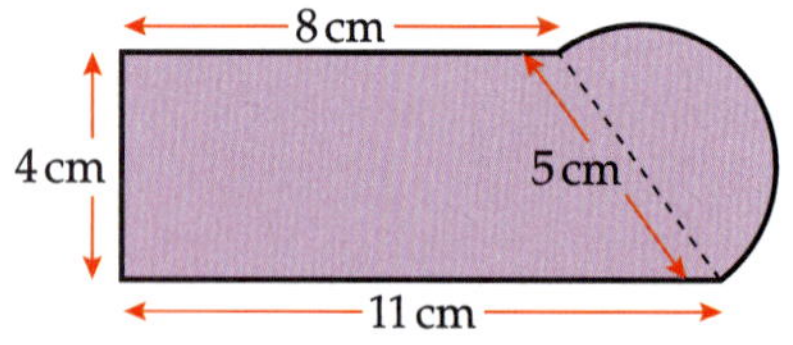

3 The diagram shows two lanes of a circular track.
Johnathan runs along the outer lane and Jeremy runs along the inner lane.

a) What distance did each boy run in one lap?

b) If Johnathan runs 10 laps and Jeremy runs 8 laps, who runs the further distance and by how many metres?

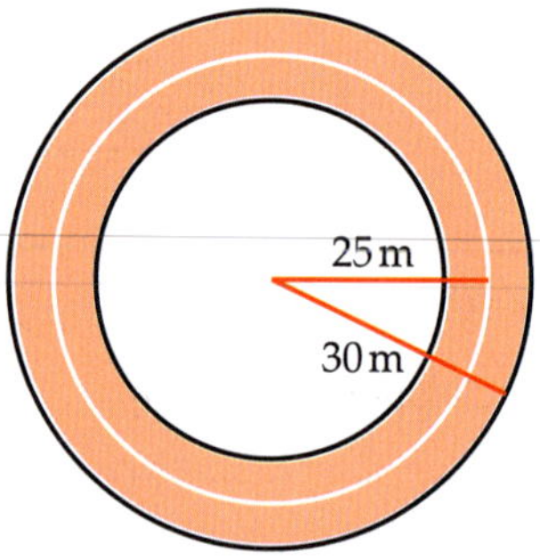

4 Aaron is making a birthday cake for his brother.
The cake is 20 cm in diameter.
When the cake is iced he will put a ribbon around the edge.
The ribbon needs to have 1.5 cm overlap.
What length of ribbon will Aaron need?

Key words
Arc
Sector
Radii
Central angle

Unit 3.2 Arc length

In this unit you will:

- use a formula to calculate the lengths of arcs of circles
- find the perimeter of shapes containing sectors of circles.

Any continuous part of the circumference of a circle is called an **arc**. The area between two radii and the arc between them is called a **sector**, and the angle between the two radii is called the **central angle**.

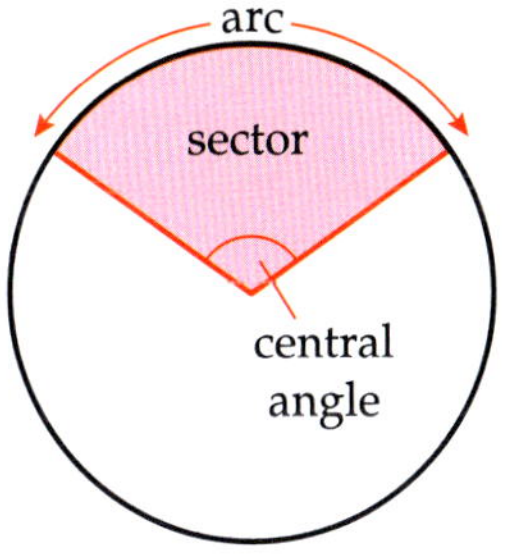

Radii is the plural of radius.

An arc is a fraction of the circumference of a circle. Since there are 360° in a full circle, the fraction is $\frac{\text{central angle}}{360°}$.

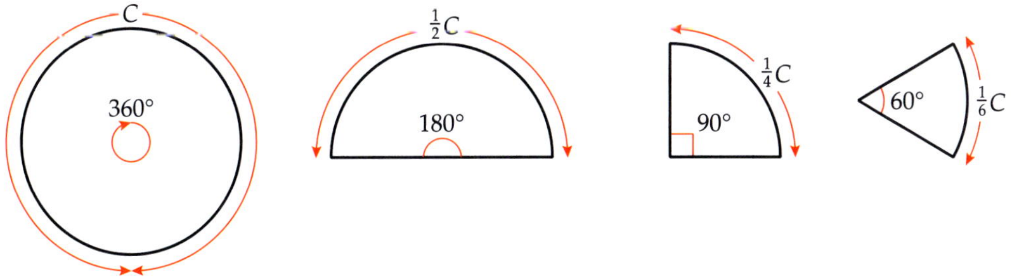

If the central angle is $x°$, then the length of the arc, L, is given by the formula $L = \frac{x}{360}C$.
Since the circumference of the circle, C, can be written as $2\pi r$, the formula may also be written as $L = \frac{x}{180}\pi r$.

Worked example

Calculate the length of an arc with central angle 50° in a circle of radius 10 cm.
Give your answer correct to one decimal place.

Length of arc $= \frac{x}{180}\pi r$

$= \frac{50}{180} \times 3.14 \times 10$

$= 8.7$ cm to one decimal place

Exercise

1 Calculate the length of each of the following arcs, using $\pi = 3.14$. Give your answers correct to one decimal place.

a)

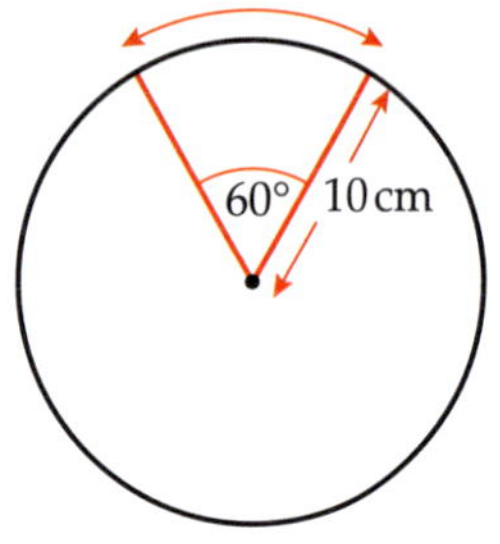

b)

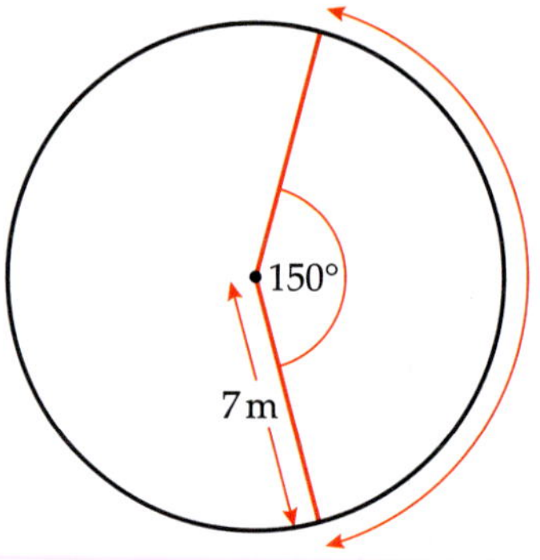

c)

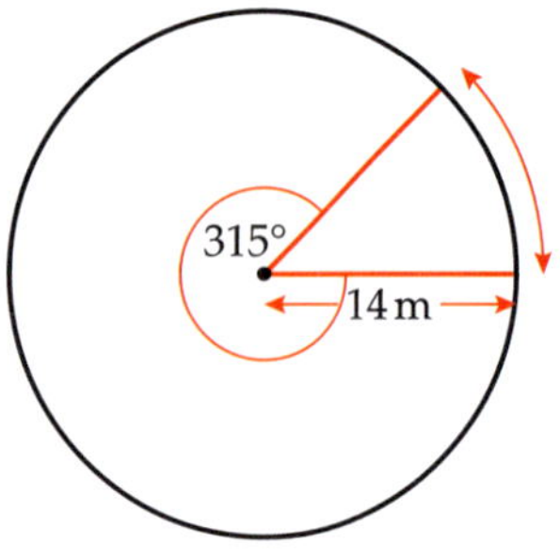

d)

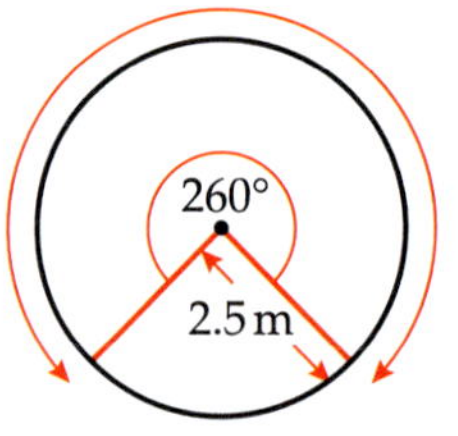

Make sure you use the correct angle in part **c)**.

2 Copy and complete these sentences:

a) The length of an arc with central angle □° is $\frac{2}{3}$ of the circumference.

b) The length of an arc with central angle 18° is □ of the circumference.

3 Calculate the perimeter of each of the shapes below. Give your answers correct to one decimal place.

a)

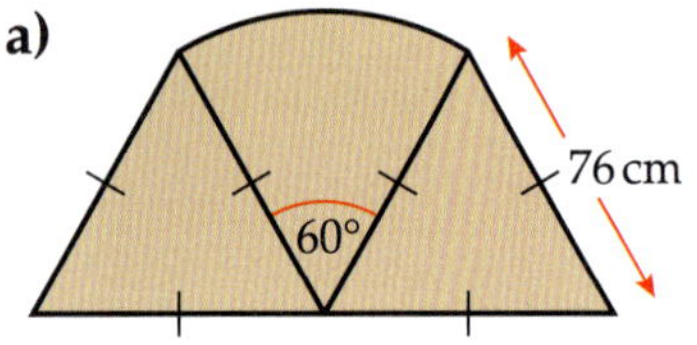

b)

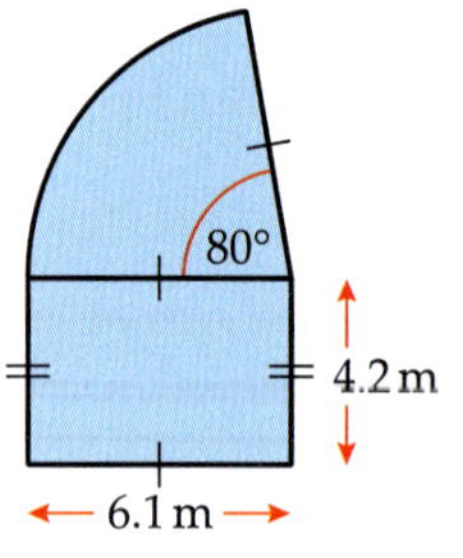

c)

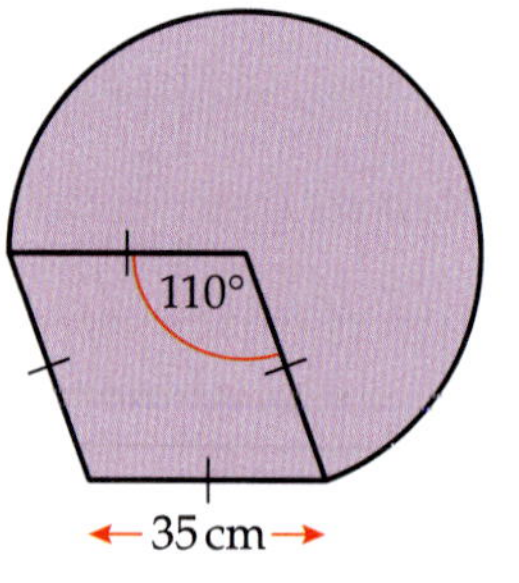

4 Calculate the length of an arc with central angle 200° on a circle of radius 9 m. Give your answer correct to one decimal place.

5 Find the central angle of an arc of length 10 mm on a circle of circumference 25 mm.

Key words
Area
Parallelogram
Trapezium
Kite

Unit 3.3 Area

In this unit you will:

- use formulae for calculating the area of polygons
- convert between different units in the metric system.

The formulae for calculating the **area** of some well-known polygons are given below.

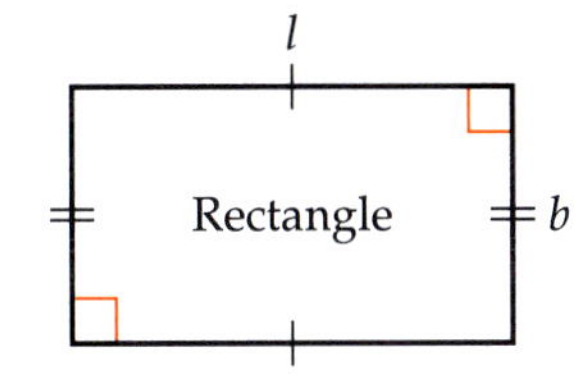

$A = bl$

Triangle

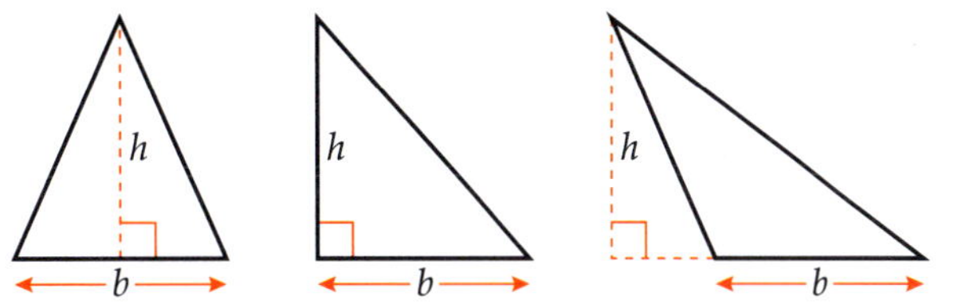

$A = \frac{bh}{2} = \frac{1}{2}bh$

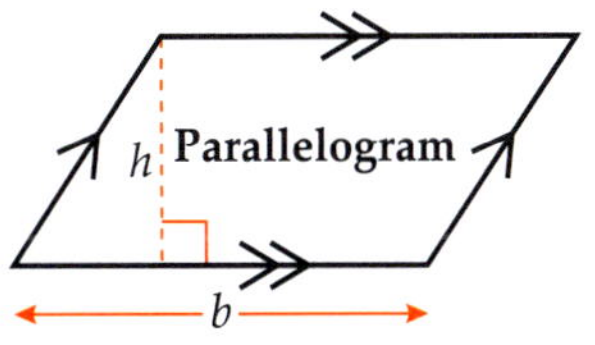

$A = bh$

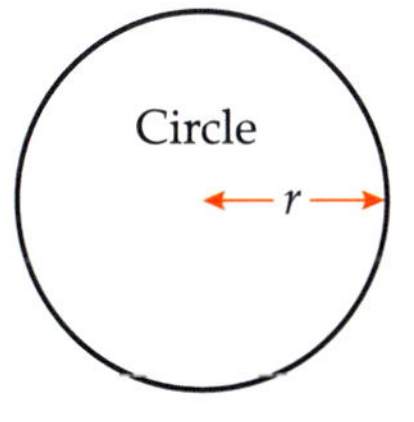

$A = \pi r^2$

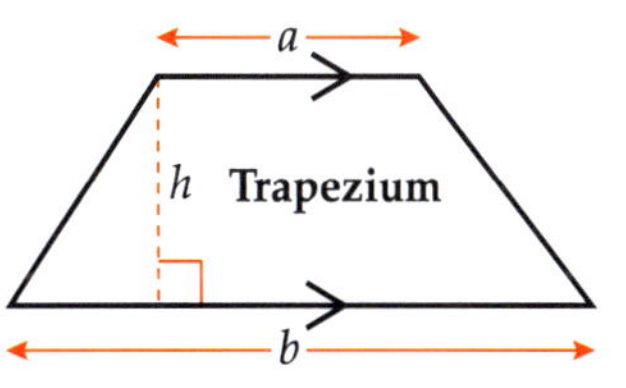

$A = \frac{h(a + b)}{2} = \frac{1}{2}(a + b)h$

The metric units of area are square metres (m^2), square centimetres (cm^2) and square millimetres (mm^2).

$1\ m^2 = 10\,000\ cm^2$ $\quad$ $1\ cm^2 = 100\ mm^2$

If you convert linear cm to mm you multiply by 10. However, if you convert square cm to square mm, you are in fact multiplying by 10×10 or 100. So, when you convert between square units, you have to double the decimal places you would normally multiply or divide by on the metric scale.

Consider a rectangle with a length of 3 cm and a width of 2 cm. The length may also be expressed as 30 mm and the width as 20 mm. Hence the area is either 6 square centimetres or 600 square millimetres.

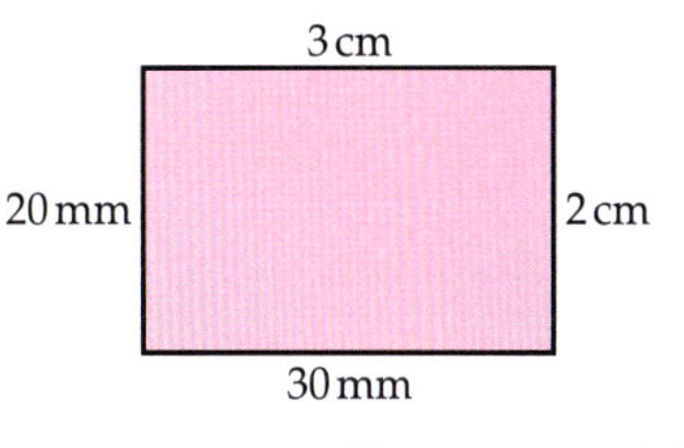

Worked examples

1. Copy and complete the following.
   a. $8\ cm^2 = \square\ mm^2$
   b. $4\ m^2 = \square\ cm^2$
   c. $340\ mm^2 = \square\ cm^2$
   d. $72\,500\ cm^2 = \square\ m^2$

a. $8\ cm^2 = 800\ mm^2$ — $8 \times 100 = 800$
b. $4\ m^2 = 40\,000\ cm^2$ — $4 \times 10\,000 = 40\,000$
c. $340\ mm^2 = 3.4\ cm^2$ — $\frac{340}{100} = 3.4$
d. $72\,500\ cm^2 = 7.25\ m^2$ — $\frac{72\,500}{10\,000} = 7.25$

2. Determine the area of this shape in

a. square metres

b. square centimetres.

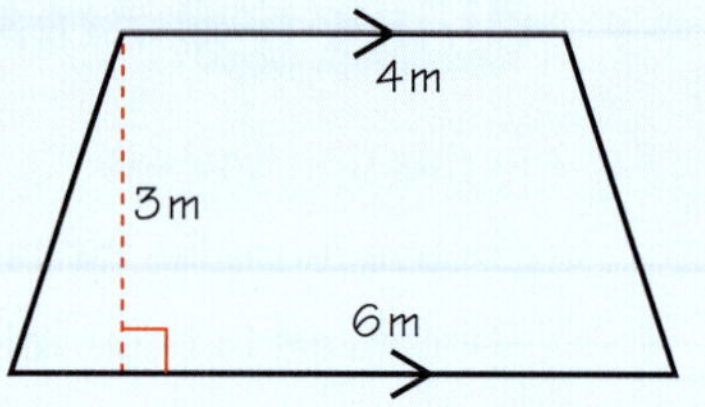

Area of the trapezium $= \frac{1}{2}(a + b)h$

$= \frac{1}{2}(4 + 6) \times 3$

$= 15\ \text{m}^2 = 15 \times 10\,000\ \text{cm}^2$

Or $\frac{1}{2}(400\ \text{cm} + 600\ \text{cm}) \times 300\ \text{cm}$

$= 150\,000\ \text{cm}^2$

Exercise

1 Convert these square centimetres to square metres.

a) $50\,000\ \text{cm}^2$ **b)** $250\,000\ \text{cm}^2$ **c)** $35\,700\ \text{cm}^2$ **d)** $8\,250\ \text{cm}^2$ **e)** $575\ \text{cm}^2$

2 Convert these square millimetres to square centimetres.

a) $500\ \text{mm}^2$ **b)** $2\,500\ \text{mm}^2$ **c)** $72\,500\ \text{mm}^2$ **d)** $8\,254\ \text{mm}^2$ **e)** $375\ \text{mm}^2$

3 Convert these square centimetres to square millimetres.

a) $5\ \text{cm}^2$ **b)** $65\ \text{cm}^2$ **c)** $2.75\ \text{cm}^2$

d) $6.7\ \text{cm}^2$ **e)** $0.22\ \text{cm}^2$ **f)** $0.3\ \text{cm}^2$

4 Convert these square metres to square centimetres.

a) $7\ \text{m}^2$ **b)** $75\ \text{m}^2$ **c)** $3.5\ \text{m}^2$ **d)** $0.5\ \text{m}^2$ **e)** $0.3475\ \text{m}^2$

5 A postage stamp measures 2.5 cm by 1.8 cm.

a) Convert the lengths to millimetres.

b) Work out the area of the stamp in

i) square centimetres

ii) square millimetres.

6 Calculate the areas of the following shapes.

a)

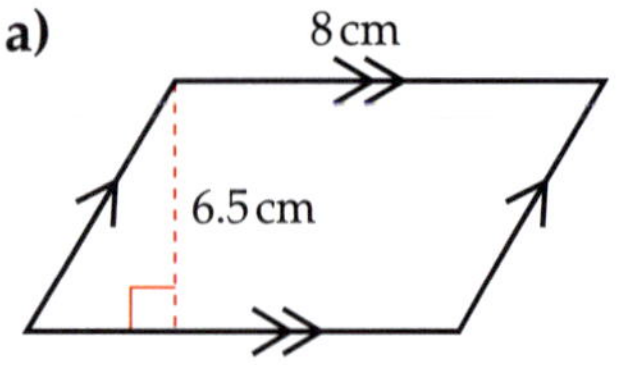

b)

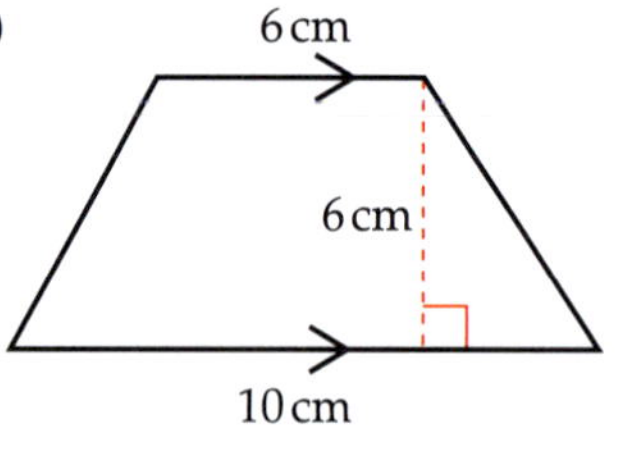

c)

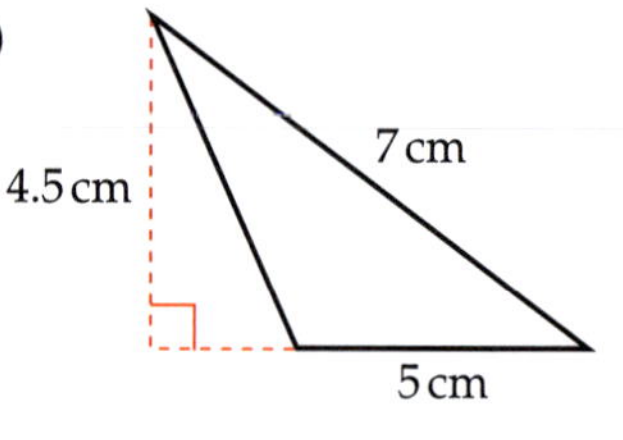

7 Calculate the area of these shapes. Show all your working.

a)

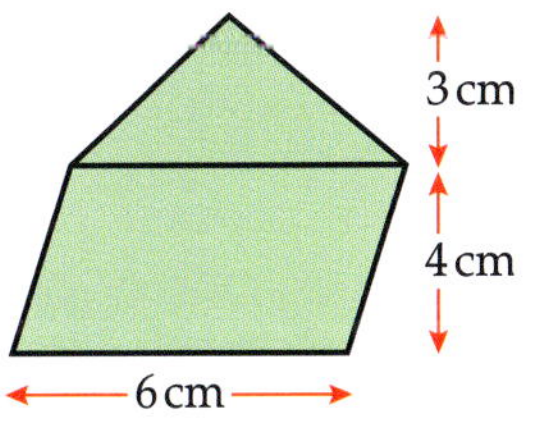

b)

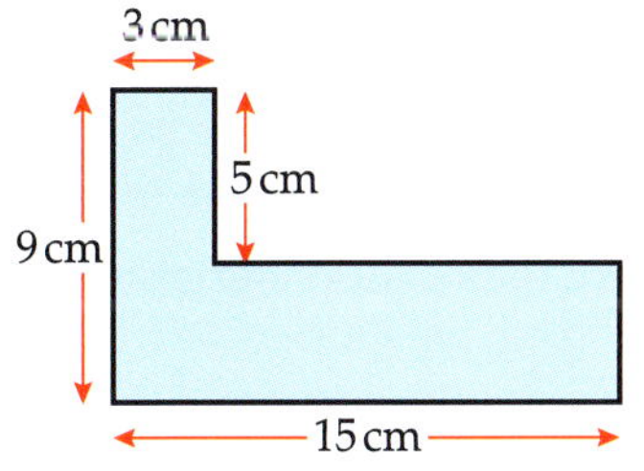

c)

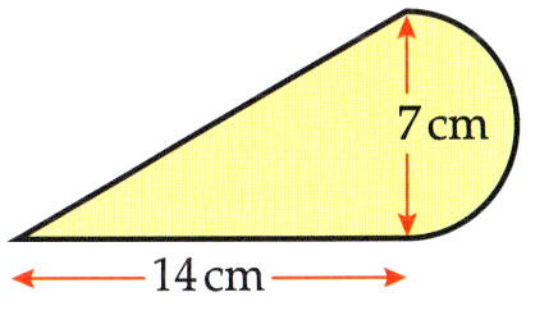

d)

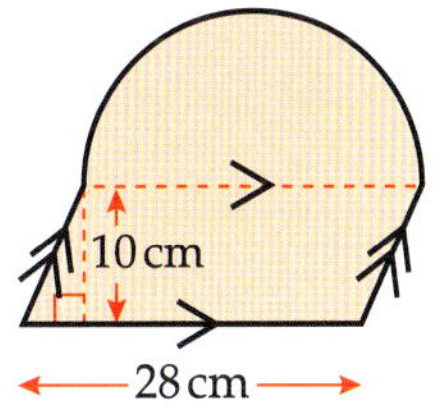

8 The piece of fabric below is used to make bandanas for a carnival band. If each bandana must be 50 cm by 40 cm, how many bandanas can be made from the fabric?

In this diagram 1 cm represents 80 cm

Scale: 1 cm = 80 cm

Investigation

9 Derive a formula for calculating the area of a **kite** where x and y are the diagonals of the kite.

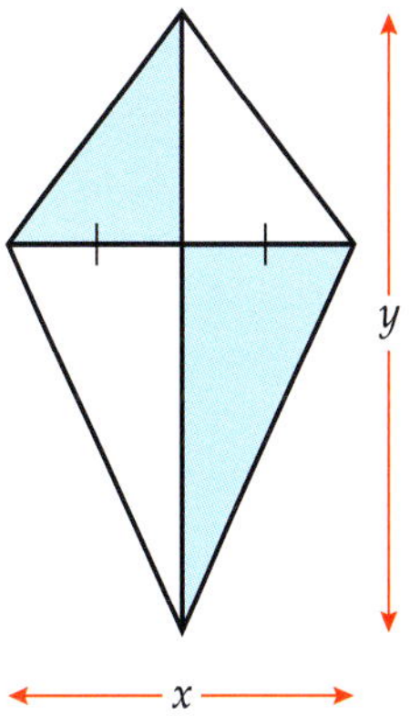

Unit 3.4 Sector area and compound shapes

In this unit you will:

- use a formula to calculate the areas of sectors of circles
- find the area of shapes containing sectors of circles.

Just as the length of an arc is a fraction of the circumference of a circle, the area of a sector is a fraction of the area of a circle. Since there are 360° in a full circle, the fraction is $\frac{\text{central angle}}{360°}$.

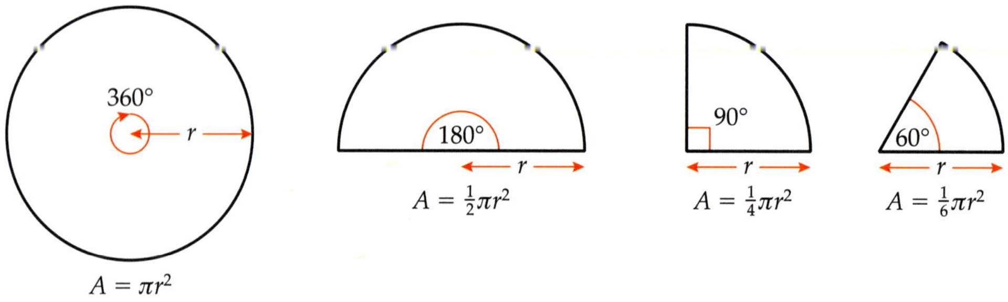

If the central angle is $x°$, then the area of the sector, A, is given by the formula $A = \frac{x}{360}\pi r^2$.

Worked examples

Calculate the area of each shape. Give your answers correct to one decimal place

a.

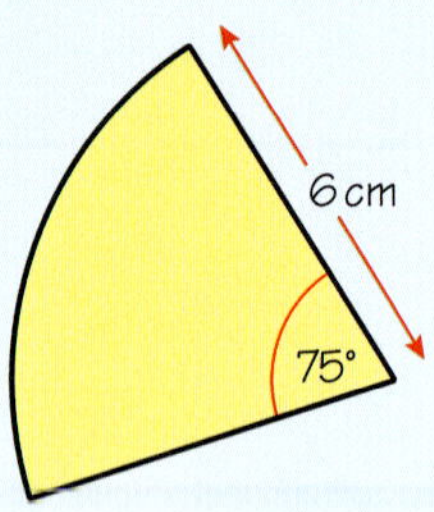

b.

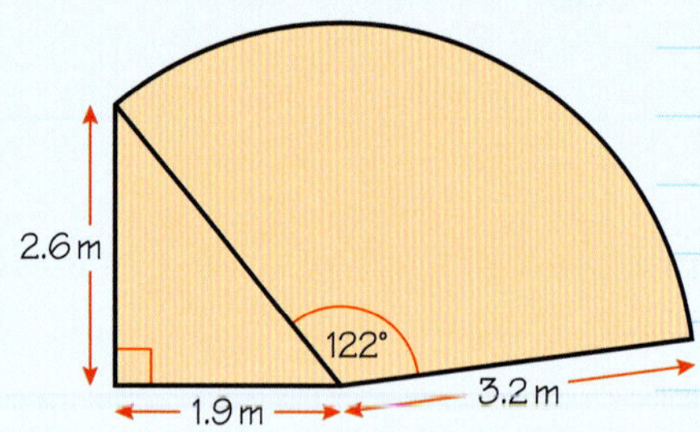

a. Area of sector $= \frac{x}{360}\pi r^2$

$= \frac{75}{360} \times 3.14 \times 6^2$

$= 23.6$ cm^2 to one decimal place

b. Area of triangle $= \frac{1}{2}bh$

$= \frac{1}{2} \times 1.9 \times 2.6$

$= 2.47$ m^2

$$\text{Area of sector} = \frac{x}{360}\pi r^2$$

$$= \frac{122}{360} \times 3.14 \times 3.2^2$$

$$= 10.896\ldots \text{ cm}^2$$

Don't round your answers in intermediate steps in a calculation. Use the full values, then round your final answer.

Total area = $2.47 + 10.896\ldots = 13.4$ cm^2 to one decimal place.

Exercise

1. Using $\pi = 3.14$, calculate the area of a sector with central angle 30° in a circle of radius 5 cm. Give your answer correct to one decimal place.

2. What fraction of each circle is shaded?

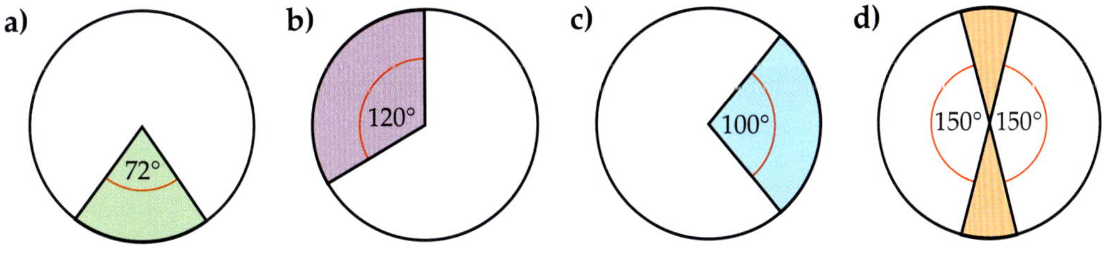

3. Calculate the area of each of the shaded sectors. Give your answers correct to one decimal place.

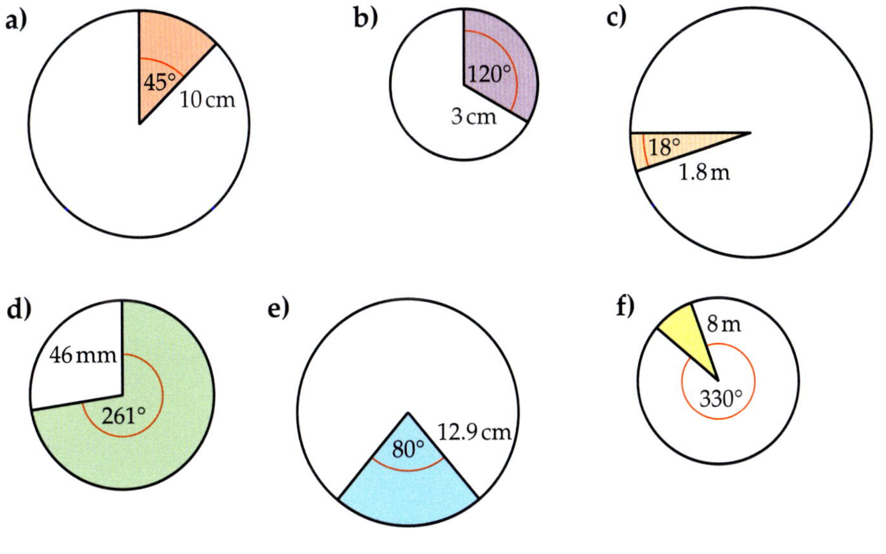

4. On a clear day the beam from Point Galera lighthouse can be seen from a distance of 22.5 km. It covers an angle of 175°. Calculate the area of the ocean which is covered by the lighthouse. Give your answer correct to one decimal place.

Unit 3.5 Volume of cuboids

In this unit you will:

- estimate and calculate the volume of a cuboid
- solve problems involving cuboids.

Key word
Volume
Cuboid
Cross-section
Formula
Prism

The **volume** of a 3-D shape is the space contained inside it. We measure volume in cubes. Examples of units of volume are cm^3 and m^3.

The volume of a litre carton of milk is 1 000 cm^3. The volume of a classroom is approximately 100 m^3.

You can think of the volume of a **cuboid** as cubes inside the shape.

Look at this cuboid:

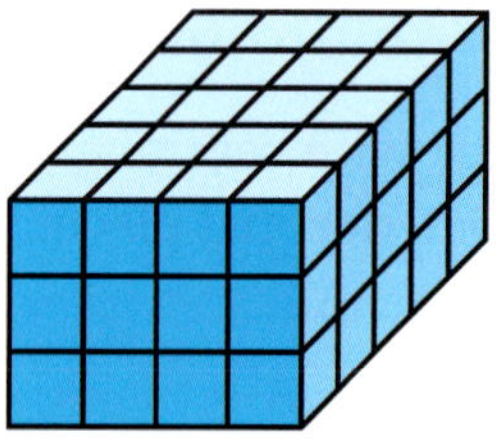

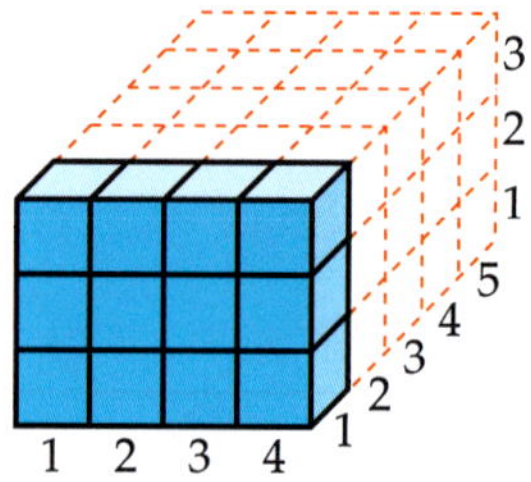

There are $4 \times 3 = 12$ cubes in each 'slice'.

The 'slice' is called the **cross-section** of the cuboid.

There are 5 'slices', each containing 12 cubes, so the total number of cubes is $5 \times 12 = 60$ cubes

The volume of the cuboid is 60 cubic units.

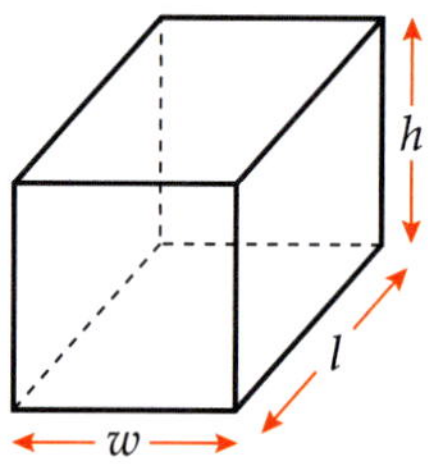

The **formula** for the volume of a cuboid of width w, length l and height h is $w \times l \times h$, written as wlh.

A cuboid is a type of **prism**. A prism has two identical faces and a uniform cross-section.

You can also think of the volume as the area of the cross section multiplied by the length.

Worked examples

1. Calculate the volume of this cuboid.

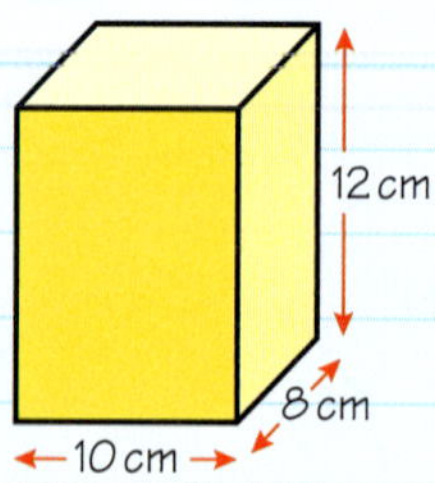

1. Volume $= w \times l \times h$
$= 10 \times 8 \times 12$
$= 960\ cm^3$

Replace w, l and h with the width, length and height of the cuboid.

2. Calculate the volume of this prism.

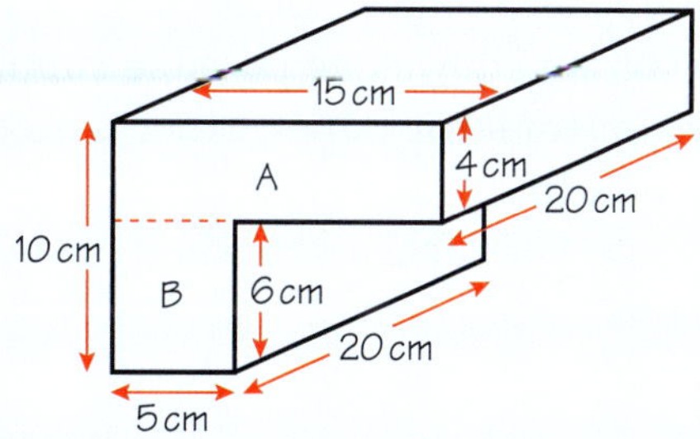

2. Volume of A $= 15 \times 20 \times 4$

$= 1200\ \text{cm}^3$

Volume of B $= 5 \times 6 \times 20$

$= 600\ \text{cm}^3$

Volume of prism $= 1200 + 600 = 1800\ \text{cm}^3$

The prism is made up of two cuboids. Calculate the volume of each cuboid and then add them together to find the total volume of the prism.

Exercise

1 Calculate the volume of each of these cuboids.

a)

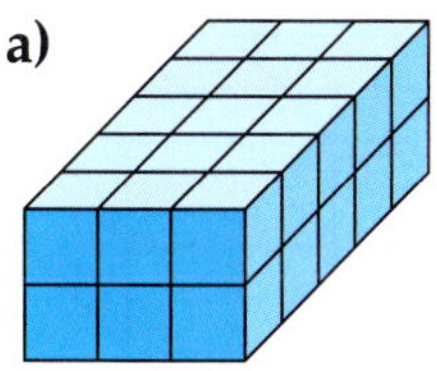

b)

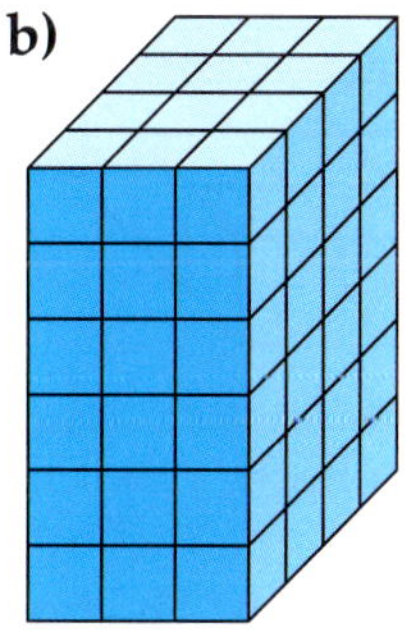

c)

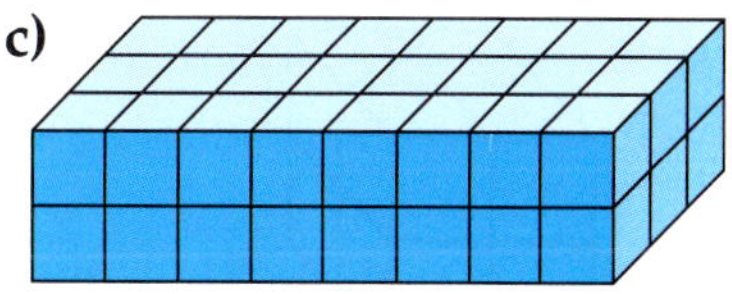

2 Calculate the volume of each of these cuboids.

a)

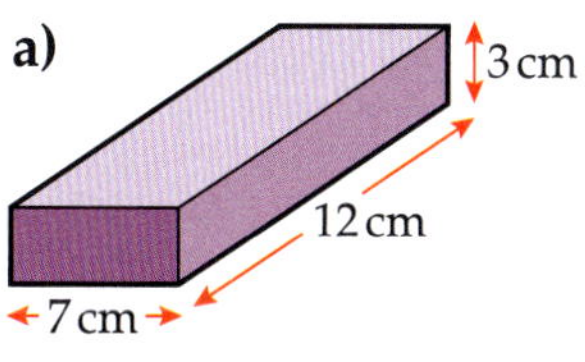

b)

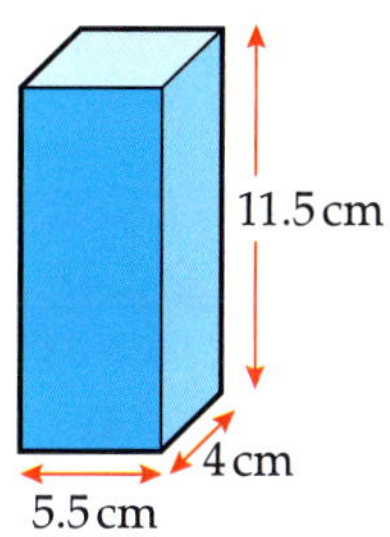

c)

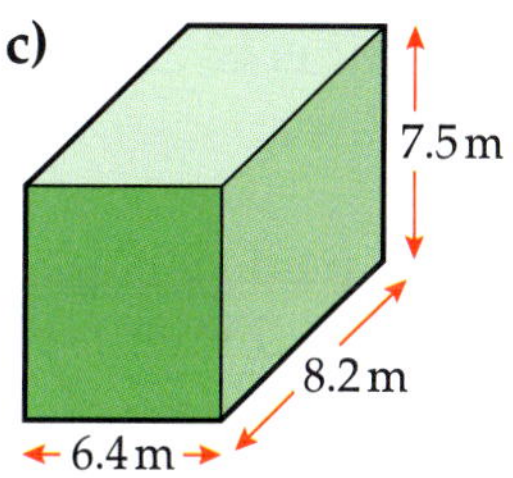

3 Calculate the volume of each of these prisms.

a)

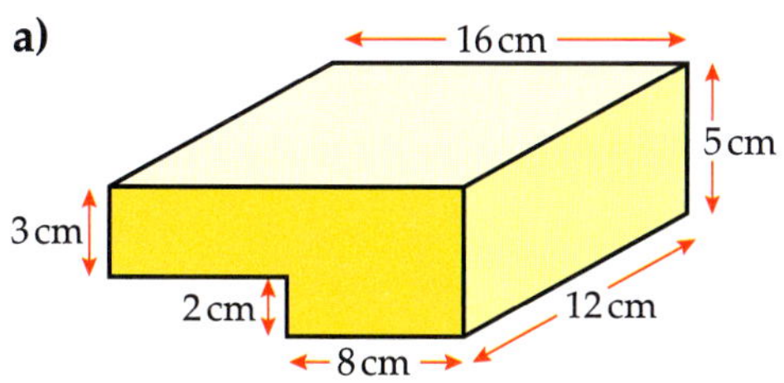

b)

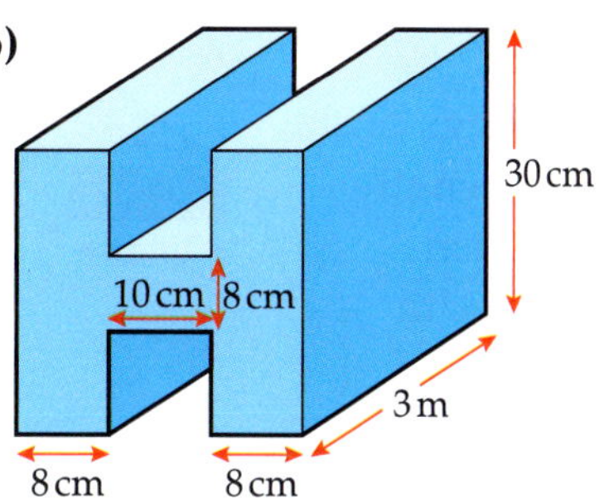

Break the prisms down into cuboids. Calculate the volume of each cuboid and then add them together to find the total volume of the prism.

4 Mr Harding has built a rectangular pond for his garden.
The pond is 1 m deep. The length is 2 m and the width is 2.5 m.
What is the volume of Mr Harding's pond?

5 A cuboid has a volume of 24 cm^3.
The diagram shows one possible cuboid.
Draw, on **isometric paper**, other possible cuboids with volume 24 cm^3.
Make sure you label all the dimensions clearly on your diagrams.

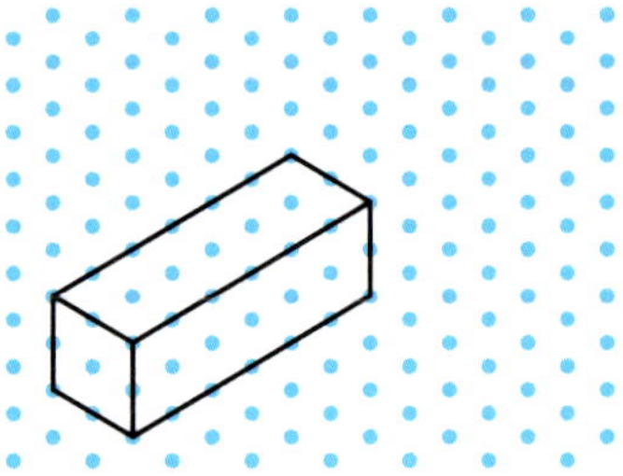

6 Find the volume of each of these shapes. Each block is 1 cm^3.

a)
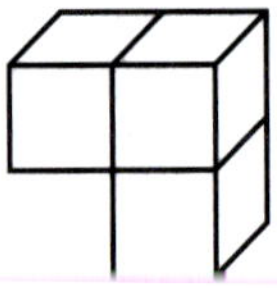

b)
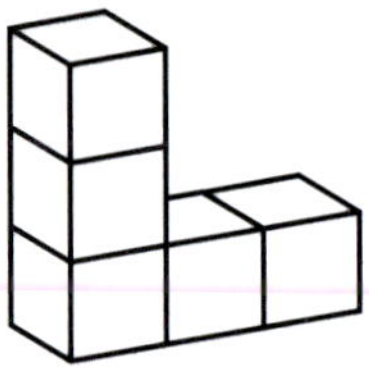

c)
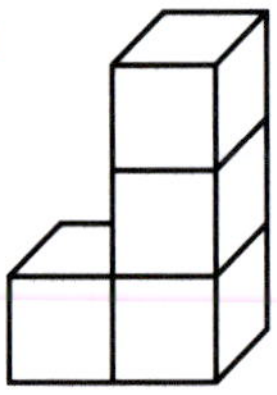

d)
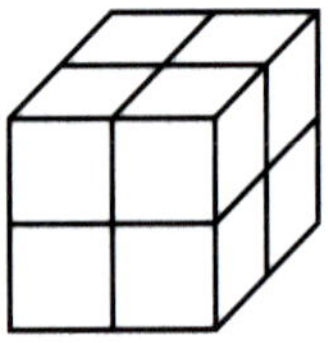

e)
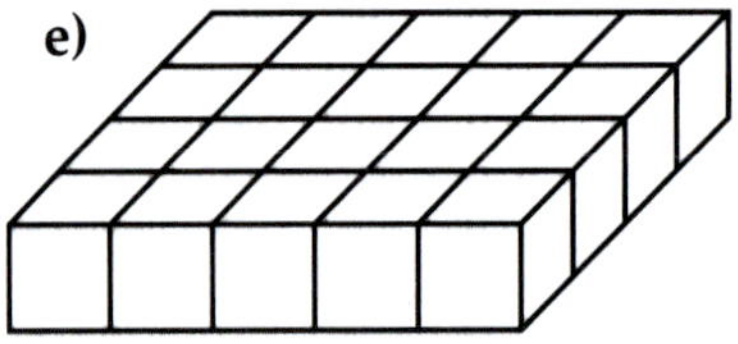

f)
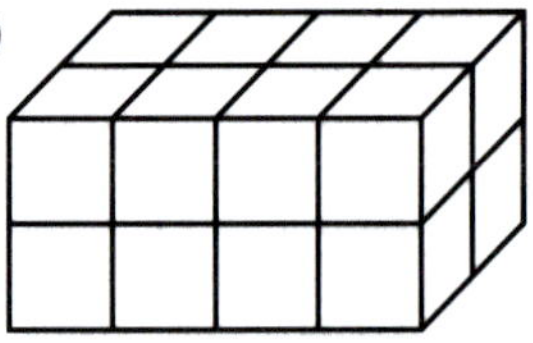

7 A room is 2.5 m long, 4 m wide and 2.2 m high. What is its volume?

8 A hymn book has a cover that is 12 cm by 9 cm. The book is 2 cm thick. What is the volume of the hymn book?

9 A chocolate bar is 16 cm long, 8 cm wide and 2 cm thick. What is the volume of 24 bars stacked together?

10 Mrs Tyler has made a large tray of macaroni pie to serve at a family dinner for six people. If she divides it equally, what volume of pie would each person receive?

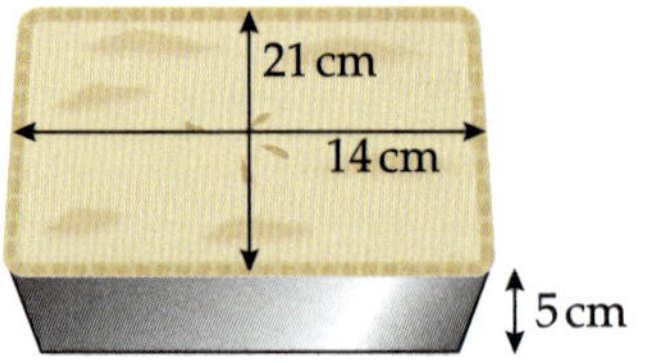

11 Boxes of greetings cards are packed in cartons.
The boxes of greetings cards have width 10 cm, length 20 cm and height 3 cm.
The cartons have width 30 cm, length 60 cm and height 45 cm.
How many boxes of greeting cards will fit into each carton?
Draw a sketch to show your answer.

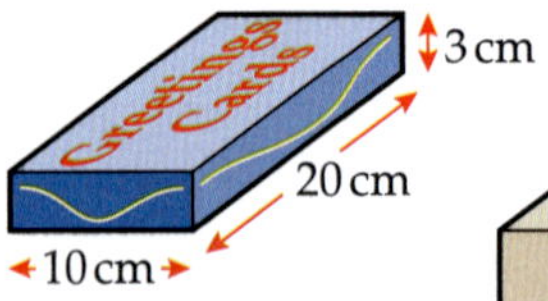

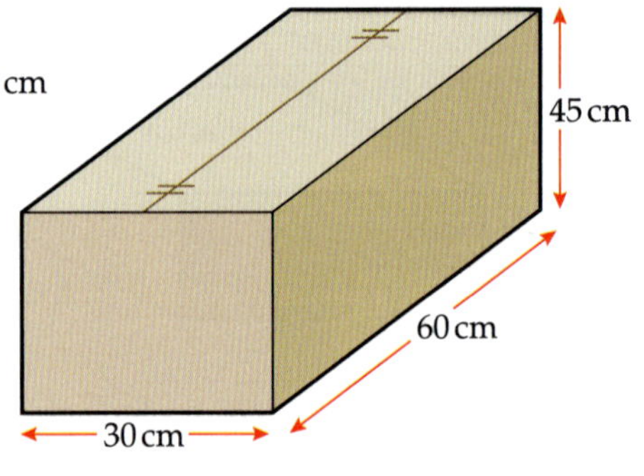

Key words
Capacity
Cubic centimetre
Millilitre
Litre

Unit 3.6 Capacity

In this unit you will:

- differentiate between capacity and volume
- compare and express the capacities of different containers.

Capacity is a measure of the amount a container can hold when full. The metric units of capacity are the **litre** (l), centilitre (cl) and **millilitre** (ml).

1 l = 100 cl
1 l = 1 000 ml
1 cl = 10 ml

Also 1 l = 1 000 **cm³**
1 ml = 1 cm³

The milk carton has a capacity of 1 litre. That is the same as 1 000 millilitres.
The medicine dropper has a capacity of about 1 millilitre. That is the same as 1 cm³.

Worked example

Arrange these containers in order of capacity from the largest to the smallest.

a

b

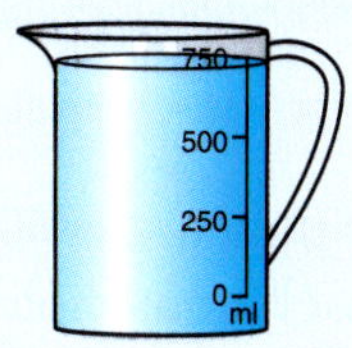

c

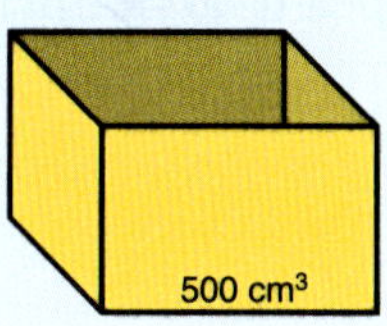

d

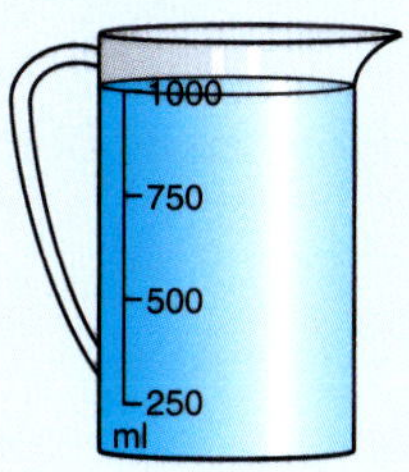

e

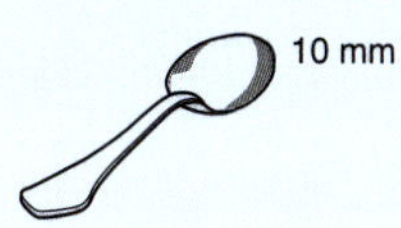

f

Order of capacity = **e, c, d, a, b, f**

Exercise

1. **a)** Write down the capacity of each container.
   b) Which containers hold less than 500 ml?
   c) Which containers hold more than 500 cm^3?
   d) How many times could you fill container **(vi)** from the barrel?
   e) How many barrels could the truck hold?

i)

ii)

iii)

iv)

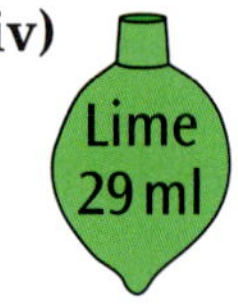

v)

vi)

vii)

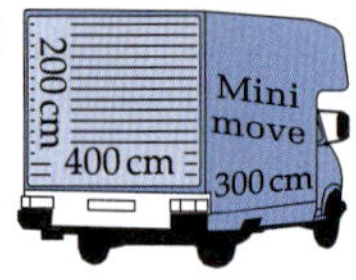

viii)

2. A carton of juice holds 250 cm^3. How many cartons are needed to fill a jug which can hold 8 litres?

3. The aquarium shown has a base width of 50 cm and a length of 40 cm. What is the height of the water in the aquarium when it contains 20 litres of water?

4. A full water tank holds 600 litres of water. How many households can get water from the tank if each household is allowed 4.8 litres?

5. The capacity of a tank is 8 litres. If the base of the tank has an area of 128 cm^2, what is the depth of the tank?

6. A bottle of milk holds 2.5 litres when full. How many bottles are needed to give thirty children 200 ml each for 5 days?

Key words
Net
Tetrahedron
Pyramid
Prism

Unit 3.7 3-D shapes

In this unit you will:

- use ruler, protractor and compasses to construct nets of 3-D shapes.

A **net** folds up to make a 3-D shape. After sketching a net, we can draw it accurately using a ruler and protractor, or a ruler and compasses, if we know enough about the 3-D shape we want to make.

Here are some examples of nets:

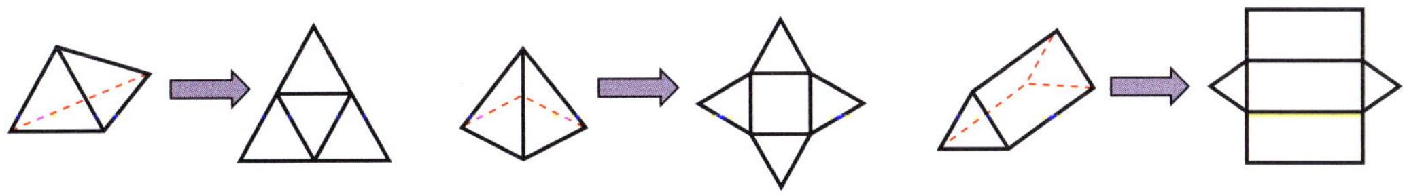

tetrahedron | square-based **pyramid** | triangular **prism**

Worked example

Construct the net for this triangular prism.

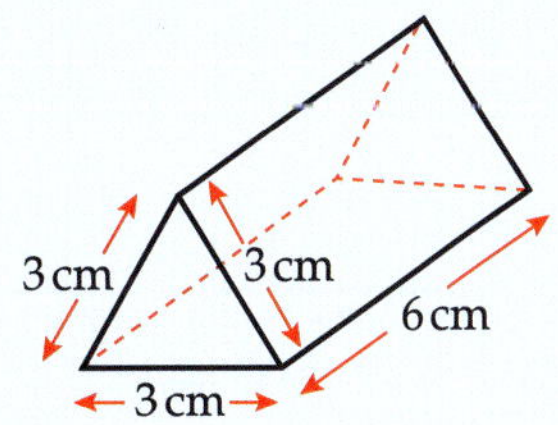

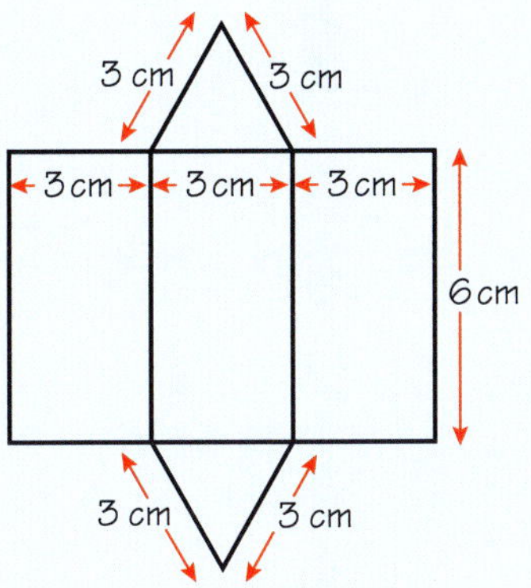

Sketch the net. We know that the faces are either equilateral triangles of side 3 cm, or rectangles of side 3 cm by 6 cm. The angles will be either 60° or right angles.

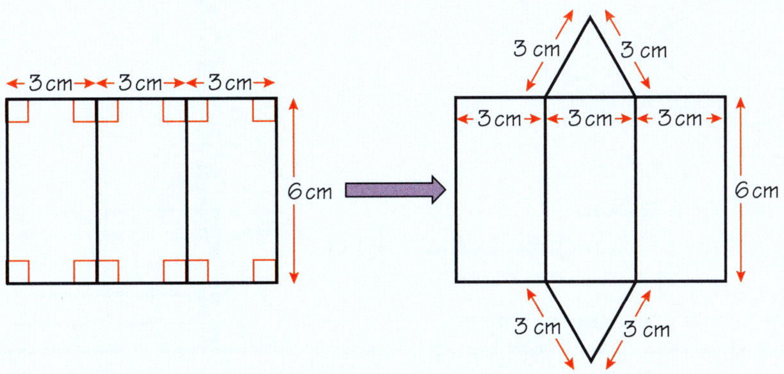

First, draw the rectangles with a protractor or a set square and ruler. Next, draw the equilateral triangles using either a protractor and ruler or using compasses opened out to 3 cm.

Exercise

1. Construct a net for this triangular prism.

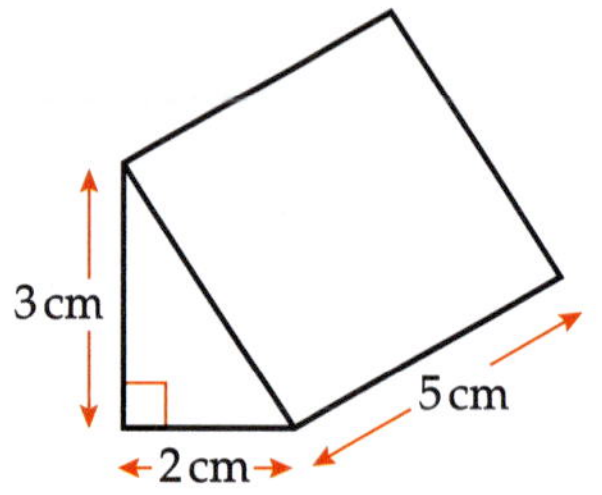

2. Construct a net for this triangular prism.

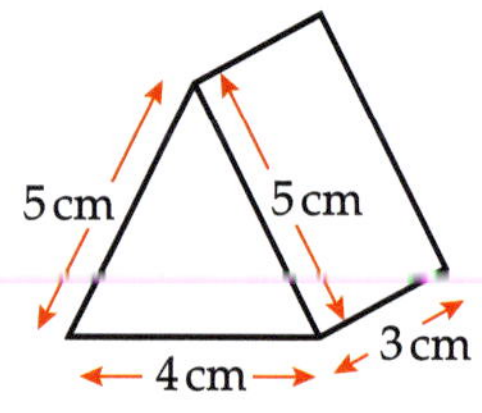

3. What 3-D shape will this net make? Copy it accurately and assemble it.

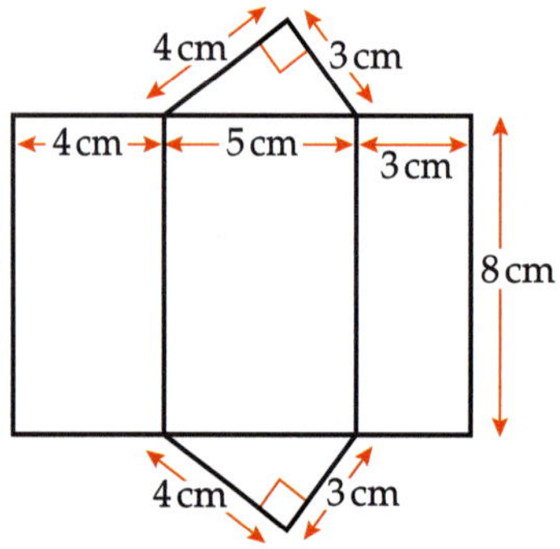

4. Construct the nets for these 3-D shapes using a ruler and compasses.

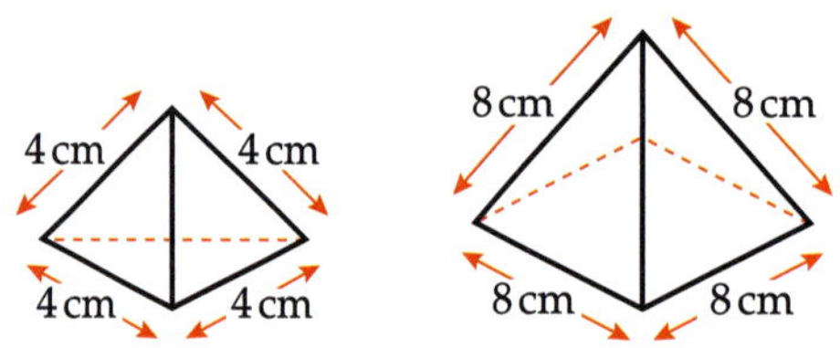

5. Construct two possible nets for this regular tetrahedron.

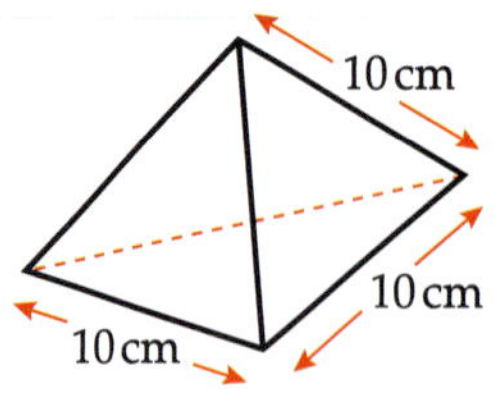

6. The diagram shows a net for a cuboid.
There are other arrangements of six rectangles that are nets for a cuboid.
Draw the other arrangements on squared paper.
Cut them out and check that they will fold to make cuboids.
Then stick them in your books and use coloured arrows to show which edges meet when the sides are folded up.

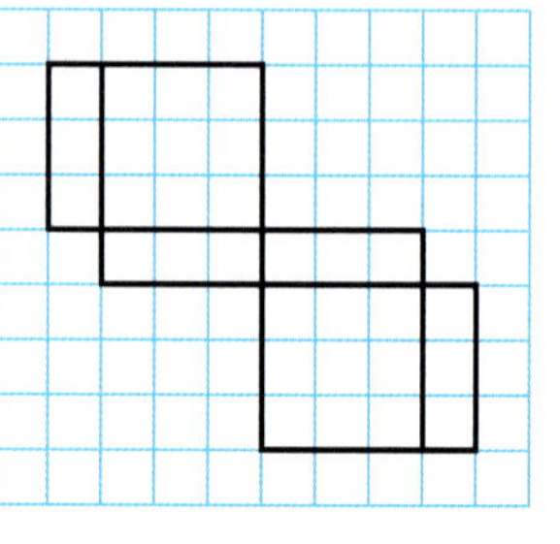

Key words
Surface area
Volume
Prism

Unit 3.8 Surface area and volume

In this unit you will:

- calculate surface area and volume of right prisms.

AThe formula for the area of any triangle is $\frac{1}{2}$(base × perpendicular height).

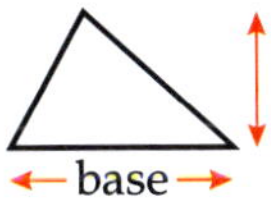

The formula for the area of a parallelogram is base × perpendicular height.

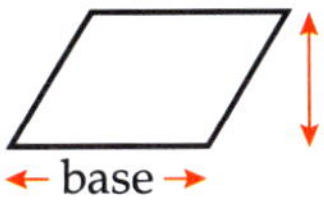

The formula for the area of a trapezium is $\frac{1}{2}(a + b)h$ where a and b are the lengths of the parallel sides and h is the perpendicular height.

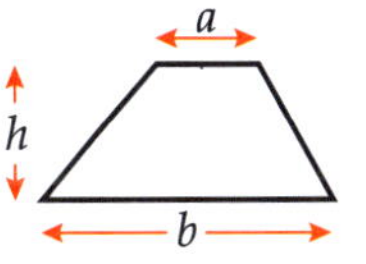

The formula for the area of a circle is πr^2, where r is the radius.

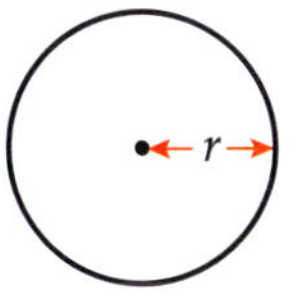

The total area of all the faces of a 3-D shape is called the **surface area**. The formula for the surface area of a cuboid is $2lw + 2lh + 2wh$, where w is the width, l is the length and h is the height.

The **volume** of a **prism** can be found by multiplying the area of the cross-section by the length.

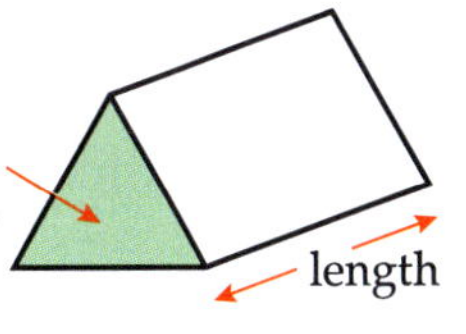

Worked example

Find:

a) the volume

b) the surface area of this prism.

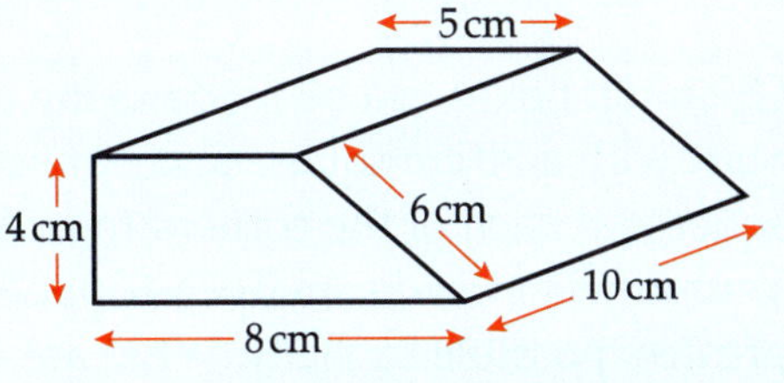

a) Area of cross-section $= \frac{1}{2}(a + b)h$

$= \frac{1}{2}(5 + 8) \times 4 \text{ cm}^2$

$= 26 \text{ cm}^2$

Volume = area of cross-section × length

$= 26 \times 10 \text{ cm}^3 = 260 \text{ cm}^3$

Use the formula for the area of a trapezium to find the area of the cross-section.

b) Surface area = area of all the faces

= top + base + left-hand side + right-hand side + cross-sections

$= (5 \times 10) + (8 \times 10) + (4 \times 10) + (6 \times 10) + (2 \times 26) \text{ cm}^2$

$= 50 + 80 + 40 + 60 + 52 \text{ cm}^2$

$= 282 \text{ cm}^2$

Add up the areas of each of the faces in turn.

Exercise

1. Find the volume and surface area of this triangular prism.

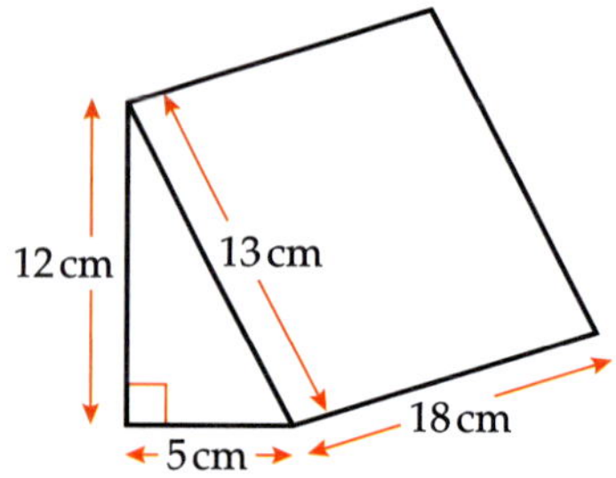

2. Find:
   a) the volumes b) the surface areas of these prisms.

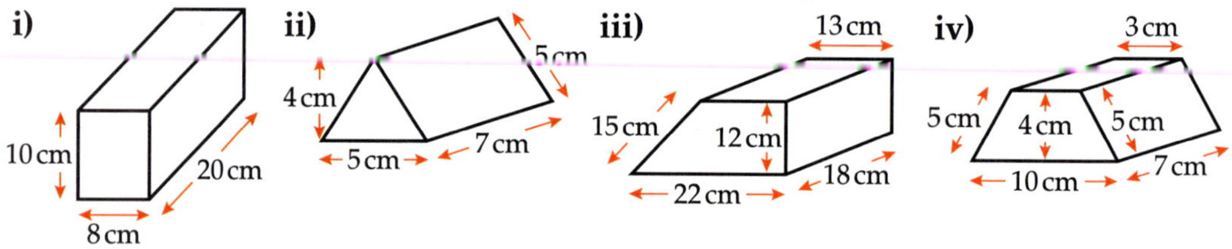

3. The diagram shows the cross-section of a pond. The length of the pond is 2.5 m.
   a) What is the volume of the pond?
   b) The owner of the pond decides to line the pond. What area needs to be lined?

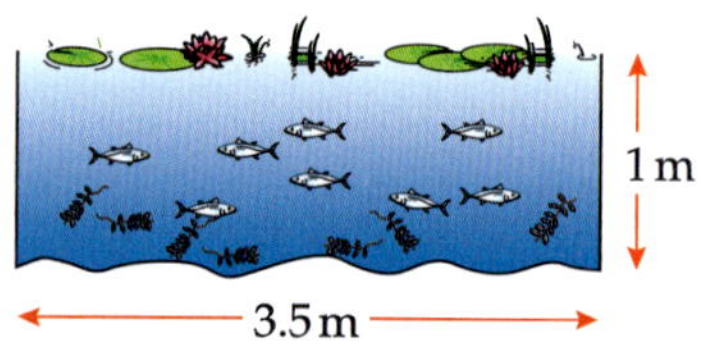

4. What is the volume of this can, to the nearest decimal place?
   Can you think of a way to calculate the surface area of the can?

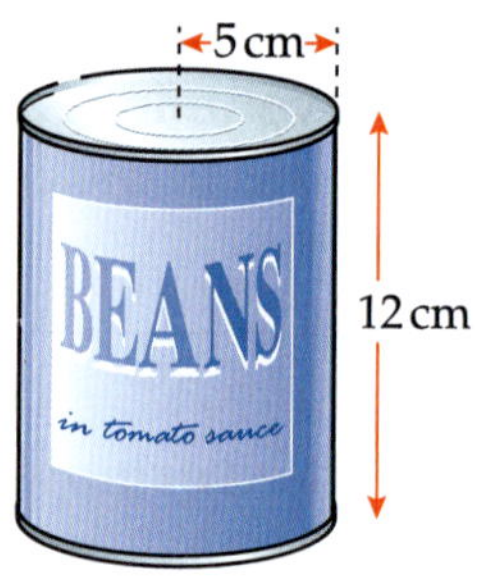

5. Open top boxes can be made using a net like this: Start with a 30 cm square and cut out identical squares at each of the corners to make the net. Your aim is to make an open-topped box with the greatest possible capacity. What are its dimensions?

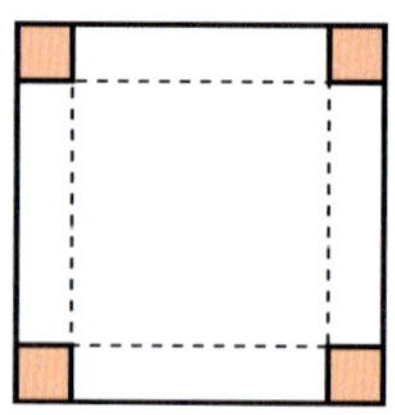

Investigations

6. Investigate the solids that you can make using eight linked cubes.
   Try to make solids with the smallest and the largest possible surface areas and draw them on isometric paper.
7. Investigate the solids that you can make using 27 linked cubes.
   Try to make solids with the smallest and the largest possible surface areas and draw them on isometric paper.
8. Investigate for your own number of linked cubes. What do you notice?

Looking back

1 Using $\pi = 3.14$, calculate the length of an arc with central angle 173° on a circle of radius 13 cm. Give your answer correct to one decimal place.

2 Calculate the perimeter of each of the shapes below. Give your answers correct to one decimal place.

a)

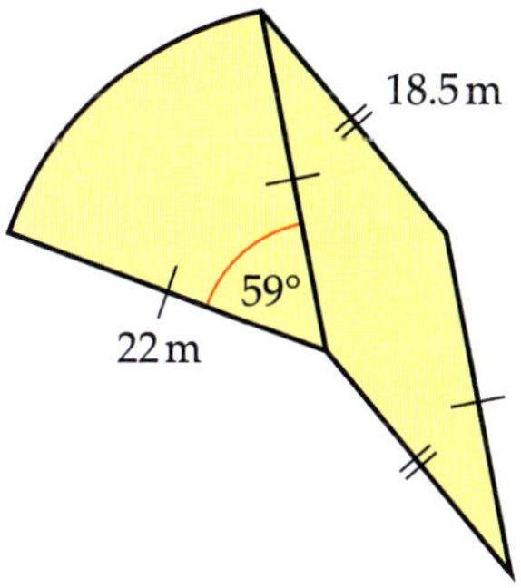

b)

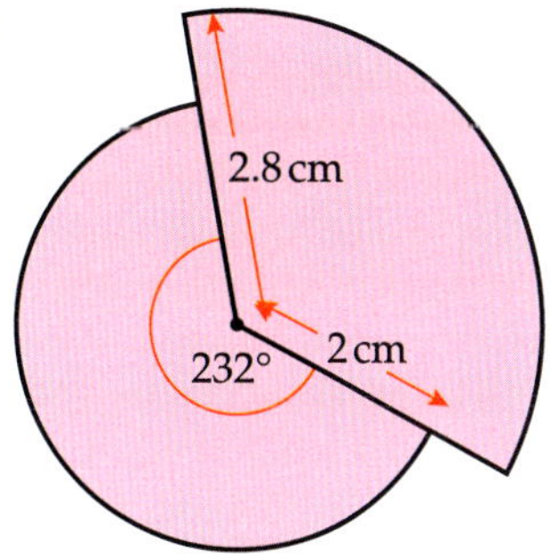

3 Calculate the area of a sector with central angle 215° on a circle of radius 8 m. Give your answer correct to one decimal place.

4 Calculate the area of each of the shapes below. Give your answers correct to one decimal place.

a)

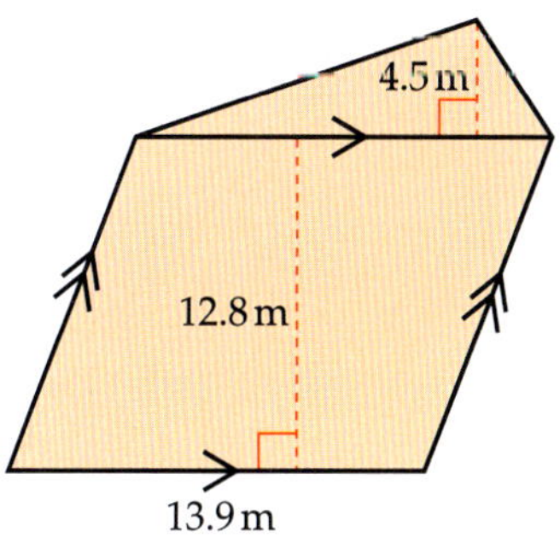

b)

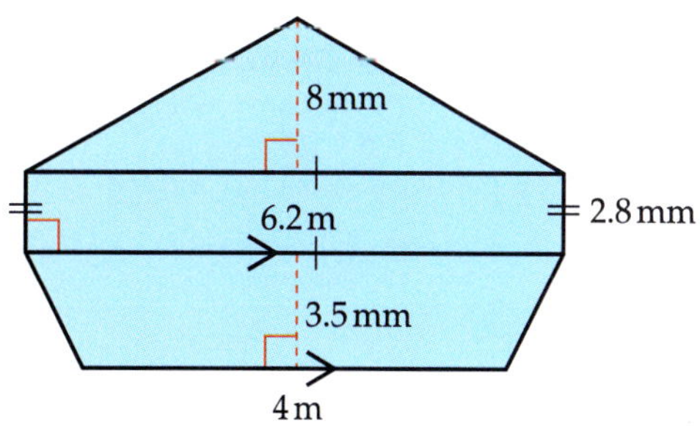

c)

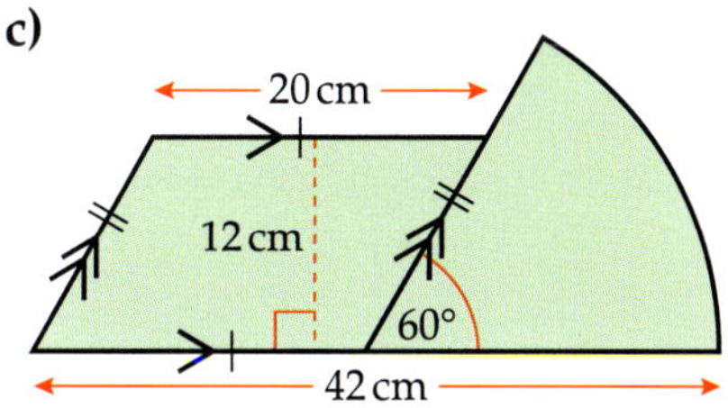

d)

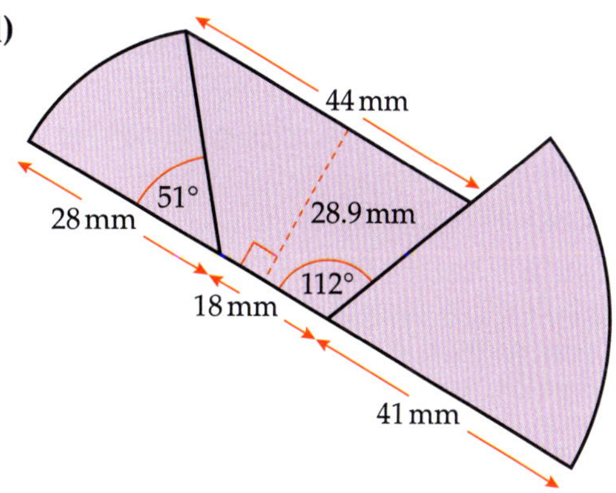

5 A block of wood is in the shape of a rectangular prism. The cross-section of the block has a length of 35 cm and a width of 40 cm. If the height of the block is 1.2 metres, what is the volume of the block?

6 The cross-section of a cuboid is a square with sides of 8.5 cm. If the length of the cuboid is 12 cm, what is its volume?

7 The volume of a cube is 125 cm^3. What is the height of the cube?

Case study - Signwriter

Graeme Morrison

I own and run a small sign shop in a village and, even though the village is quite small, I am normally fairly busy. My business often involves using quotations and so I have to measure the surface area of the sign to base the quotation on. I need to be careful with my calculations or the project could end up costing me money!

Let me first, however, explain what the art of signwriting entails. There are not many true sign "writers" left. Computers and technology have changed the way signs are made. I suppose one could call the new generation sign makers rather than writers.

In my job, the first step would be a call from a customer. I would then set up an on-site meeting with them to discuss what signage they want. At the meeting (which has now cost me money in petrol), I will measure the space in which the sign will be erected. The space could be a steel plate to which we will apply adhesive vinyl, a wall that will be painted or a glass shop front. I need to use maths to first measure the space and then to convert the area into square metres.

Back in the office I will design the sign on a computer and from there calculate how much material I will need. I will also need to figure out how much petrol is going to cost me. Some projects may be as far as 40 km away and may need several days work with the travelling that it entails. Any extra labour needed will also need to be calculated.

So let's do some maths and work out a quote.

We have a large agricultural expo annually near to the village and I do a lot of work for a large petroleum company who is the major sponsor and advertiser.
They want two double sided signs mounted in frames, measuring $1.8\text{ m} \times 6\text{ m}$ and attached to the top of a 4 m high wall.

We will use white coated steel. The steel comes in some standard sizes, $1.8 \times 900\text{ mm}$ being one of them.
We need to know how many sheets we will need:
$6\text{ m} \div 900\text{ mm} = 6\text{ m} \div 0.9\text{ m} = 6.6$
I will need 7 sheets per side and will trim a piece off to get to 6 m.

Coated steel of this size costs \$150 per sheet so 4 signs, each using 7 sheets will cost: $\$150 \times 7 \times 4 = \$4\,200.00$.

Now we need to work out how much adhesive vinyl we will need:
$1.8\text{ m} \times 6\text{ m} = 10.8\text{ m}^2$
$4\text{ signs} \times 10.8\text{ m}^2 = 43.2\text{ m}^2$

My price for cut and applied vinyl is \$250 per m^2:
$\$250 \times 43.2\text{ m}^2 = \$10\,800$

The steel for the structure that will hold the sign has to be added. To do this I need to calculate how much steel I need. After drawing up a plan I can determine how much steel I need. This would be in metres.

Steel also comes in 6 m lengths, so I need to work my waste in. Steel is \$20 per m, we need 30 m per frame and we are making 2 frames:

$\$20 \times 30\text{ m} \times 2 = \$1\,200.00$

Now we have to cost the following:

- welding rods and electricity
- the salary for the guy who works for me
- the chemicals to clean the steel frame
- the undercoat and paint and thinners to paint it
- the drill bits and rivets and rawl bolts to assemble it
- the extra guys I will need to employ for a few days to erect the sign
- the truck with crane I will need to hire to transport and lift the sign onto the wall.

Maths, maths and more maths!
As you can see, everything around me in the running of a business involves maths. In addition to each job, I have to work out salaries, petrol costs, rent, electricity and water costs. The list is endless!

Key words
Like terms
Collecting
Simplified
Expand
Unknown

Chapter 4

Algebraic expressions and identities

Algebra is used to make generalisations and to help in solving problems. In this chapter you will learn how to expand and factorise algebraic expressions, including algebraic fractions and quadratics.

Unit 4.1 Simplifying expressions

In this unit you will:

- multiply an unknown value over a bracket
- simplify algebraic expressions.

$3a$ and $4a$ can be added together to give $7a$. These terms are called **like terms** because they contain the same letter. Like terms can be combined – this is called **collecting** like terms.

Algebraic expressions can be **simplified** by collecting like terms:

For example: $3a + 4b - 2a + 2b - 3ab$
$= 3a - 2a + 4b + 2b - 3ab$
$= a + 6b - 3ab$

$a \times b$ is written as ab.
$a \times b$ and ab are **equivalent expressions**.

To simplify expressions it is sometimes necessary to remove (or **expand**) the brackets. The number outside the brackets is multiplied by **every value inside the brackets**.

For example: $3(7x - 4)$
$= 3 \times 7x - 3 \times 4$
$= 21x - 12$

We can expand brackets with an **unknown** value outside the bracket in the same way:

For example: $b(b + 4)$
$= b \times b + b \times 4$
$= b^2 + 4b$

We always write the number before the letter.

Remember, we write $b \times b$ as b^2.

Worked examples

1. Simplify the following expression: $x(2x + 5) - 12 + 3x$
2. Decide which two of the following expressions have the same value.
   a) $4(3x - 3)$ **b)** $12x + 3(x - 1)$ **c)** $2(4x + 1) + 4x - 14$

1. $x(2x + 5) - 12 + 3x$
$= x \times 2x + x \times 5 - 12 + 3x$
$= 2x^2 + 5x - 12 + 3x$
$\mathbf{= 2x^2 + 8x - 12}$

First, expand the bracket.

Now simplify the expression.
Remember we can only add terms together that have exactly the same letters. The terms x^2 and x are **not** like terms.

2. **a)** $4(3x - 3)$
$= 4 \times 3x - 4 \times 3$
$= 12x - 12$

b) $12x + 3(x + 1)$
$= 12x + 3 \times x + 3 \times 1$
$= 12x + 3x + 3$
$= 15x + 3$

c) $2(4x + 1) + 4x - 14$
$= 2 \times 4x + 2 \times 1 + 4x - 14$
$= 8x + 2 + 4x - 14$
$= 12x - 12$

Expressions **a)** and **c)** have the same value.

Don't forget the order of operations: multiplication and division before addition and subtraction.

Exercise

1 Expand the brackets in the expressions below:

a) $4(3x + 5)$ **b)** $6(7x - 3)$
c) $n(x + 2)$ **d)** $a(a + 4)$
e) $b(b - 2)$ **f)** $r(3r^2 - 5)$
g) $y(3y + z - 3)$ **h)** $2p(p^2 + r + 2)$

Remember, $r \times r^2 = r^3$

2 Simplify the following algebraic expressions:

a) $3(x + 4) - 2x$ **b)** $5(x - 2) + 12$
c) $x(x + 5) + 4x$ **d)** $5x + 6(x - 3)$
e) $2 + x(x^2 + 1) - 2x$ **f)** $3x^4 + x^3(2x - 3) - 7$

See Example 1.

3 Decide which of the following expressions are identical:

a) $4(3x - 5)$ **b)** $10x + 2(x - 3)$
c) $5(2x - 2) - 2x - 10$ **d)** $8x + 2(2x + 3) - 12$

See Example 2.

4 In a magic square, the sum of the expressions in each row, column and diagonal is the same.
Show that this square is a magic square.

$a(3a - 3)$	$3a^2 + 3a - 4$	$3a^2 + 4$
$3a^2 + 3a + 4$	$3a^2$	$3(a^2 - a) - 4$
$3a^2 - 4$	$a(3a - 3) + 4$	$3a^2 + 3a$

5 **a)** Describe in words the relationship between the width and height of the blue rectangle.
b) Describe in words the relationship between the width and height of the red rectangle.

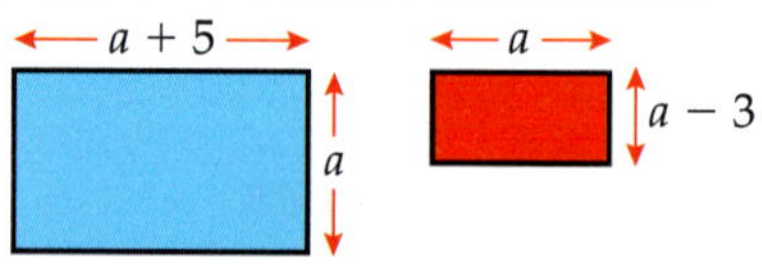

c) Write down an algebraic expression for finding the area of the blue rectangle. (Do not expand the brackets.)
d) Write down an algebraic expression for finding the area of the red rectangle.

The blue rectangle has a piece the size of the red rectangle cut out of it as shown:

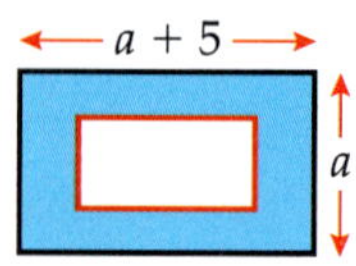

e) Write down an expression for the area of the new blue shape.
f) Simplify your answer to part **e)** by expanding the brackets.
g) If $a = 4$ cm what is the area of the new blue shape?

Key words

Common denominator

Unit 4.2 Algebraic fractions

In this unit you will:

- add and subtract algebraic fractions.

Before we can add or subtract fractions, we must find their **common denominator**.

One way to find the common denominator is to multiply the denominators together.

For example: $\frac{2}{7} - \frac{1}{10}$

We could choose $7 \times 10 = 70$ as the common denominator for this pair of fractions.

$= \frac{2 \times 10}{7 \times 10} - \frac{1 \times 7}{10 \times 7}$

If we multiply the bottom of a fraction by a number, we must do the same to the top of a fraction.

$= \frac{20}{70} - \frac{7}{70}$

Now we simplify the fractions.

$= \frac{13}{70}$

When we have an unknown instead of a number in the fraction, we use **exactly the same method** as shown in the examples below.

Worked examples

1. Simplify the following: $\frac{1}{g} + \frac{3}{5}$

2. Simplify the following: $\frac{x}{2} - \frac{1}{3}$

1. $\frac{1}{g} + \frac{3}{5} = \frac{1 \times 5}{g \times 5} + \frac{3 \times g}{5 \times g}$

The common denominator will be $g \times 5 = 5g$.

We must multiply the numerator and the denominator by the same number.

$= \frac{5}{5g} + \frac{3g}{5g}$

Simplify the fractions.

$= \frac{5 + 3g}{5g}$

2. $\frac{x}{2} - \frac{1}{3} = \frac{x \times 3}{2 \times 3} - \frac{1 \times 2}{3 \times 2}$

$= \frac{3x}{6} - \frac{2}{6}$

The common denominator is 6.

$= \frac{3x - 2}{6}$

Exercise

1 Find the value of each of the following:

a) $\frac{2}{3}-\frac{1}{7}$ **b)** $\frac{4}{11}+\frac{2}{5}$ **c)** $\frac{9}{10}-\frac{2}{3}$ **d)** $\frac{1}{9}+\frac{6}{7}$ **e)** $\frac{5}{12}-\frac{1}{10}$ **f)** $\frac{1}{3}+\frac{1}{2}-\frac{3}{5}$

2 Simplify each of the following:

See Example 1.

a) $\frac{1}{x}+\frac{1}{2}$ **b)** $\frac{2}{m}+\frac{3}{4}$ **c)** $\frac{9}{p}-\frac{1}{5}$ **d)** $\frac{8}{9}-\frac{2}{y}$ **e)** $\frac{1}{2}+\frac{3}{j}$ **f)** $\frac{9}{a}-\frac{9}{10}$

3 Simplify each of the following:

See Example 2.

a) $\frac{a}{3}+\frac{4}{5}$ **b)** $\frac{b}{7}-\frac{3}{9}$ **c)** $\frac{c}{9}-\frac{4}{5}$ **d)** $\frac{9}{12}-\frac{d}{7}$ **e)** $\frac{1}{5}+\frac{e}{11}$ **f)** $\frac{f}{3}-\frac{1}{3}$

4 Simplify each of the following:

a) $\frac{x}{2}+\frac{y}{3}$ **b)** $\frac{a}{5}-\frac{p}{3}$ **c)** $\frac{q}{7}+\frac{w}{4}$

d) $\frac{3}{a}+\frac{1}{b}$ **e)** $\frac{4}{c}-\frac{5}{d}$ **f)** $\frac{11}{e}-\frac{12}{f}$

g) $\frac{3}{w}+\frac{x}{2}$ **h)** $\frac{m}{5}-\frac{9}{q}$ **i)** $\frac{8}{7}+\frac{p}{q}$

5 Aaron has two identical birthday cakes. He shares one equally with his classmates at school. There are x people in his class. He shares the second cake equally with his family. There are y people in his family.

a) How much of the first cake will Aaron eat?

b) How much of the second cake will Aaron eat?

c) How much cake does he eat altogether?

d) If $x = 15$ and $y = 4$, what fraction of a whole cake does he eat altogether?

6 A colony of bacteria is bred in a laboratory. The colony is split into two equal groups, A and B.

a) If there are x bacteria cells altogether, how many are there in each group?

Group A is put into the fridge; group B is left at room temperature. After two days the group in the fridge has doubled in number and the group left at room temperature has halved.

b) How many cells are there now in group A?

c) How many cells are there now in group B?

The groups are put back together.

d) How many cells are there altogether?

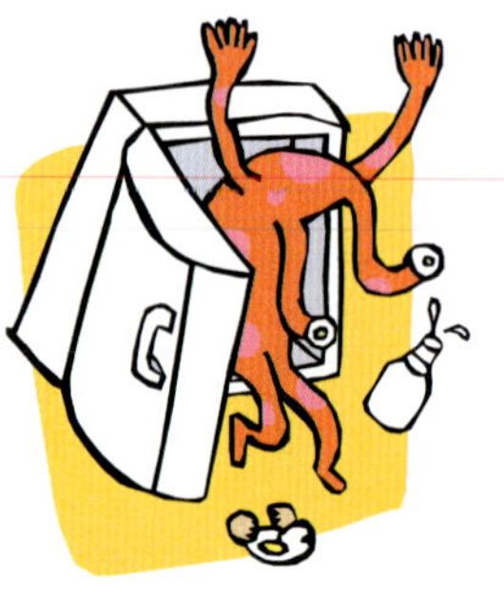

7 Class 9A are asked to add together a pair of fractions.
Some of the answers given are shown. Decide who has given the correct answer and describe what the others did incorrectly.

Alexandra: $\frac{1}{t}+\frac{1}{q}=\frac{2}{t+q}$ James: $\frac{1}{t}+\frac{1}{q}=\frac{tq}{tq}$

Carla: $\frac{1}{t}+\frac{1}{q}=\frac{q+t}{qt}$ Glyn: $\frac{1}{t}+\frac{1}{q}=\frac{1}{t+q}$

Key words
Expand
Factorise
Highest common factor

Unit 4.3 Factorising

In this unit you will:

- factorise algebraic expressions.

We have learnt how to **expand** expressions involving brackets, such as $3(2x + 5)$, by multiplying the value outside the bracket by each of the terms inside the bracket.

$$3(2x + 5) = 3 \times 2x + 3 \times 5$$
$$= 6x + 15$$

To **factorise** an expression, we work the opposite way, making a simple algebraic expression into an algebraic expression involving brackets.

In order to factorise an algebraic expression, we find the **highest common factor (HCF)** of all the terms in the algebraic expression.

For example: $10x + 25$

The highest common factor of $10x$ and 25 is 5.

So we know that: $10x + 25 = 5(\square + \square)$

To find the terms to place inside the bracket, we must **divide** $10x$ **and** 25 by 5 to give: $10x + 25 = 5(2x + 5)$

Always check your answer by multiplying out the brackets.

Worked examples

1. Copy and complete
   a) $4a + 6 = 2(\square + 3)$ b) $18x - 6 = \square(3x - 1)$ c) $3x^2 + 2x = x(\square + \square)$
2. Factorise these algebraic expressions:
   a) $21z + 14$ b) $3b^4 - 2b$

1. a) $4a + 6 = 2(\square + 3)$
   $4a + 6 = 2(2a + 3)$
   The highest common factor of $4a$ and 6 is 2.
   To find the missing term, we must divide $4a$ by 2.
   b) $18x - 6 = \square(3x - 1)$
   $18x - 6 = 6(3x - 1)$
   The HCF of $18x - 6$ and $3x - 1$ is 6.
   c) $3x^2 + 2x = x(\square + \square)$
   $3x^2 + 2x = x(3x + 2)$
   We must divide $3x^2$ and $2x$ by x to find the terms inside the bracket. $3x^2 \div x = 3x$ $2x \div x = 2$

2. a) $21z + 14 = 7(\square + \square)$
   The highest common factor of $21z$ and 14 is 7, so we know that the number 7 appears outside the bracket.
   $21z + 14 = 7(3z + 2)$
   To find the terms inside the bracket, we must divide $21z$ and 14 by 7, this gives us $3z$ and 2.
   b) $3b^4 - 2b$
   $3b^4 - 2b = b(\square - \square)$
   The highest common factor of $3b^4$ and $-2b$, is b so we know that b appears outside the bracket.
   $3b^4 - 2b = b(3b^3 - 2)$
   To find the terms inside the bracket, we must divide $3b^4$ and $-2b$ by b. This gives us $3b^3$ and -2.

Exercise

1 Copy and complete the following:

See Example 1.

a) $6a + 9 = 3(2a + \square)$ **b)** $5(3b - \square) = 15b - 20$

c) $32c - 16 = 16(2c - \square)$ **d)** $18d + 24 = 6(\square + 4)$ **e)** $11(\square + 5) = 11e + 55$

f) $24f - 60 = 12(\square - 5)$ **g)** $20(\square - 3) = 40g - 60$ **h)** $2h^2 + 5h = h(2h + \square)$

2 Copy and complete the following:

a) $10a + 25 = \square(2a + 5)$ **b)** $22c - 110 = \square(c - 5)$ **c)** $\square(5e + 9) = 20e + 36$

d) $\square(2f - 3) = 2f^2 - 3f$ **e)** $\square(e - 4) = e^2 - 4e$ **f)** $\square(3m^2 + m) = 3m^3 + m^2$

3 Factorise the following algebraic expressions:

See Example 2.

a) $3x + 6$ **b)** $4q + 8$ **c)** $5r - 20$ **d)** $10p + 120$

e) $6n + 9$ **f)** $10m - 25$ **g)** $21w - 70$ **h)** $8 + 12x$

i) $24 + 80t$ **j)** $77 - 11y$

4 Factorise the following algebraic expressions:

a) $3a^2 + 5a$ **b)** $4b^2 - 7b$ **c)** $c + 9c^2$ **d)** $3d + 4d^3$

e) $7e^5 + e$ **f)** $2f^3 - f$ **g)** $15g - 17g^3$ **h)** $h^3 + h^1$

5 The area of a rectangle is $2x^2 + 4x$.

a) Factorise this expression.

b) Using your answer to part **a)**, suggest possible lengths for the sides.

To find the area of a rectangle, you multiply the length by width. What two things are you multiplying together here?

6 The sum of three consecutive integers is 24.

Integers are whole numbers.

a) If the first integer is n, write down expressions for:

i) the second integer in terms of n. **ii)** the third integer in terms of n.

b) Write down and simplify an algebraic expression for the sum of the three integers.

c) Factorise your answer to part **b)**.

d) Find the three integers.

e) Repeat, but this time let the middle number be x.

Investigation

7 Look at your answer to **Q6** part **c)**.

a) Find the value of the expression when:

i) $n = 1$ **ii)** $n = 2$ **iii)** $n = 3$ **iv)** $n = 4$

b) All your answers to part **a)** are a multiple of which number?

c) Why is this?

d) Write an expression for the sum of four consecutive integers and factorise it.

e) These values will all be multiples of which number?

f) Investigate for the sum of 5, 6, 7, … consecutive integers. Describe any patterns.

Key words
Factorising
Coefficient
Algebraic expression
Quadratic expression

Unit 4.4 Factorising quadratic expressions

In this unit you will:

- factorise quadratic expressions.

Consider $(x + 3)(x + 8) = x^2 + 11x + 24$
You will observe that the coefficient of x, $11 = 3 + 8$
The constant term, $24 = 3 \times 8$

Investigate $(x - 6)(x + 3) = x^2 - 3x - 18$.
Observe: the coefficient of x, $-3 = -6 + 3$
The constant term, $-18 = -6 \times 3$

Generally, if we have to multiply two binomials of the form $(x + a)(x + b)$, the result will be $x^2 + (a + b)x + ab$.
An expression of this form is called a **quadratic expression**. $(a + b)$ is the coefficient of x and ab is the constant term.

We can immediately write the quadratic expressions that represent the following:
$(x + 2)(x + 3) = x^2 + (2 + 3)x + 2 \times 3 = x^2 + 5x + 6$
$(x + 2)(x - 4) = x^2 + (2 + -4)x + 2 \times -4 = x^2 - 2x - 8$
$(x - 4)(x - 8) = x^2 + (-4 + -8) + -4 \times -8 = x^2 - 12x + 32$

We can now factorise a quadratic expression or represent the quadratic expression as a product of two binomials.

$x^2 + 8x + 15$ is a quadratic expression and factorises to give $(x + 3)(x + 5)$.

Note that the coefficient of x is $8 = 3 + 5$ and the constant term is $15 = 3 \times 5$

Worked example

1. Factorise $x^2 + 9x + 14$
2. Factorise $x^2 - 2x - 15$
3. Factorise $x^2 + 2x - 15$
4. Factorise $x^2 - 8x + 15$

1. $x^2 + 9x + 14$
   The coefficient of x is 9 and the constant term is 14.
   $9 = 2 + 7$
   $14 = 2 \times 7$
   Therefore $x^2 + 9x + 14 = (x + 2)(x + 7)$
2. $x^2 - 2x - 15$
   The coefficient of x is -2 and the constant term is -15
   $-2 = -5 + 3$
   $-15 = -5 \times 3$
   Therefore $x^2 - 2x - 15 = (x - 5)(x + 3)$

3. $x^2 + 2x - 15$

The coefficient of x is 2 and the constant term is -15

$2 = 5 + (-3)$

$-15 = 5 \times -3$

Therefore $x^2 - 2x - 15 = (x + 5)(x - 3)$

4. $x^2 - 8x + 15$

The coefficient of x is -8 and the constant term is 15.

$-8 = -3 + (-5)$

$15 = -3 \times -5$

Therefore $x^2 - 8x + 15 = (x - 3)(x - 5)$

Exercise

1. Factorise the following

a) $x^2 + 11x + 28$ **b)** $x^2 + 9x + 18$ **c)** $x^2 - 9x + 20$

d) $x^2 - 6x - 7$ **e)** $x^2 + x - 30$

2. Write expressions for these shapes.
How many different expressions can you find for each?

a)

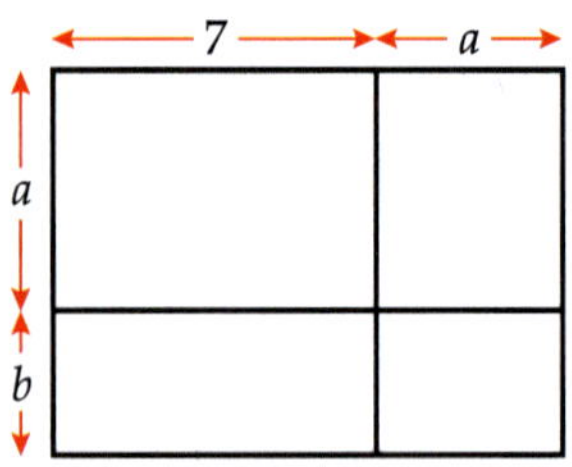

b)

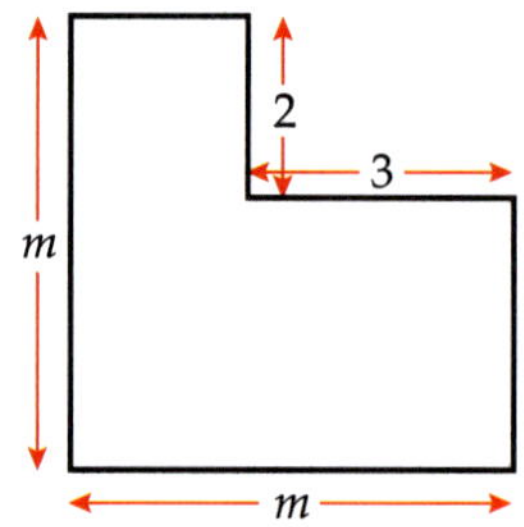

c)

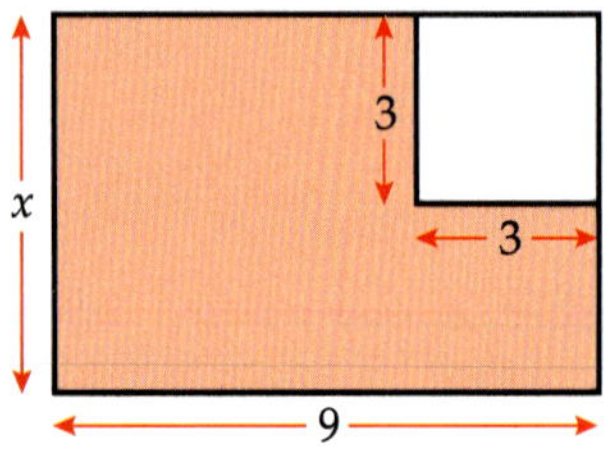

3. The length of a rectangle is 5 cm longer than the width. If the area is 36 cm^2, what are the dimensions of the rectangle?

Let x = width.

Looking back

1 Expand the brackets in these expressions:

a) $2(5x + 3)$ **b)** $5(7x - 1)$ **c)** $y(y + 3)$ **d)** $a(a^2 + 3b - 4)$

2 Simplify:

a) $\frac{1}{n} + \frac{1}{3}$ **b)** $\frac{5}{7} - \frac{3}{x}$ **c)** $\frac{4}{a} + \frac{7}{b}$ **d)** $\frac{2}{5} + \frac{p}{q}$

3 Factorise these expressions:

a) $5x + 15$ **b)** $28n - 63$ **c)** $3a^2 - 5a$ **d)** $\frac{m^2}{2} + 3m$

4 Factorise the following

a) $e^2 + 7e + 12$ **b)** $x^2 + 11x + 18$ **c)** $a^2 + 5a - 14$ **d)** $g^2 - 6g + 9$

Investigation

5 **a)** Find the value of $(x + 5) - (x - 5)$ when:

i) $x = 1$ **ii)** $x = 2$ **iii)** $x = 12$

iv) $x = {}^{-}6$ **v)** $x = 0$

b) Describe what you notice.

c) Explain why this is.

d) Write another algebraic expression, which will give you similar answers to parts **b)** and **c)**.

R1 Revise and consolidate 1

1 Complete the following calculations, giving your answer in its lowest terms.

a) $\frac{3}{4} + \frac{2}{5}$ **b)** $\frac{8}{9} - \frac{5}{6}$ **c)** $1\frac{3}{5} + 2\frac{1}{3}$ **d)** $3\frac{2}{3} - 1\frac{3}{4}$

2 Evaluate:

a) $\frac{3}{5} \times \frac{20}{21}$ **b)** $\frac{5}{6} \div \frac{5}{9}$ **c)** $1\frac{3}{4} \div 2\frac{5}{8}$ **d)** $\frac{5}{6} \times \frac{4}{15} \div \frac{2}{3}$

3 True or False?

a) Multiplying by 0.1 is equivalent to dividing by 10.

b) Dividing by 0.6 is equivalent to dividing by 0.1 and then dividing by 6.

c) Dividing by 0.01 is equivalent to multiplying by 10.

d) $0.3 \times 0.4 = 1.2$

e) $0.34 \div 0.02 = 1.7$

4 A cube is made with a wire frame for all the edges and coloured perspex for its faces. It has a side length of $5\frac{1}{2}$ cm.

a) Find: **i)** the length of wire needed for the frame

i) the length of wire needed for the frame

ii) the area of perspex needed for the cube

iii) the volume of powder it can hold.

b) If the cube comes $\frac{7}{8}$ full and has a 5% filling error, what is the smallest amount of powder it may hold? Give your answer to the nearest cm^3.

5 Write as a power of 10

a) 1 million **b)** 1000 **c)** 0.01

6 Write as a decimal number

a) 10^2 **b)** 10^{-1} **c)** 7.2×10^3

7 Give the decimal value of:

a) 4×10^6 **b)** 2.7×10^{-2} **c)** $5 \div 10^3$ **d)** $9.3 \div 10^{-1}$

8 Using the laws of indices, simplify the following:

a) $4a^2 \times 2a^3$ **b)** $10b^4 \div 2b^3$ **c)** $\frac{14t^2}{7t^3}$ **d)** $3m^3 \times 4m^2 \times 2m$

9 Find the value of each expression if $a = 5$

a) a^3 **b)** $3a^2 - 4$ **c)** $\frac{4a^2}{10a}$ **d)** $2a(3a + a^2)$

10 **a)** Using $\frac{22}{7}$ as an approximation for π, find the area of a circle with a radius of 7 cm.

b) Using 3.14 for π, find the area of a circle with a diameter of 6 cm. Give your answer correct to 1 d.p.

11 If the circumference of a bicycle wheel is 282.6 cm, what is its diameter? (Use 3.14 as an approximation for π.)

12 **a)** Convert the following fractions into decimals.

i) $\frac{5}{9}$ **ii)** $\frac{14}{27}$ **iii)** $\frac{18}{45}$ **iv)** $\frac{22}{7}$

b) Which are recurring decimals?

13 A metal pipe has an outside diameter of 6 cm. The inside radius is 22 mm.

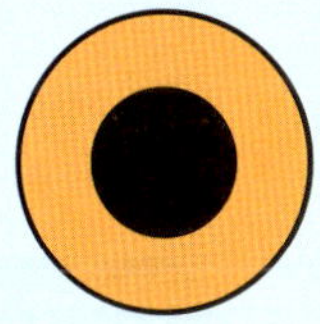

a) Find the area of metal used on the cross-section, giving your answer in mm^2. (Use 3.14 for π)

b) The pipe is 1.3 m long. What is the volume of metal in the pipe in cm^3?

c) Round your answer **i)** to the nearest hundred cm^3 **ii)** to 5 s.f.

d) 8.4 cm^3 of metal has a mass of 10 g. Find the mass of the pipe, to the nearest 10 g.

14 **a)** Simplify the following expressions.

i) $3m \times 10^5 \times 2p \times 10^3$ **ii)** $4p^3 \times 10^{-2} \div 2p^2 \times 10^{-5}$

b) If $m = 3$ and $p = \frac{1}{2}$, find the value of the two expressions.

15 Factorise the following algebraic expressions.

a) $8a + 12 = 4(2a + \square)$ **b)** $3x + 6$ **c)** $15y - 20$

d) $48 + 12x$ **e)** $a^2 + 2a$ **f)** $3t + 4t^2$

16 Simplify

a) $\frac{1}{x} + \frac{2}{3}$ **b)** $\frac{d}{2} - \frac{5}{6}$ **c)** $\frac{x}{3} + \frac{y}{4}$ **d)** $\frac{1}{a} - \frac{3}{b}$ **e)** $\frac{4}{r} + \frac{p}{6}$ **f)** $\frac{a}{b} + \frac{2}{c}$

17 Simplify the following fractions:

a) $2\frac{2}{3} + 3\frac{3}{4} - 2\frac{5}{6}$ **b)** $\frac{3a^2}{b} + \frac{4ab}{a^2}$ **c)** $3\frac{1}{3} \times 2\frac{1}{4} \times 1\frac{3}{5}$

d) $\frac{2a^2b}{b^3} \times \frac{3b^2}{a^2} \times \frac{a^2b}{6a}$ **e)** $3\frac{3}{4} \div 2\frac{1}{2} - 1\frac{1}{6}$ **f)** $\frac{y}{x} \div \frac{2}{x} + \frac{x}{y}$

18 **a)** Which of the following are recurring decimals?

i) 0.42 **ii)** $0.\dot{7}$ **iii)** $3.\dot{1}2\dot{1}$ **iv)** 6.33

b) Write the following recurring decimals using the 'dot' notation.

i) 0.1111… **ii)** 0.232323… **iii)** 0.3444…

c) Convert your decimals in part **b)** into fractions.

19 Simplify the following, giving your answer as a single power where possible.

a) $8a^2 \times 3a^7$ **b)** $15b^3 \div 5b^6$ **c)** $(2^3)^4$

d) $t^{-2} \times t^3 \times t^{-2}$ **e)** $\frac{12\,a^2b^4c}{4b^{-2}c}$

20 Expand the following:

a) $(a + 3)(a + 6)$ **b)** $(b - 5)(b + 3)$ **c)** $(c - 4)(2c - 3)$ **d)** $(3d - 4)^2$

21 Factorise the following expressions.

a) $3x + 9y$ **b)** $4x^2 + 6x$ **c)** $5x^3 + 7x^2$ **d)** $2x^3 + 10x^4 + 4x^2$

Chapter 5

Modelling real-life situations using equations

Key words
Solve

In this chapter you will revise the work on line graphs and intersections of lines that you covered last year. You will learn how to solve linear and quadratic equations, and how to apply your knowledge to real-life problems.

Unit 5.1 Using graphs to solve equations

In this unit you will:

- solve simultaneously equations graphically.

All points on a line will **solve** the equation for that line. Therefore where two lines intersect, the coordinates will solve both equations.

Worked example

Solve the following equations graphically, $2y = 4x - 3$ and $y = \frac{1}{2}x$.

Remember to substitute three values for x to find the value of y for each equation, then plot the points and draw the graphs.

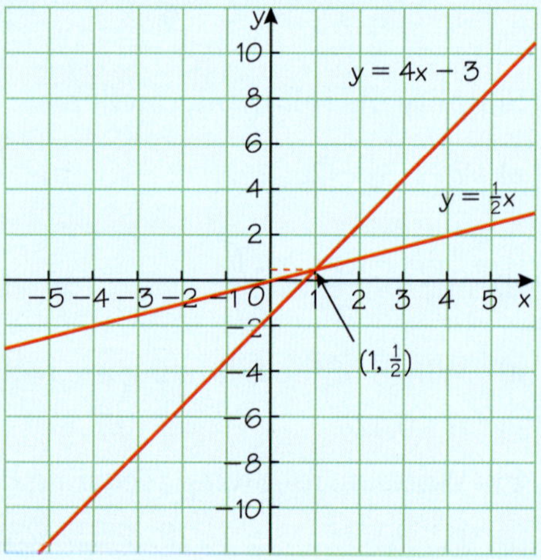

The graphs intersect at $(1, \frac{1}{2})$ so $x = 1$ and $y = \frac{1}{2}$ are the values that satisfy both equations.

Remember to check your result

Equation 1 LHS: $2y = 2 \times \frac{1}{2} = 1$

RHS: $4x - 3 = 4 \times 1 - 3 = 1$

LHS = RHS

Equation 2 LHS: $y = \frac{1}{2}$

RHS: $\frac{1}{2}x = \frac{1}{2}(1) = \frac{1}{2}$

LHS = RHS

Exercise

1 Draw graphs and use them to solve the following pairs of simultaneous equations.

a) $3x + 1 = y$ and $x + 5 = y$

b) $x - 2 = y$ and $2x = y$

c) $6 - 4x = y$ and $y = 3x + 1$

d) $y = \dfrac{x + 4}{2}$ and $y = 3 - x$

e) $\dfrac{x}{3} - 4 = y$ and $3 - 2x = y$

f) $4x - 3 = y$ and $-3x = y$

2 Write the equation(s) that the labelled points on the graph will solve.

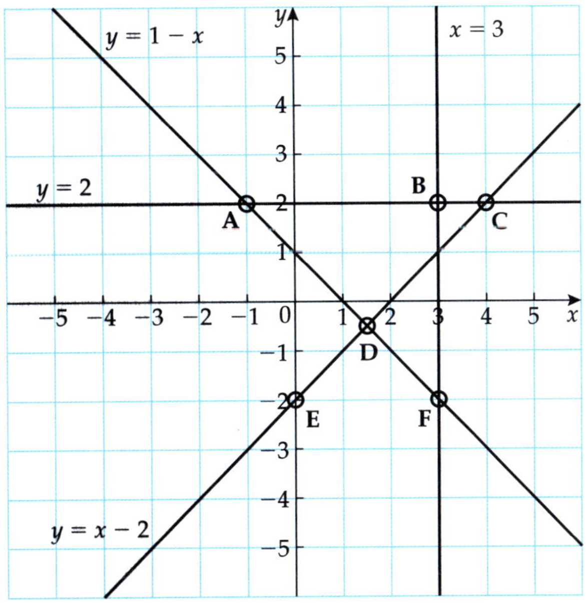

3 On the same grid, draw the graphs of $y = 5 - x^2$ and $y = x + 3$ and find their solutions from the graph.

Unit 5.2 Solving linear equations by elimination or substitution

Key words
Simultaneous
Elimination
Substitution

In this unit you will:

- learn how to solve simultaneous equations by elimination and substitution
- determine the best method for solving pairs of simultaneous equations.

Mrs Green buys 8 bananas and 3 apples for \$14.97 and Mrs Jones buys 8 bananas and 4 apples for \$17.96.

Let 1 represent \$1.

$8b + 3a = 14.97$ Equation 1
$8b + 4a = 17.96$ Equation 2

We can subtract equation 1 from equation 2:

$$(8b + 4a) - (8b + 3a) = 17.96 - 14.97$$

$a = 17.97 - 14.97$
$= 2.99$

Substitute 2.99 for a in equation 1:

$8b + 3 \times 2.99 = 14.97$
$8b = 14.97 - 3 \times 2.99$
$= 14.97 - 8.97$
$= 6$
$b = \frac{6}{8} = 0.75$

The same result would be obtained if we substituted in equation 2:

$8b = 17.96 - 4 \times 2.99$
$= 17.96 - 11.96$
$= 6$
$b = 0.75$

In the example above, we used two conditions to find the cost of an apple and the cost of a banana. We formed two equations and we needed both equations to determine that an apple cost \$2.99 and a banana cost 75c. These equations are called **simultaneous** equations because they must be satisfied at the same time or simultaneously.

We removed or **eliminated** the variable b and got a simpler equation from which we knew immediately that an apple costs \$2.99. When we solve simultaneous equations like this we are using the **elimination method**.

We can also solve simultaneous equations by **substituting** or replacing one of the variables with an equivalent algebraic expression or value. We will investigate first where the coefficient of one of the variables in one of the equations is 1.

$x + y = 12$ **1**
$x - y = 6$ **2**

from equation 2

$x - y + y = 6 + y$
$x = 6 + y$

Add y to both sides.

We can now use this expression for x in equation 1:

$$6 + y + y = 12$$
$$6 + 2y = 12$$
$$2y = 12 - 6 = 6$$
$$y = 3$$

Substitute 3 for y in equation 1:

$$x + 3 = 12$$
$$x = 12 - 3 = 9$$

Check in equation 2:

$$x - y = 6$$
$$9 - 3 = 6 \quad \checkmark$$

The substitution and elimination methods can be used at any time to solve a pair of simultaneous equations. However, you must always attempt to determine the method that is most appropriate. One guiding principle is the substitution method is most appropriate when one of the coefficients of one of the variables is 1.

Worked example

Solve the equations: $4x + 3y = 8$, $2x + 4y = 14$

$4x + 3y = 8$	**1**	
$2x + 4y = 14$	**2**	Multiply equation 2 by 2:
$4x + 8y = 28$	**2**	Subtract 1 from 2:
$4x + 8y - (4x + 3y) = 28 - 8$		
$5y = 20$		
$y = 4$		Substitute in equation 1:
$4x + 3 \times 4 = 8$		
$4x = 8 - 12 = -4$		
$x = -1$		Check in equation 2:
$2 \times (-1) + 4 \times 4 = 14$ ✓		

Exercise

1. Solve the following simultaneous equations
   - **a)** $x + y = 15, x - y = 3$
   - **b)** $x + y = 3, -x + y = 17$
   - **c)** $x + y = 3, x + 2y = 5$
   - **d)** $4x + 5y = 2, 2x + 3y - 4$
   - **e)** $x + y = 5, 4x - 3y = 6$
   - **f)** $5x - 2y = -1, -7x + 3y = 2$
   - **g)** $3x - y = -3, x + 2y = 4$
   - **h)** $1 + 4x = 2y, 3y - 4 = 8x$

2. Look at the simultaneous equations in the worked example. Obtain an expression for x from one of the equations and substitute in the other equation. Find the values for x and y. Which method is easier for this pair of simultaneous equations – the method of elimination or the method of substitution?

Unit 5.3 Graphing quadratic functions to obtain roots

In this unit you will:

- recognise quadratic functions
- plot graphs of quadratic functions
- find roots of quadratic functions using graphical methods.

Key words
Quadratic function
Parabola
Root
y-intercept

A quadratic function is a function where the highest power of x that appears is x^2. An equation of a quadratic function will always look like:

$$y = ax^2 + bx + c$$

where a, b and c can be different numbers but a is not zero.

The graph of a quadratic function is a special curve called a **parabola**.

The **roots** of a quadratic function can be found from its graph. These are the points where the graph crosses the x-axis.

The roots occur when $y = 0$.

There may be one, two or no roots for a quadratic function. The ***y*-intercept** of a quadratic graph is at $y = c$.

When $x = 0$.

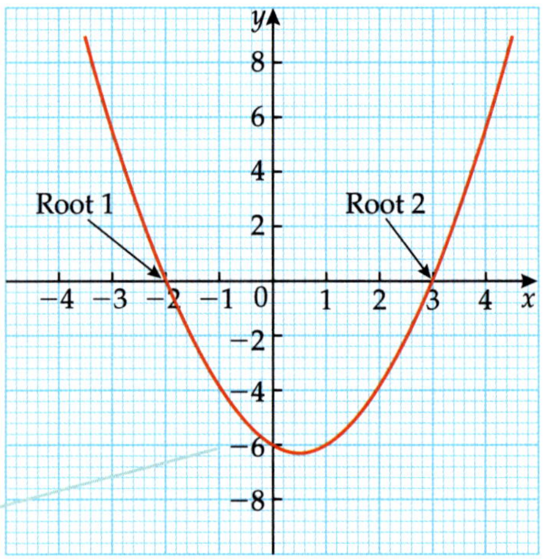

Worked example

Plot the graph of $y = 2x^2 + 3x - x^2 - 2x - 6$, for $-4 \leqslant x \leqslant +4$.

What are the roots for this graph?

$y = 2x^2 + 3x - x^2 - 2x - 6$

$y = x^2 + x - 6$

Simplify the equation.

x	−4	−3	−2	−1	0	1	2	3	4
$y = x^2 + x - 6$	6	0	−4	−6	−6	−4	0	6	14

Draw a table of values for the quadratic function.

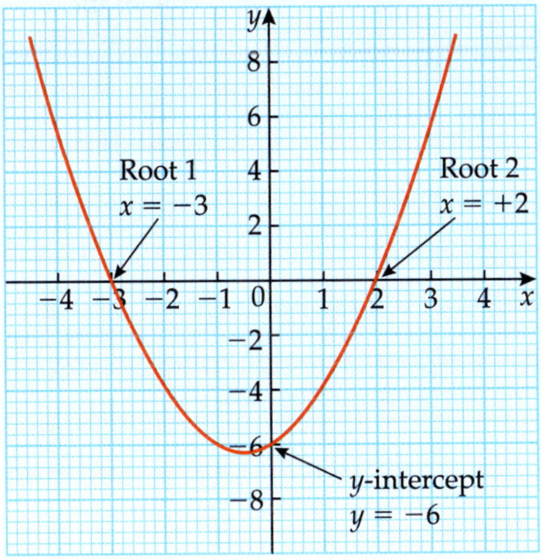

Exercise

1 Which of the following are quadratic functions?

a) $y = 4x^3 + x - 3$

b) $y = 8 + \frac{x^5}{x^3} + (-2)(3x)$

c) $y = x + 3x^2 + 5x^3$

d) $y = 3x^2 + 3x + 5 - 2x^2 + 3x$

e) $y = 4 - x$

f) $y = x(x + 2)$

g) $y = x^2 - 4$

h) $y = \frac{1}{4x^2}$

i) $y = x^6 - x^4$

2 **a)** Complete this table for three different quadratic functions.

x	−3	−2	−1	0	1	2	3
$x^2 + 1$							
x^2							
$x^2 - 1$							

b) Draw the graph of each function on the same set of axes.

c) How many roots does each function have?

3 For each quadratic graph in question **1**, draw its graph, showing values from $-4 < x < +4$.

4 **a)** Draw the graph of $x^2 - 8x - 17$ for $-1 \leqslant x \leqslant +5$.

b) Draw the line of symmetry of this graph. What is the value of x for this line?

Investigation

Draw a table for the quadratic functions $y = x^2$, $y = 2x^2$, $y = 4x^2$ and $y = \frac{1}{2}x^2$.
Include at least the values from $-4 \leqslant x \leqslant +4$.
Now draw these four graphs on the same set of axes. Can you describe the effect of different a values on the graphs of quadratic functions?

Unit 5.4 Solving quadratic equations by factorisation

Key words
Quadratic equation
Parabola
Roots
Factorise

In this unit you will:

- factorise quadratic equations
- determine roots from factorised quadratic equations
- plot graphs of quadratic equations from the factorised form of the equation
- write factorised quadratic equations using graphs.

A **quadratic equation** has the form $y = ax^2 + bx + c$. The graph of this equation is a **parabola**. The graph crosses the x-axis when $y = 0$ or

$$ax^2 + bx + c = 0$$

The points where the graph crosses the x-axis are called the **roots** and they can be found by solving this quadratic equation.

The roots are sometimes also called the solutions.

This can be done by drawing the graph or by factorising the quadratic equation, the same way as you factorised algebraic expressions (see Unit 4.3, Factorising).

To solve a quadratic equation, arrange it so that it is of the form $ax^2 + bx + c = 0$. Factorise the quadratic equation so that it contains two separate factors, each involving x.

If $a \times b = 0$ then either $a = 0$, or $b = 0$.

The roots of a quadratic equation are found by equating each factor in x to zero.

For example: $x^2 = x$ (highest power is x^2, therefore equation is quadratic)
$x^2 - x = 0$ (re-arrange equation to form $ax^2 + bx + c = 0$)
$x(x - 1) = 0$ (x is the HCF. This is the factorised quadratic equation.)

Check your factorising by multiplying out the brackets.

For this equation to be true, either

$$x = 0$$

or $\quad x - 1 = 0$

So the roots are $x = 0$ or $x = 1$.

Remember: a quadratic function may have one, two or no roots!

Factorising a quadratic equation can reveal information about the graph. The roots of a quadratic equation show where the function crosses the x-axis.

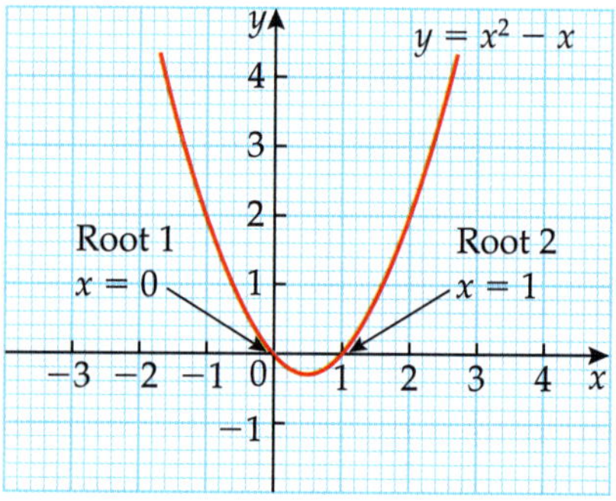

Worked examples

1. Draw the graph for the equation $y = 3x + x^2 - x$. Clearly label any roots.
2. a) Write the factorised form of the quadratic equation shown on the graph.
   b) What is the expanded form of this quadratic equation?

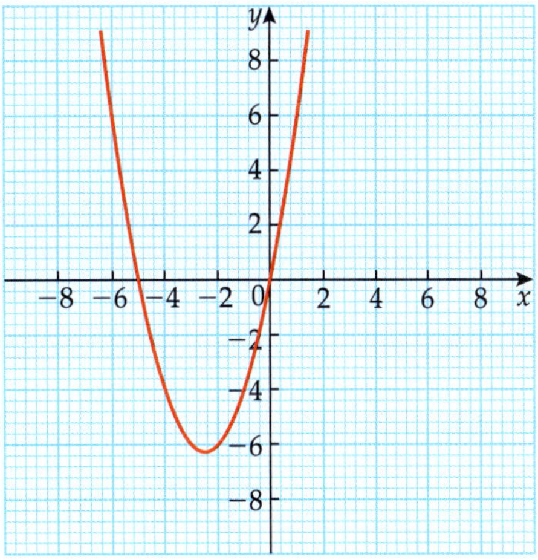

1. $y = 3x + x^2 - x$. The highest power is x^2 so the graph will be a parabola.
   $y = x^2 + 2x$ Simplify the equation.
   $0 = x^2 + 2x$ Rearrange the equation to the form $ax^2 + bx + c = 0$.
   $0 = x(x + 2)$ Factorise the equation.

   From this quadratic equation, either $x = 0$ or $x + 2 = 0$, therefore $x = 0$ and $x = -2$ are the roots.
   The graph is shaped like a parabola, passing through the x-axis when $x = 0$ and $x = -2$.
   The y-intercept is when $y = 0$.

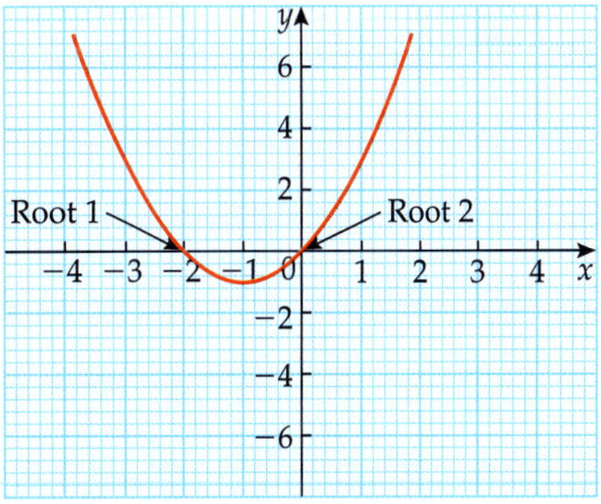

2. a) From the graph, the roots occur at $x = -5$ and $x = 0$. (Identify the roots from where the graph intercepts the x-axis).
   So,
   $y = (x - 0)(x + 5)$ ((x − root) = 0 for each factor of the qudratic equation)
   $y = x(x + 5)$
   Check the equation is correct by calculating 2 points (which are not intercepts) for the graph.

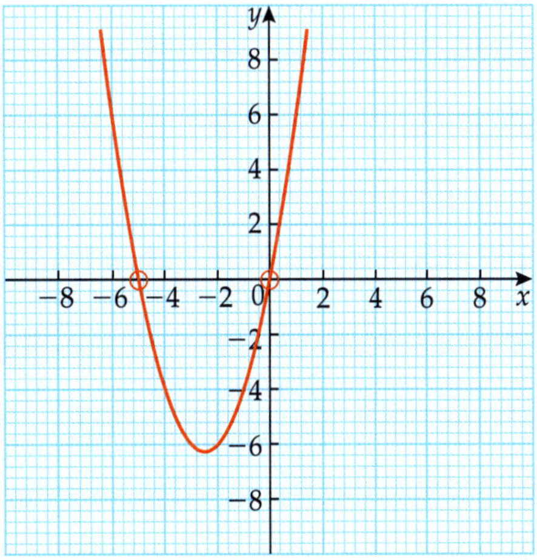

for $x = -1$, $y = -1\,(-1 + 5) = -4$ The point $(-1, -4)$ lies on the graph.
for $x = -2$ $y = -2\,(-2 + 5) = -6$ The point $(-2, -6)$ lies on the graph.

b) $y = x(x + 5)$

$y = x^2 + 5x$ (expand out the brackets)

(Check: From the graph, the y-intercept is 0. From the expanded equation $c = 0$.)

Exercise

1 Factorise these quadratic equations.

a) $y = 2x^2 - 6x - x^2 + 9$

b) $y = 3x^2 + 3x - 7x^2 - 15x + 5x^2$

c) $y = -11x + 3x^2 + \frac{x^2}{x} - 2x^2 + x$

d) $y = 3 + 2x^2 + 2 \times 5x - 3 - x^2$

2 Find the roots of the following quadratic equations, by factorising or using other methods.

a) $y = x^2 - 2x$

b) $2 + 5x = 2 - x^2 + y$

c) $y - 3x + 2x^2 = 3x^2 - 4x$

d) $y = x^2 - 1$

What are the possible values of x when $y = 0$?

3 Sketch the graphs of the following equations. Label any roots or intercepts.

a) $y = x^2 + 6x$

b) $y = x^2 + 1$

c) $9x^2 - 2x + 5 = 5 - y + 10x^2$

4 Write factorised and expanded quadratic equations for the following graphs.

a)

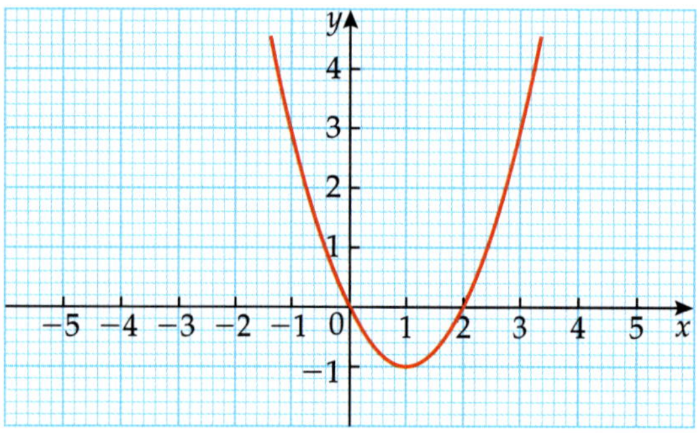

b)

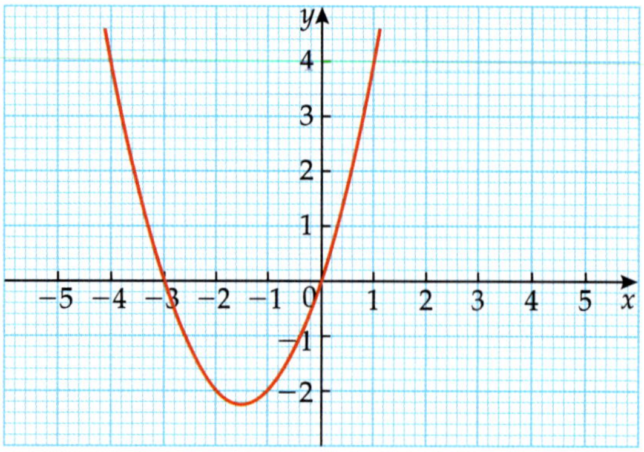

c) 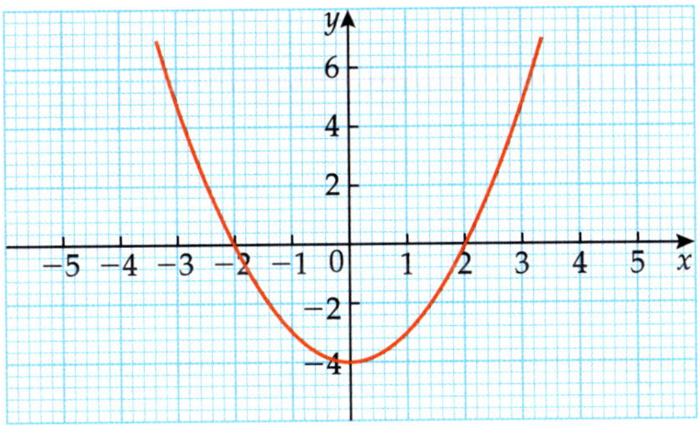

Looking back

1 Draw graphs of these equations. Use each graph to solve the pairs of simultaneous equations.

a) $y = 2x + 9,\ y = 4x + 5$

b) $y = 3 - x,\ y = 5 + x$

c) $\frac{x}{3} + 2 = y,\ 5y - 3x = -6$

2 Emma buys 4 pens and 7 notebooks for \$47.80. Jay buys 5 pens and 10 notebooks for \$64.00. How much does 1 pen and 1 notebook cost?

3 For each of the following quadratic functions, write a table and plot the graph using $-3 \leqslant x \leqslant +3$. What are the roots for each function?

a) $x^2 - 11x + 10$ **b)** $x^2 - 3x - 10$ **c)** $x^2 - 8x + 12$ **d)** $x^2 + 3x$

4 Draw graphs of the following equations, labelling the x-intercepts.

a) $y = x(x - 4)$ **b)** $y = x(x - 3)$ **c)** $y = x^2 + 4x$

d) $y = 2x + x^2$ **e)** $y = x^2 - 9x$

Chapter 6

Interest, depreciation and investment

Key words
Profit
Loss
Selling price
Cost price

This chapter deals with the calculations, formulae and skills that people use on a daily basis to manage their money affairs in the real world. You will solve problems related to profit and loss. You will revisit loans and interest rates and learn how to calculate compound interest. You will also learn about utility bills, depreciation and stocks and shares.

Unit 6.1 Profit and loss

In this unit you will:

- review actual and percentage profit and loss calculations
- solve problems related to cost price, selling price, profit and loss.

People who buy and sell goods for a living rely on making money from the sale of the goods. The amount of money they make from sales is called their **profit**. If the person selling the goods does not make money, we say the goods are sold at a **loss**. The price a shopkeeper pays for his or her goods is the **cost price** of the goods. The shopkeeper then marks up the goods to cover his or her expenses (transport, storage, rent and salaries) and sells them to the public at a higher price, called the **selling price**. The difference between the cost price and the selling price is the profit (if it is a positive amount) or loss (if it is a negative amount).

$$\text{Percentage profit} = \frac{\text{actual profit}}{\text{cost price}} \times 100\%$$

$$\text{Percentage loss} = \frac{\text{actual loss}}{\text{cost price}} \times 100\%$$

Shopkeepers have to decide how much profit they are going to make and what prices they are going to charge for their goods. To do this, they have to perform some calculations.

For example, to calculate the selling price when you want to make a profit, you have to calculate the required profit then add it to the cost price.

Selling price = cost price + profit

Similarly when you want to sell goods at a loss, you have to calculate the required loss then subtract it from the selling price.

Selling price = cost price − loss

When you know the selling price and the percentage profit, you can work out the cost price of an article. In this case, you have to remember that the profit is a percentage of the cost price – it is not a percentage of the selling price!

For example, if an article was sold for \$120 with a 20% profit, what was its cost price?
Cost percentage = 100%
Profit percentage = 20% of cost price
Selling percentage = cost percentage + profit percentage
= 100% + 20% = 120%

So, selling price is 120% of the cost price.

120% of cost price = \$120

$$100\% \text{ of the cost price} = \frac{\$120}{120} \times 100 = \$100$$

This means the item cost \$100.

Worked examples

1. A shopkeeper buys an item for \$400 and sells it for \$450. What is her percentage profit?
2. Mr Jones buys a used car for \$12 000. He sells it after six months for \$10 000. Calculate the percentage lost on the sale.
3. A stallholder at the market buys T-shirts for \$10 and sells them at a 25% profit. What is the selling price of the T-shirts?
4. Find the selling price of crates of coconuts bought fresh for \$150 and sold at a loss of 10%.
5. A travel agent sells air-tickets at a special last minute rate of \$440 and loses 12% on the price. What was the cost price of each ticket?

1. Profit = selling price − cost price
   = \$450 − \$400
   = \$50

   $$\text{Percentage profit} = \frac{\text{profit}}{\text{cost}} \times 100 = \frac{\$50}{\$400} \times 100 = 12.5\%$$

   Her percentage profit is 12.5%.

2. Loss = cost price − selling price
   = \$12 000 − \$10 000
   = \$2 000

   $$\text{Percentage loss} = \frac{\$2\,000}{\$12\,000} \times 100 = 16.7\%$$

   He loses 16.7% on the sale.

3. Cost price = \$10

   $$\text{Profit} = 25\% \text{ of } \$10 = \frac{25}{100} \times 10 = \$2.50$$

   Selling price = cost price + profit
   = \$10 + \$2.50
   = \$12.50

   He sells them for \$12.50.

4. Cost price = \$150

   $$\text{Loss} = 10\% \text{ of } \$150 = \frac{10}{100} \times \$150 = \$15$$

   Selling price = Cost price − loss
   = \$150 − \$15
   = \$135

   They are sold for \$135.

5. Cost percentage = 100%, Loss percentage = 12%

Selling percentage = cost percentage − loss percentage = 100% − 12% = 88%

Selling price = 88% of cost price = $440

So $100\% = \frac{\$440}{88} \times 100 = \$500$

The cost price of each ticket was $500.

Exercise

1. Find the actual profit and the percentage profit in each of the following.
   - **a)** an item costs $40 and is sold for $45
   - **b)** an item costs $50 and is sold for $55
   - **c)** an item costs $1.75 and is sold for $1.80
   - **d)** an item costs 50 cents and is sold for 75 cents

2. Calculate the percentage loss in each of the following.
   - **a)** cost price $500, selling price $400
   - **b)** cost price $75, selling price $60
   - **c)** cost price $5.00, selling price $3.50
   - **d)** cost price $2.99, selling price $0.50

3. A woman buys a hand of 100 bananas for $25. She sells them all for 50 cents each. Calculate the percentage profit or loss she made on the sale.

4. An item is bought for $450 and sold at a 20% profit. What was its selling price?

5. What is the selling price of an item bought for $400 and sold at a 33% profit?

6. A student bought books for university for $500. After using them for a year, he sold them at a loss of 15%. At what price did he sell them?

7. Mrs Hamilton makes table cloths. Her total costs for ten large table cloths were $246.50. At what price should she sell ten table cloths if she wants to make a 15% profit?

8. Find the cost price of each of the following items.
   - **a)** selling price $360, profit 20%
   - **b)** selling price $460, profit 15%
   - **c)** selling price $867, loss 15%
   - **d)** selling price $1 200, loss of $33\frac{1}{3}\%$

Investigation

9. What is a reasonable profit?

   Work in groups. Discuss various home industries in your area. Consider how much work people put into making craft items and discuss how you think they should calculate their profit margins. Report back to the class and discuss your ideas.

   What is discount and how is it calculated?

Key words
Rate
Simple interest

Unit 6.2 Simple interest

In this unit you will:

- solve problems related to simple interest.

When you invest money in a bank (or other financial institution) the bank pays you interest on your investment. Similarly, when you borrow money in the form of a loan, the bank charges you interest for the use of the money.
The letter I is used to represent interest. The letter P is used to represent the original amount of money invested or borrowed (the principal amount).
The amount of interest depends on the length of time for which the money is invested or borrowed. T is used to represent time. Time is normally calculated in years.
Banks pay or charge interest at different **rates** of interest. This is normally given as a percentage per year (% pa) and is represented by the letter R.

You have already worked with the concept of **simple interest**.

The following formula can be used to calculate simple interest: $I = \frac{PTR}{100}$

You can manipulate this formula to find P, T and R given the other variables.
The final amount you earn or pay is the principal plus the interest: $A = P + I$

Worked examples

1. \$300 is borrowed at 10% pa simple interest. How much interest is paid in two years?
2. How long will it take for \$500 invested at a simple interest rate of 8% pa to amount to \$620?
3. A businesswoman borrows \$16 000 and repays her loan at the end of 2 years by repaying \$20 000. What rate of simple interest was she charged?

1. 10% × \$300 = \$30 pa — Calculate the interest paid per year.
   2 × 30 = \$60 — Multiply this by the number of years.
   In 2 years, \$60 is paid in interest.
2. Interest = amount − principal
   = \$620 − \$500
   = \$120
   Interest rate = 8%, so $\frac{8}{100} \times \$500 = \$40$ — \$40 is earned in one year.
   $\frac{\$120}{\$40} = 3$ years
   So it will take 3 years to earn \$120 interest.
3. Interest = amount − principal
   = \$20 000 − \$16 000 = \$4 000
   She paid \$2 000 per year (the loan was for 2 years).
   $\frac{\$2\,000}{\$16\,000} \times \frac{100}{1} = 12.5\%$ The interest rate was 12.5%.

Exercise

1. Calculate the simple interest on
   a) \$400 invested for a year at a rate of 3% pa
   b) \$550 invested for 5 years at a rate of 9% pa
   c) \$720 invested for $2\frac{1}{2}$ years at a rate of 12% pa
   d) \$800 invested for 8 years at a rate of 11% pa
   e) \$50 000 invested for 10 years at a rate of 13.5%.

2. \$1 250 is invested at a simple interest rate of 5% per annum. How long will it take for this amount to reach \$1 562.50?

3. Mrs King invests \$1 800 for 5 years. She receives \$630 in simple interest. What is the percentage interest rate per annum?

4. A bank offers a simple interest rate of 8.5% pa over 3 years on loans to small business owners. Calculate
   a) the amount of interest charged
   b) the monthly repayment due on each of the following loans
   i) \$10 000 ii) \$28 000 iii) \$55 500.

5. A bank charges 12.9% interest on loans paid back in four years or less and 13.5% interest on loans paid back over five to seven years. The following table shows the monthly repayments required on a loan of \$8 000 taken over different periods of time.

Period of loan (time)	1	2	3	4	5	6	7
Monthly repayment (\$)	714.16	379.96	269.17	214.22	183.67	162.29	147.28

   a) Calculate the total amount payable over each period.
   b) Calculate the total interest payable over each period.
   c) Give one advantage of paying the loan back quickly.
   d) Give one advantage of paying the loan back over a longer period.

Investigation

6. Find out what the following terms mean:
   - guarantor
   - surety
   - usury
   - ombudsman.

Key words
Compound interest
Period
Principal

Unit 6.3 Compound interest

In this unit you will:

- **learn how to calculate compound interest using a formula**
- **solve problems related to compound interest.**

When you worked with simple interest, all of the interest was calculated on the original amount borrowed (the **principal**). However, this is seldom what happens in the real world. When banks lend you money, the interest charged or paid each time **period** is added to the principal. This means that the principal increases all the time, so that the interest calculated in the next time period will be greater. This type of interest is known as **compound interest**.

> ***Caribank Loans***
> Applying for a Caribank Loan is simple and fast. We approve loans quickly, usually the same day, so there's no waiting and wondering. Our Caribank Loans offer great benefits to the borrower and allow you the financial flexibility to get on with your plans:
>
> - Competitive interest rates
> - Customized repayment terms to fit your budget
> - Monthly payments charged directly to your deposit account on the date that suits you
> - No arrangement fee
>
> Contact your Caribank branch today to learn more about Caribank loans, including current rates, and to get a free quote.

When you invest money, compound interest works in your favour, as the principal investment grows and you earn more interest each year. At a compound interest rate of 15%, your original investment will double every five years!
However, when you borrow money, the opposite happens – the interest due is added to the principal every period and the debt in fact grows.

It is very time consuming to calculate compound interest by multiplication and addition, so we use a formula to make the calculation easier.

The formula is:

$$A = K\left(1 + \frac{r}{100}\right)^n,$$

where A is the total repayment, including the principal (investment or debt) and the interest,
K is the principal,
r is the interest rate,

r is divided by 100 because interest rates are always given as a percentage.

n is the number of interest periods over which the principal is borrowed.

In compound interest, n is not always given as the number of years over which the principal is invested or borrowed because the interest may be compounded quarterly or monthly.

So, for example, if you borrow money for five years, and the interest is calculated and added to the principal every four months then there are three interest periods for every year of the loan, or 15 interest periods altogether – in this case n would be 15 and not 5.

Worked examples

1. Delia borrows \$9 000 from Caribank to pay for her studies. The bank offers an interest rate of 21% pa, compounded annually. The loan must be repaid within 3 years.
   a. How much must Delia repay altogether?
   b. How much interest did she end up paying on her loan?
2. James also borrows \$9 000, but he uses a finance company. The company offers him an interest rate of 21% compounded half yearly. His loan also has to be repaid in three years.
   a. How much does James repay altogether?
   b. How much interest did James end up paying on his loan?
   c. What accounts for the difference in interest between these two loans?

1. a. $A = K\left(1 + \frac{r}{100}\right)^n$

$= 9\,000\left(1 + \frac{21}{100}\right)^3$

$= 9\,000(1.21)^3$

$= 9\,000 \times 1.772$

$= \$15\,948$

She must repay \$15 948.

b. \$15 948 − \$9 000 = \$6 948

She paid \$6 948 interest.

2. a. $A = K\left(1 + \frac{r}{100}\right)^n$

$= 9\,000\left(1 + \frac{10.5}{100}\right)^6$

$= 9\,000(1.105)^6$

$= 9\,000 \times 1.8204$

$= \$16\,383.60$

James repays \$16 383.60

b. \$16 383.60 − \$9 000 = \$7 383.60

He paid \$7 383.60 interest.

c. The interest in James' loan was calculated and added to his debt more often, increasing it and causing him to pay more interest each subsequent period.

Exercise

1. Maria borrows $27 000 from her uncle to buy a car. He charges her 12% pa over two years, compounded annually. How much money must she repay her uncle?

2. Mrs Kelly gets a gift of $750 from her niece in Miami. She invests it in a fixed deposit account at the bank. The bank offers 7% interest compounded every six months. How much interest will Mrs Kelly have earned
   - **a)** after 1 year
   - **b)** after 5 years
   - **c)** after 10 years?

3. Mr Jones has no money and he has to pay for his daughter's school books. He decides to take a loan from a micro-lending company which charges 19% interest per annum, compounded every four months. He borrows $500 for a year.
   - **a)** How much does Mr Jones owe the micro-lenders after a year?
   - **b)** Mr Jones can't afford to pay the loan back at the end of the year. He uses his bonus to pay back the principal ($500). The company agrees that he can pay off the rest of the loan over four years, but the rate increases to 23%, compounded every four months. How much does Mr Jones owe the company after four years?
   - **c)** How much interest will the micro-lenders have earned from Mr Jones after four years?
   - **d)** Is this fair? Discuss this in groups and prepare an advice sheet for people considering borrowing money.

Unit 6.4 Depreciation

In this unit you will:

- learn about how the value of an item can change over time
- calculate depreciation based on given rates.

Key words
Value
Depreciation

Some items you can buy increase in **value** over time.
The value of houses often increases every year.

Most items you can buy will decrease in value over time. This is called **depreciation**. The value after a certain period of time will depend on the rate of depreciation. The rate of depreciation is usually given as a percentage.

This second-hand car is one year old.
It is worth \$102 000. The same car new costs \$120 000.
The car has depreciated in value by \$18 000. You can write this as a percentage of the original cost.

$$\text{Depreciation in one year} = \frac{\$18\,000}{\$120\,000} \times 100 = 15\%$$

The car has depreciated by 15% in one year. If the car continues to depreciate at the same rate in the second year it will decrease in value by $\$102\,000 \times \frac{15}{100} = \$15\,300$.
After two years the car will be worth $\$102\,000 - \$15\,300 = \$86\,700$.
You can calculate depreciation more quickly using this formula:

$$V = C\left(1 - \frac{r}{100}\right)^t$$

where V is the value of the item
C is the original cost of the item
r is the rate of depreciation as a percentage
and t is the number of time periods that have passed.

So to find the value of the car after three years you use the values:
$C = \$120000$
$r = 15\%$
$t = 3$
The value is:

$$V = \$120\,000\left(1 - \frac{15}{100}\right)^3 = \$73\,695$$

Worked example

A computer costs \$3600 new and its value depreciates by 31% every year. Calculate its value after 4 years. Give your answer correct to the nearest dollar.

$$V = C\left(1 - \frac{r}{100}\right)^t$$

$$= \$3600\left(1 - \frac{31}{100}\right)^4$$

$= \$816$ to the nearest dollar

Exercise

1 A motor car costs \$92 000 new and its value depreciates by 12% every year.
  a) Calculate its value after 1 year.
  b) Calculate its value after 3 years.

2 An armchair costs \$895 new and its value depreciates by 8% every year.
  a) Calculate its value after 1 year.
  b) Calculate its value after 3 years.
  c) After how many years is the armchair worth less than half its original value?

3 Mr Brown buys a second-hand fishing boat for \$38 500. After a year he sells it for \$32 000. Calculate the depreciation in value of the fishing boat in one year. Give your answer as a percentage correct to one decimal place.

4 A digital camera costs \$2600. After two years it is worth \$1664. Calculate the rate of depreciation per year of the camera.

5 Leo buys a new bicyle for \$900. It depreciates in value by 20% every year. Leo says that this means his bicycle will be worthless after 5 years.
  a) Explain why he is wrong.
  b) Calculate the value of Leo's bicycle after 5 years. Give your answer correct to the nearest dollar.

Investigation

6 Use the internet to research depreciation. Find out the value of second-hand items on an internet auction site like *eBay* and compare it to the cost of the items new. Make sure you are comparing like for like. What types of item keep their value well and what types of items depreciate more quickly? You could compare:

- DVDs
- Cars
- Computer equipment
- Clothing
- CDs
- Books

Unit 6.5 Investments

In this unit you will:

- learn how money can be invested in bank accounts
- investigate how stocks and shares can be used to generate returns.

Key words
Investment
Stock
Share
Principal
Stock Exchange

An **investment** is a way of using money to make a profit. The original amount of money invested is called the **principal**. You can invest money in a number of different ways:

Banks and building societies
When you invest a sum of money in a bank or building society account, you will be paid interest. This is usually calculated as a percentage and compounded daily, monthly or yearly. Banks and building societies are a very safe way of investing money because it is very unlikely that your investment will go down.

Stocks and shares
Stocks and shares are certificates which represent ownership of a business or corporation. When you invest in stocks and shares, you are buying a share of one or more businesses. If the value of those businesses increases, then the value of your share increases. However, if the value of the businesses decreases, then the value of your share decreases. Your investment will be worth less than your principal.

To invest successfully in stocks and shares you must be able to predict which businesses will do well.

Shares are traded at the **Stock Exchange**, which is located in Nicholas Tower in Port of Spain.

Each share in a business is usually only worth a small amount of money, say \$30. In order to invest larger sums of money you must buy lots of shares.

On 4 January 2005 shares in Trinidad Publishing Company were trading at \$10 each. The cost of 500 shares would have been \$5000. A year later, the same shares were trading at \$12.70 each. 500 shares were then worth \$6350. By investing \$5000 you would have made a profit on your investment of \$1350. This is a profit of 12.7%.

Worked example

Shares in a mining company are trading at \$5 each. Mr Browne buys 2500 shares.

a. What is the value of Mr Browne's investment?

b. The share price increases to \$5.20. How much profit has Mr Browne made?

c. Calculate Mr Browne's profit as a percentage of his principal.

a. $\$5 \times 2500 = \$12\,500$

b. Value of investment after increase $= \$5.20 \times 2500 = \$13\,000$

Mr Browne has made \$500 profit.

c. $\frac{\$500}{\$12\,500} \times \frac{100}{1} = 4\%$

Exercise

1 Mrs Toussaint buys 600 shares in a banking company, costing \$48 each.

a) What is the value of Mrs Toussaint's investment?

b) After one year, the share price has increased to \$54. Calculate Mrs Toussaint's profit, as an amount of money and as a percentage of her principal.

2 An investment company decided to buy shares in a property business. The shares cost \$40 and the company invested \$100 000.

a) How many shares did the company buy?

b) One year later the share price was \$38. How much was the company's investment now worth?

c) Calculate the percentage change in the value of the investment, and state whether it is a profit or a loss.

3 Lewis and James each want to invest \$1000 on the stock exchange. Lewis invests his money in a company with a share price of \$25. James invests his money in a company with a share price of \$50. Which one of these sentences is true? Explain your answer.

- James's investment is worth twice as much as Lewis's.
- James's investment is worth half as much as Lewis's.
- James's investment is worth the same as Lewis's.

4 A company's share price decreases by 5% each year for two years. Find the value of a \$26 000 investment at the end of each year.

5 Mr Solomon invests \$31 500 in the stock exchange. He divides his investment equally between Scotiabank shares costing \$30 each and Flavorite Foods shares costing \$4.50 each. After one year the price of one Flavorite Foods share has increased by 50 cents, whilst the price of one Scotiabank share has decreased by \$1.20. Calculate the new value of Mr Solomon's investment.

Project

You can visit the Trinidad and Tobago stock exchange on the internet at http://stockex.co.tt.

Click "Listed Securities" to see a list of all the companies that have shares available to buy. Each company has a three letter abbreviation. You can find the share price of any company by clicking "Latest Daily Report" and looking for the right abbreviation.

- Explore the stock exchange website. Find out about different companies that are listed, and what their share prices are.
- Imagine you have \$1000 to invest. Choose a company to invest in, and work out how many of their shares you can buy at the current trading price.
- In one month look at the share price of your company. Is your 'investment' worth more or less than \$1000? Compare your results to other members of your class. What can you say about investing money in stocks and shares?

Looking back

1. A wristwatch costs $2400 new and its value depreciates by 15% every six months.
   a) Calculate its value after six months.
   b) Calculate its value after 2 years.
   c) After how long is the wristwatch worth less than one third of its original value?

2. A motorbike costs $18 000. After two years it is worth $16 245. Calculate the rate of depreciation per year of the camera.

3. Mrs Allen buys 4000 shares in an electricity company, costing $21 each.
   a) What is the value of Mrs Allen's investment?
   b) The share price drops to $17.01. How much of a loss has Mrs Allen made?
   c) Calculate the percentage by which Mrs Allen's investment has depreciated.

Case study – Pre-primary school principal and teacher St Joseph's Educare Centre

Ms Loretta Bowen-Williams

Qualifications/Experience:

Diploma in Pre-Primary Education (Miami College of Education)

Montessori Teaching Diploma (London)

Ten years teaching experience

I run a small private pre-primary school in a large town. We have seventy children in all and five staff members. I currently rent the premises from a local businessman – they are very lovely and spacious with large grounds for the children to play in.

My work has two elements. As the principal I am responsible for the administration and day-to-day running of the school. I spend about three hours a day doing this part of my job. I have to make sure that the teachers are doing their jobs and that they are paid. I also report to the board of trustees and spend time fund-raising for the school. I have to make sure that the financial records are kept up-to-date and that fee payments are entered and receipts issued. We charge fees on a monthly basis and I have to send out accounts to parents. We charge extra for outings, food and additional equipment. I also have to pay the school accounts – telephone, electricity, rent and daily expenses such as food and equipment that we buy.

Once I have dealt with the financial aspect of the job, I have to keep records for the Ministry of Education. This involves attendance records for both pupils and teachers – when a teacher takes a leave of absence, we have to complete a form and record the details in a register. We keep an overall attendance record for the pupils too. I also deal with the ordering of equipment and books and make appointments to see parents if necessary. I keep a diary and find that it helps to manage my time efficiently.

In addition to my work as principal, I also teach art and drama for two hours each day. I really love this part of my job as it allows me to have contact with the children. I plan lessons in advance and I try to collect all the materials that I will need for the lessons before I get to school. For example, this week we are making templates and painting on fabric and I have to work out what I will need for each class. For each group of four children I need a set of four plastic paint containers and four brushes as well as four pairs of scissors, rulers and thin plastic for making the templates. I also need at least a metre of plain fabric for each group. I have three classes with twelve to sixteen children in each, so it takes a little organisation to get it all sorted out. I make up a basket or box with the equipment for each class and keep it all together.

Some lessons require much more planning than others. For example, when we built models out of drinking straws, I had to draw money from school funds to buy straws and plasticine before the lessons. However, it is all worth it when the children are creative and they make beautiful artworks.

Chapter 7

Pythagoras' theorem

Many right-angled triangles are found in our environment and their distinctive properties can be used to calculate lengths or distances, and angles or directions. Trigonometry is a branch of mathematics which deals with the relationship between the sides and angles of triangles. In this chapter you will explore the properties of right-angled triangles, the relationship of the sides that Pythagoras discovered and the three basic trigonometric ratios (tangent, sine and cosine). Then you will apply the information to solve real-world problems involving triangles.

Key words
Hypotenuse
Pythagoras' theorem
Squares
Square roots

Unit 7.1 Pythagoras' theorem

In this unit you will:

- describe Pythagoras' theorem
- use Pythagoras' theorem to calculate sides of right-angled triangles.

Pythagoras was a Greek mathematician and philosopher. He discovered that in any right-angled triangle, the **square** of the **hypotenuse** equals the sum of the squares on the other two sides. This is known as **Pythagoras' theorem**.

We can draw grids on each side and count the squares or square the lengths of the sides to illustrate his theorem.

The hypotenuse is the longest side of a right-angled triangle. It is always opposite the right angle.

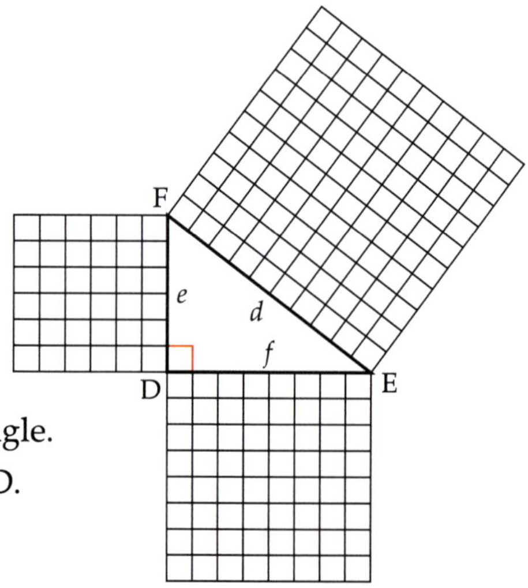

Side d is the hypotenuse; it faces the right angle D.

$e^2 + f^2 = 6^2 + 8^2 = 36 + 64 = 100$ square units

$d^2 = 100$ square units

$d = \sqrt{100} = 10$ units

$d^2 = e^2 + f^2$

We use this information to calculate the lengths of sides of any right-angled triangle.

Worked example

Find the length of a and b in the following diagrams.

a.

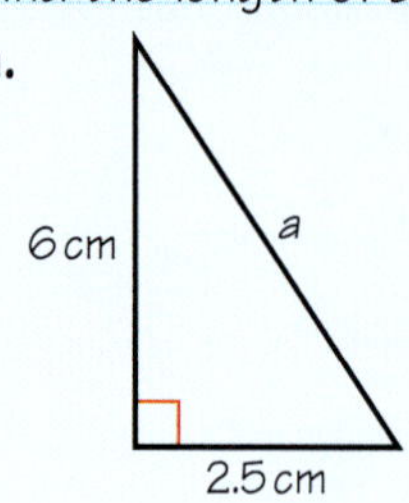

b.

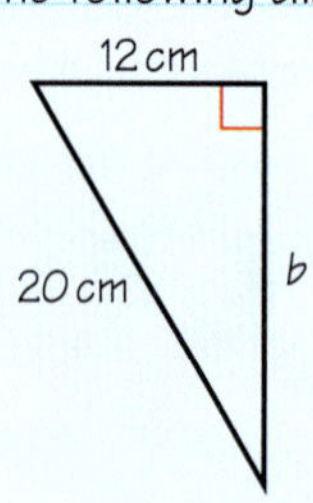

a. $a^2 = 6^2 + 2.5^2 = 36 + 6.25 = 42.25$

$a = \sqrt{42.25} = 6.5$ cm

b. $b^2 + 12^2 = 20^2$

$b^2 = 20^2 - 12^2 = 400 - 144 = 256$

$b = \sqrt{256} = 16$ cm

Exercise

Review **squares** and **square roots** before you do these problems.

1. Calculate the lengths of the unknown sides in the following figures.

a)

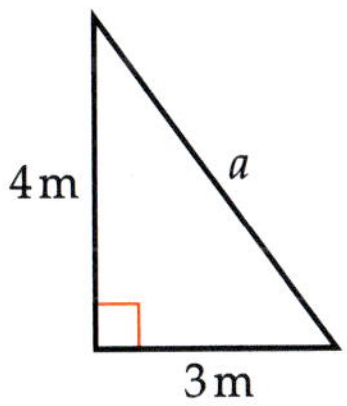

b)

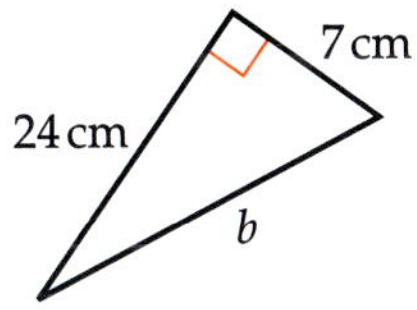

c)

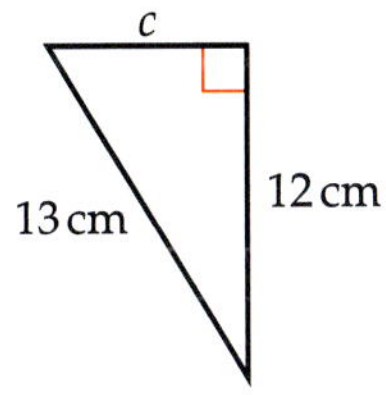

d)

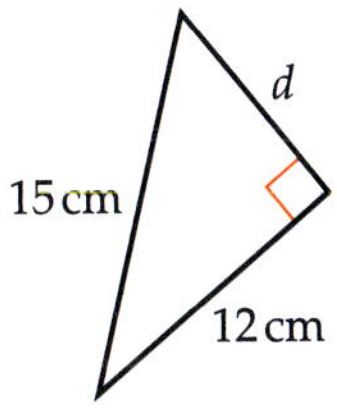

e)

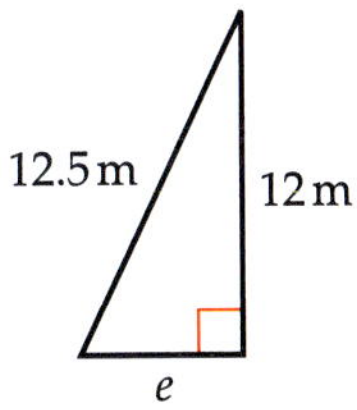

2. The sides of a square are 10 m. Find the length of a diagonal.

3. The sides of a rectangle are 15 cm by 20 cm. What is the length of a diagonal?

4. What is the length of the sides of a rhombus if the diagonals are 7 cm and 24 cm?

5. Which of the following measurements are incorrect for right-angled triangles? Explain your answers.

a)

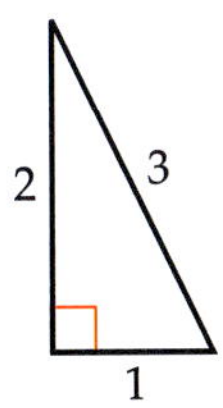

b)

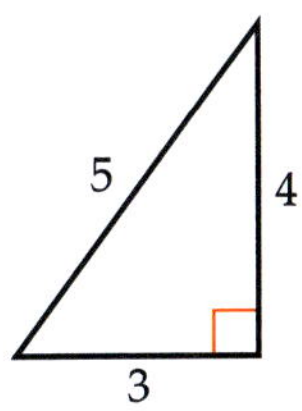

c)

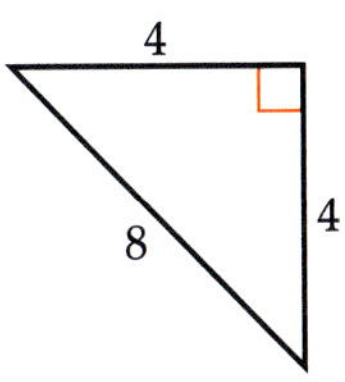

d)

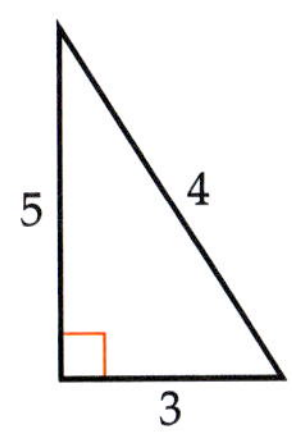

e)

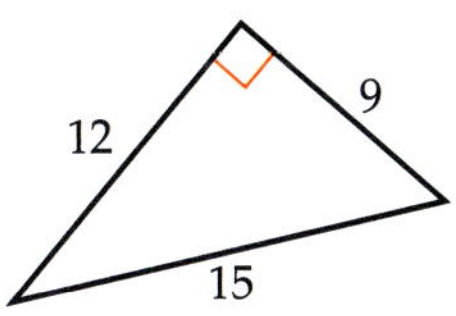

f)

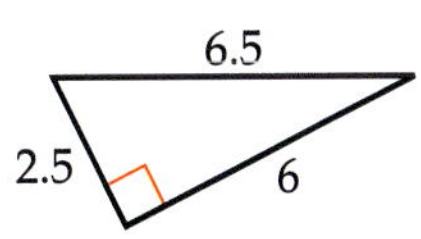

6 **a)**

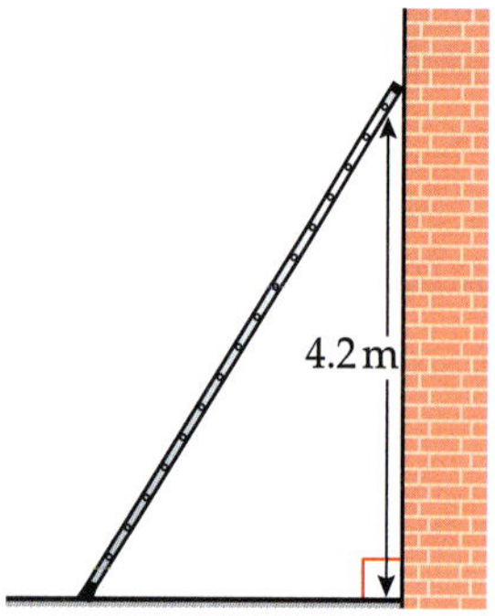

What length of ladder is needed to reach a window 4.2 m high, if the base is 3 m from the wall?

b)

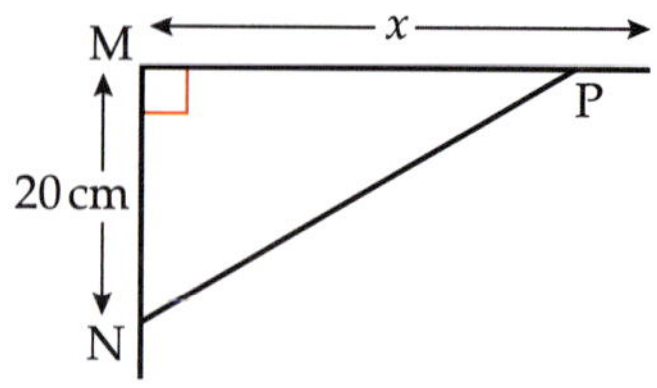

NP = 28 cm

What is the maximum shelf width (x), if it can extend 4 cm beyond the end of the support NP?

Give your answers to 3 s.f.

7 A ladder is 8 m long and Mark needs to go on top of a wall 6 m high. The top of the ladder is almost in line with the top of the wall.

a) Use this information to make a sketch of a right-angled triangle, putting in the appropriate measurements.

b) Calculate the approximate distance between the wall and the foot of the ladder.

8 The diagram shows four flag poles at the vertices of a square. The longest distance between two of the feet of the flag poles is $\sqrt{50}$ metres.

a) Sketch the triangle that can be used to find the shortest distance between two of the flag poles.

b) Calculate this shortest distance.

c) If a pole is 3 metres high, what is the approximate length of the wire that joins the top of one pole to the foot of another

i) along a side

ii) across the middle of the square?

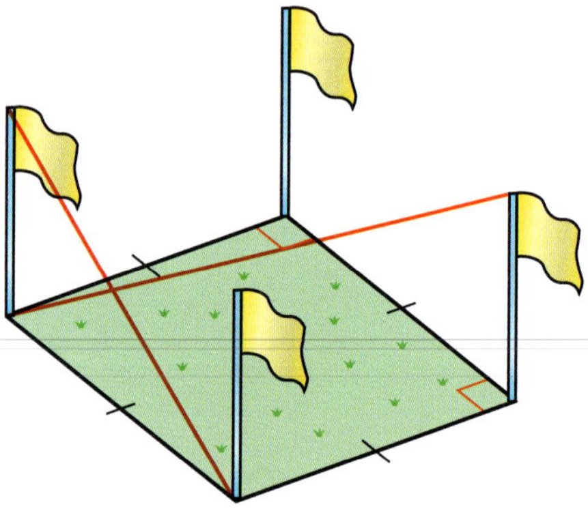

Investigation

9 Do research to find out more about Pythagoras and his contributions to mathematics.

There are sets of three numbers called Pythagorean triples. Find some and explain them to a friend.

Key words
Hypotenuse
Opposite side
Adjacent side
Tangent
Degrees
Minutes

Unit 7.2 The tangent ratio

In this unit you will:

- determine the tangent of an acute angle
- use a calculator or table to find an angle, given the tangent ratio and vice versa
- use the tangent ratio to calculate angles or sides of a right-angled triangle.

In any right-angled triangle the **hypotenuse** is the longest side. It faces the right angle. The side facing an angle is the **opposite side**. There are two **adjacent sides** to an angle. But the hypotenuse already has its special name. In the diagram below

AC is the hypotenuse, it faces the right angle B.

AB is adjacent to $\angle A$ and BC is opposite to $\angle A$.

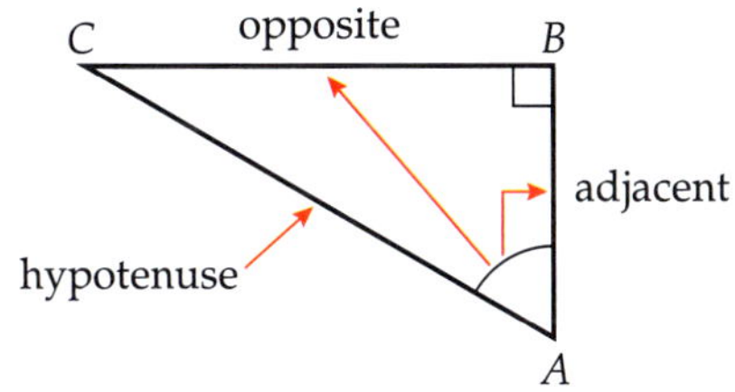

The ratio of the opposite side to the adjacent side gives a fraction that is called the '**tangent** of the angle'.

For example, the tangent of $\angle P = \dfrac{QR}{QP} = \dfrac{3}{10} = 0.3$.

That is, $\tan P = 0.3$.

In short tan P

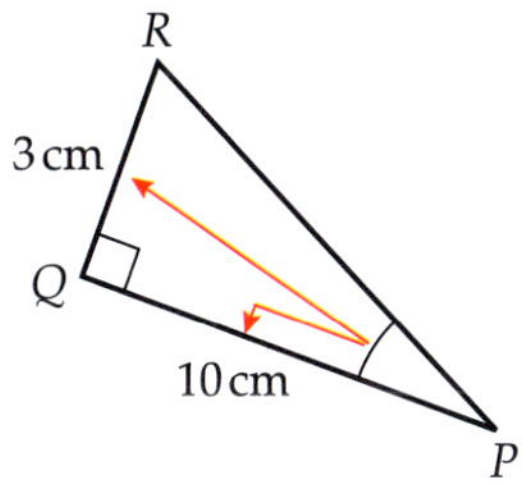

Angles are measured in **degrees**. There are 60 **minutes** in 1 degree.

We can use a calculator or a tangent table to find the value of the angles.

Using tables

The tables sometimes give the values as degrees and fractions of degrees and sometimes as degrees and minutes. Look in the body of the table for the number or the one closest to what you need e.g. 0.300.

Minutes / Degrees / Degrees	0′ 0.0°	6′ 0.1°	12′ 0.2°	18′ 0.3°	24′ 0.4°	30′ 0.5°	36′ 0.6°	42′ 0.7°	48′ 0.8°	54′ 0.9°
0	0.000	0.002	0.004	0.005	0.007	0.009	0.011	0.012	0.014	0.016
1	0.018	0.019	0.021	0.023	0.025	0.026	0.028	0.030	0.031	0.033
2	0.035	0.037	0.038	0.040	0.042	0.044	0.045	0.047	0.049	0.051
3	0.052	0.054	0.056	0.058	0.059	0.061	0.063	0.065	0.066	0.068
4	0.070	0.072	0.073	0.075	0.077	0.079	0.081	0.082	0.084	0.086
5	0.088	0.089	0.091	0.093	0.095	0.096	0.098	0.100	0.102	0.103
9	0.158	0.160	0.162	0.164	0.166	0.167	0.170	0.171	0.173	0.175
10	0.176	0.178	0.180	0.182	0.184	0.185	0.187	0.189	0.191	0.192
15	0.268	0.270	0.272	0.274	0.275	0.277	0.279	0.281	0.283	0.285
16	0.287	0.289	0.291	0.292	0.294	0.296	0.298	0.300	0.302	0.304
81	6.31	6.39	6.46	6.56	6.61	6.70	6.77	6.86	6.94	7.03
87	19.1	19.7	20.5	21.2	22.0	22.9	23.9	24.9	26.0	27.3
89	57.3	63.7	71.6	81.9	95.5	114	143	191	287	573

That is $\angle P \approx 16.7°$ or $16°42' \approx 17°$

Using the scientific calculator

Input [.] [3], then [inverse], then [tan]. The answer is 16.699.
Calculators differ so learn how to use your calculator efficiently.

Worked examples

1. What is the tangent of

a. 15° **b.** 10.5° **c.** 87° 24′?

Look at the purple squares in the tan table.

2. Find the angle that has a tangent of

a. 0.158 **b.** 6.460 **c.** 0.277.

Look at the blue squares in the tan table.

3. In the following diagrams, find

a. ∠A

b. MN.

a.

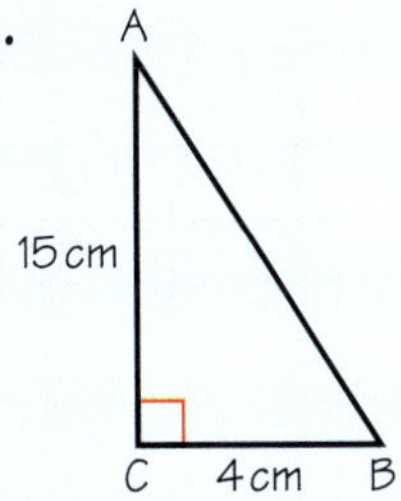

b.

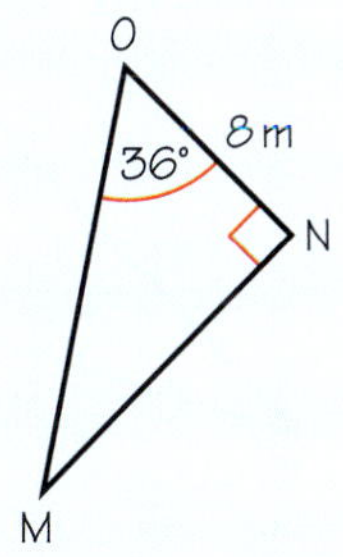

1. Using the table

a. 15° in the first column, under 0° in the 2nd column '0.268'

b. 10° in the first column, under 0.5° in 8th column '0.185'

c. 87° in the first column, under 24′ or 0.4° in the 7th column '22.0'. $24' = \left(\frac{24}{60}\right)^\circ = 0.4^\circ$.

Using the calculator

a. Input **15** then **tan** or **tan** then **15**

Remember calculators function differently.

b. As for **a.** but input 10.5.

c. For 87° 24′ you may have to change it to 87.4°.

2. a. 9°

b. 81.2° or 81° 12′

c. 15.5° or 15° 30′

See tables, or use the calculator as above.

3. a. $\tan A = \frac{\text{opp}}{\text{adj}} = \frac{4}{15} = 0.267$

So ∠A = 15°.

b. $\tan 36^\circ = \frac{ON}{MN} = \frac{8}{MN}$

So $0.727 = \frac{8}{MN}$

$MN = \frac{8}{0.727} = 11$ m.

Transposing.

Exercise

1. Find the tangent of the following angles.
   a) 19° **b)** 74° **c)** 25° **d)** 60° **e)** 59° **f)** 45°

2. Find the values of the following angles if the tangents of the angles are
   a) 0.719 **b)** 63.7 **c)** 1.402 **d)** 0.845 **e)** 0.007.

3. Calculate the tangent of the following angles.
   a) 11.5° **b)** 47.8° **c)** 55.9° **d)** 25.6° **e)** 7°54′

4. Work out the angles marked a, b, c in the following diagrams.

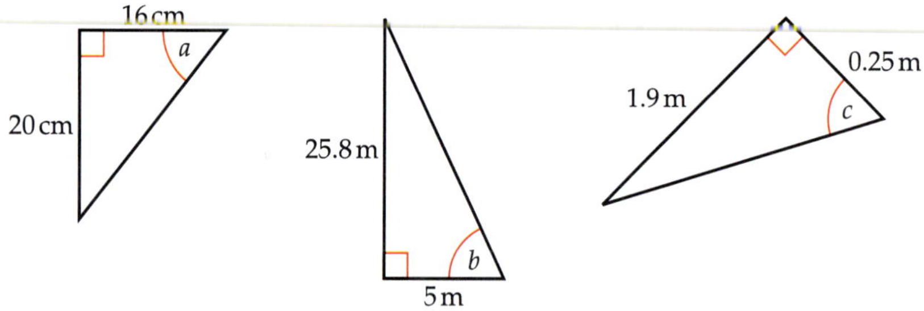

5. Calculate the sides marked n, p, q in the following diagrams.

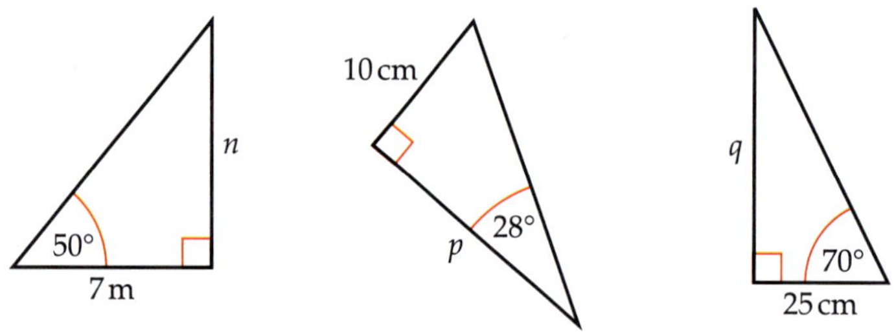

Key words
Sine
Cosine

Unit 7.3 The sine and cosine ratios

In this unit you will:

- **distinguish between the sine and cosine of an acute angle**
- **use tables or calculators to find the sine or cosine of an angle**
- **use tables or calculators to find an angle, given the sine or cosine**
- **determine the use of sine or cosine to find the values of angles or sides of right-angled triangles.**

Now we will examine ratios that involve the hypotenuse.

The **sine** of an angle, in a right-angled triangle, is the ratio $\frac{\text{opposite}}{\text{hypotenuse}}$

The **cosine** of an angle, in a right-angled triangle, is the ratio $\frac{\text{adjacent}}{\text{hypotenuse}}$

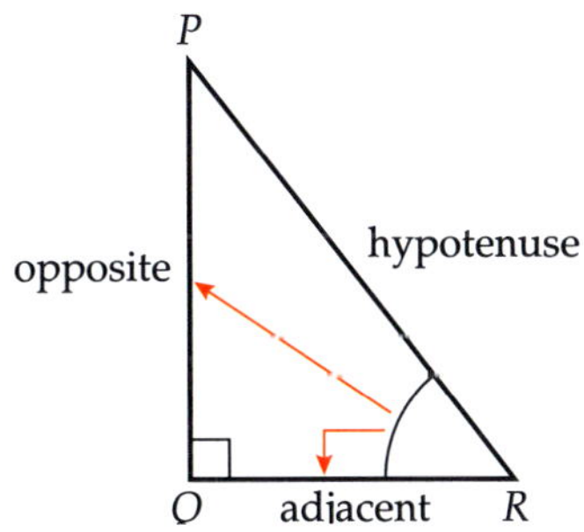

sine $\angle R = \frac{PQ}{PR}$ or $\frac{r}{q}$, cosine of $\angle R = \frac{QR}{PR}$ or $\frac{p}{q}$

Remember the lowercase letter represent the side opposite the angle with the corresponding uppercase letters e.g. $\angle R$ faces side r.

Practice using your tables and calculators to find the sine and cosine of angles. These tables are used the same way as you use the tangent tables. You input sin or cos instead of tan on the calculator.

Worked examples

1. Calculate the value of $\angle R$ and $\angle S$ in the figures.

a.

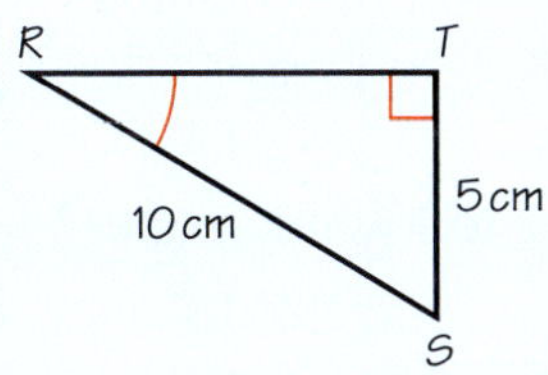

b.

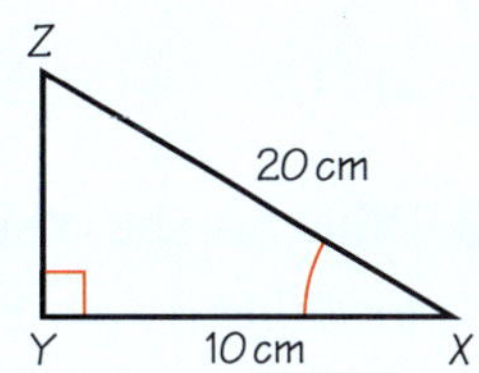

1. **a.** $\sin \angle Z = \frac{\text{opposite}}{\text{hypotenuse}} \left(\frac{O}{H}\right)$

In short, sin Z.

i.e. $\sin Z = \frac{5}{10} = 0.5$

$\angle Z = 30°$

b. cosine $\angle X = \dfrac{\text{adjacent}}{\text{hypotenuse}} \left(\dfrac{A}{H}\right)$

In short, cos X

i.e. $\cos X = \dfrac{10}{20} = 0.5$

$\angle X = 60°$

2. Use the sine or cosine ratios to calculate the unknown sides.

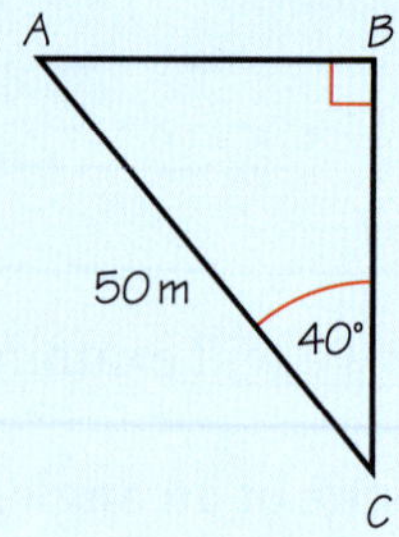

2. $\sin C = \dfrac{O}{H} = \dfrac{AB}{AC}$

$\sin 40° = \dfrac{c}{50}$ Substituting the values.

So $0.643 = \dfrac{c}{50}$ Simple equation.

$50 \times 0.643 = c$

$c = 32.2$ m

$\cos C = \dfrac{A}{H} = \dfrac{BC}{AC}$

$\cos 40° = \dfrac{a}{50}$ Substituting the values.

So $0.766 = \dfrac{a}{50}$ Simple equation.

$50 \times 0.766 = a$

$a = 38.3$ m

Exercise

1. Use a table or calculator to find the sine of
   a) 50° **b)** 45° **c)** 72° **d)** 60° **e)** 88° **f)** 90°.
2. Use a table or calculator to work out the sine of
   a) 35.4° **b)** 27.5° **c)** 72°48′ **d)** 61°12′ **e)** 8°30′.
3. Use a table or calculator to find the cosine of
   a) 17° **b)** 5° **c)** 59° **d)** 90° **e)** 30° **f)** 45°.
4. What are the cosines of the following angles?
   a) 70.1° **b)** 5.9° **c)** 57°18′ **d)** 36°24′ **e)** 45°42′

5 Use the sine or cosine ratio to find the values of the angles *a* to *e* in the following figures.

a)

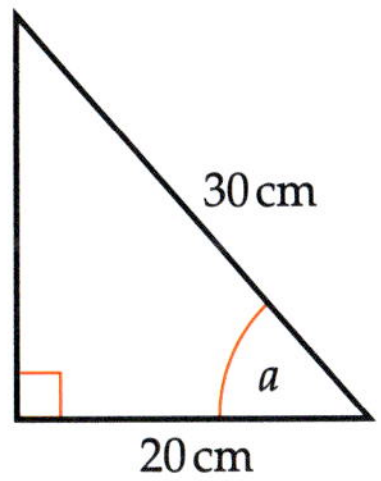

b)

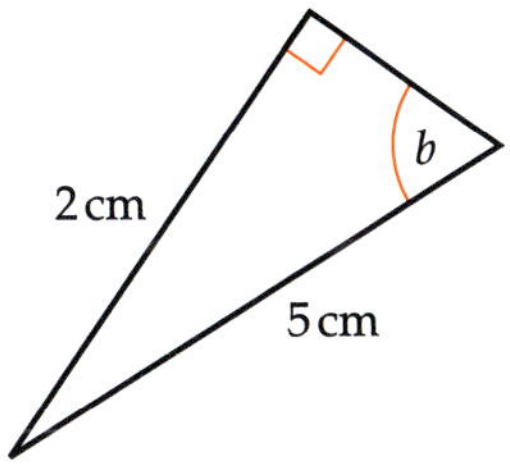

c)

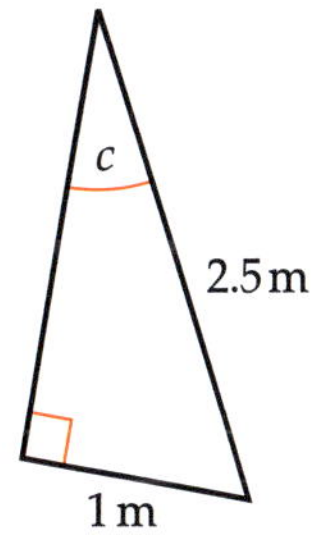

d)

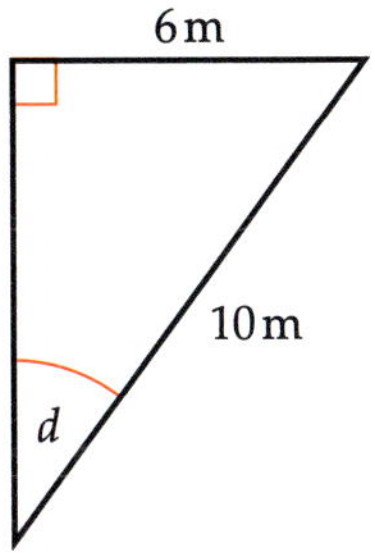

e)

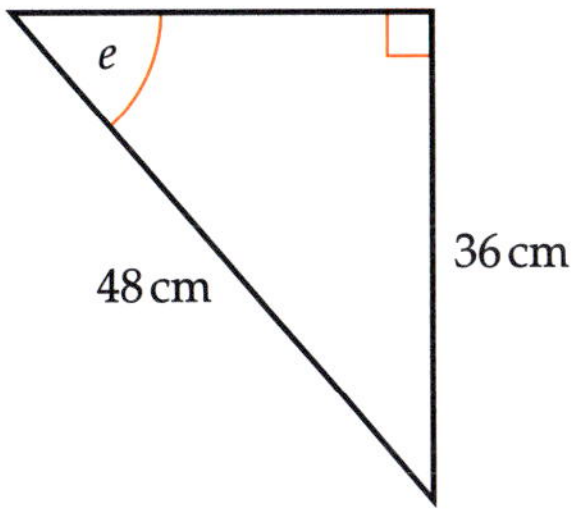

6 Use the sine or cosine ratio to calculate the value of the sides *p* to *t* in the following diagrams.

a)

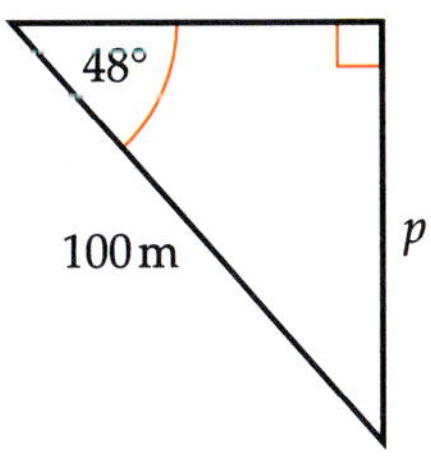

b)

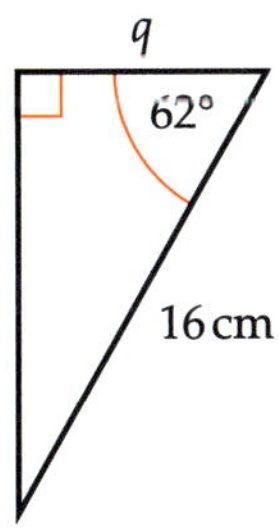

c)

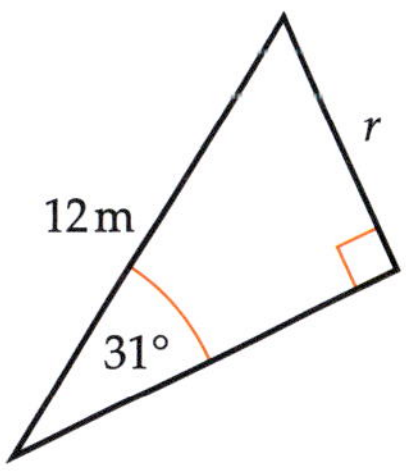

d)

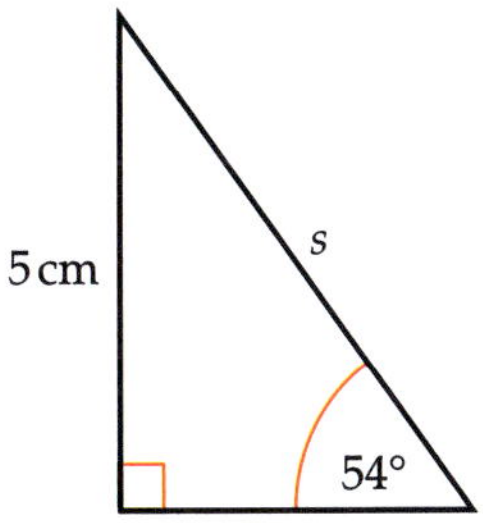

e)

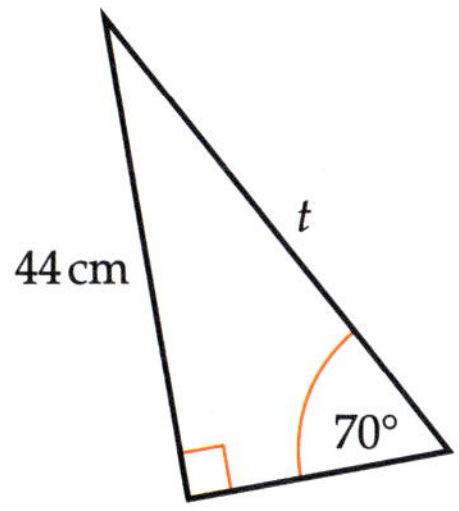

7 The two equal sides of an isosceles triangle are 15 cm. Two angles are 45° each. Do a sketch and put in the given information.

a) What is the length of the line of symmetry?

b) Work out the length of the third side.

8 The slant height of a cone is 8 cm. The angle between the radius and slant height is 65°.

a) Calculate the radius.

b) Find the vertical height.

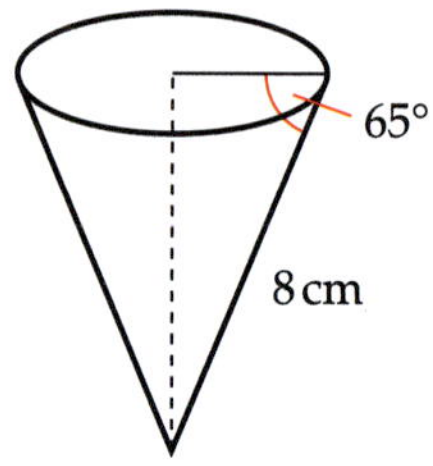

Investigation

9 Compare the three sets of values, tangent, sine and cosine, for angles from 0° to 90°. Which ones increase or decrease?

10 Compare the sine and cosine values of 60°, 30°, 45°, 15° and 75°. Describe any pattern you observe and use the pattern to check other sets of angles.

11 What is the tangent of 45°? Explain this result to a classmate.

12 Explain the relationship between the tangent of an angle and the tangent of a line to a classmate.

Key words
Angle of elevation
Angle of depression
Bearings

Unit 7.4 Using trigonometry to solve real-life problems

In this unit you will:

- **determine when and how to use a specific trigonometric ratio and/or Pythagoras' theorem.**

Let us summarise what we know about right-angled triangles, using triangle ABC, right-angled at B.

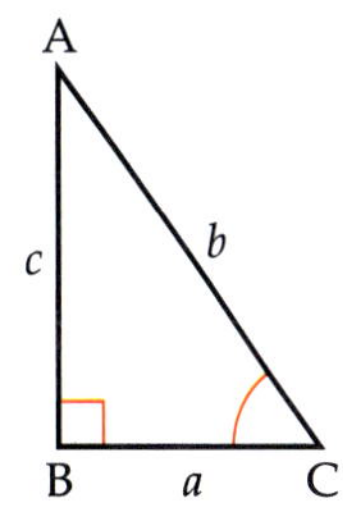

$b^2 = a^2 + c^2$

$\sin C = \frac{c}{b}$ $\quad\quad$ $\cos C = \frac{a}{b}$ $\quad\quad$ $\tan C = \frac{c}{a}$

S is $\frac{O}{H}$ $\quad\quad$ C is $\frac{A}{H}$ $\quad\quad$ T is $\frac{O}{A}$

Let us review what we know about **bearings** and other angles in real life.

- Bearing is always taken from the north line.
- **Angles of elevation** and **depression** are always taken from the line of sight or eye level.

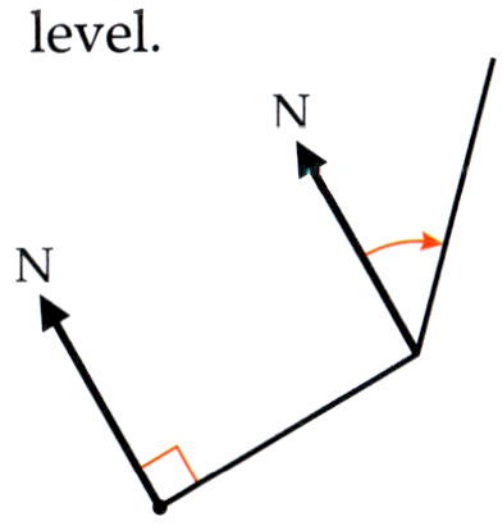

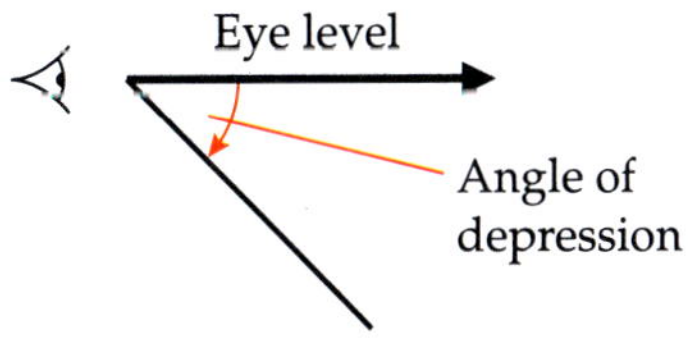

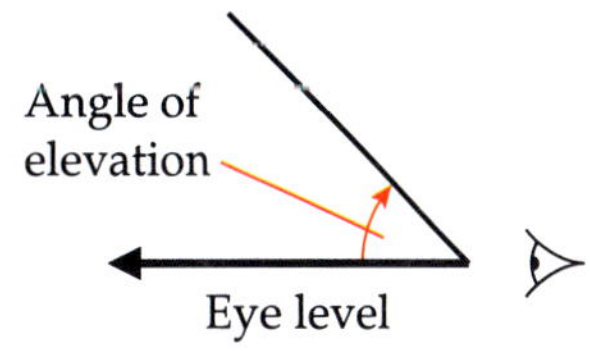

Heights, widths and depths in real life are usually measured perpendicularly.

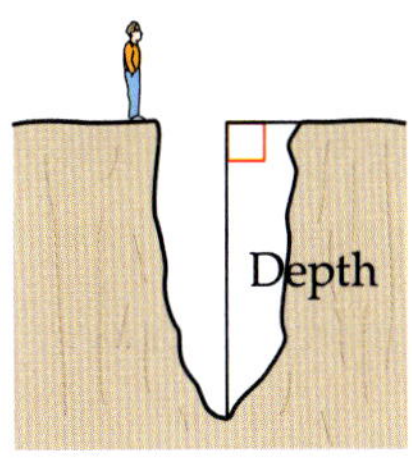

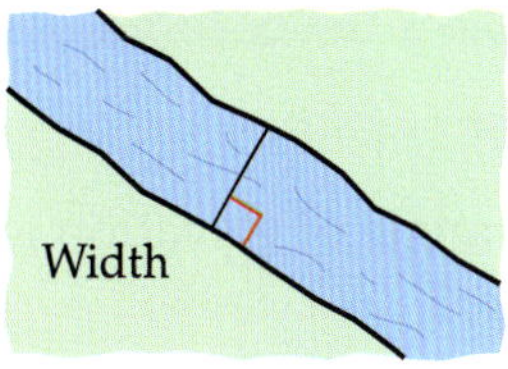

Worked example

A ferocious dog is tied to a post that is 5 metres from a path. The length of its chain is 6 m.

a. Calculate the length of the side of the path that may not be safe from the dog (to the nearest metre).

b. What is the angle through which the dog would turn to cover this distance?

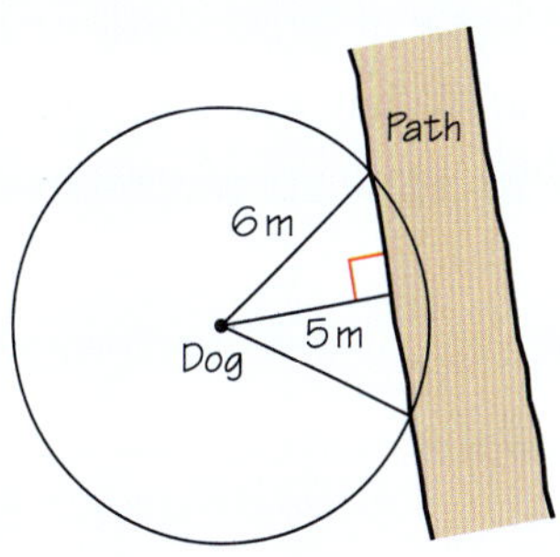

The dog can cover a circle if not restricted anymore. Part of the path is within his reach because his chain is longer than the distance he is tied from the path.

Let the required diagram be the isosceles triangle ABC.

The length BC is the distance along the path and $\angle A$ is the angle that the dog turns.

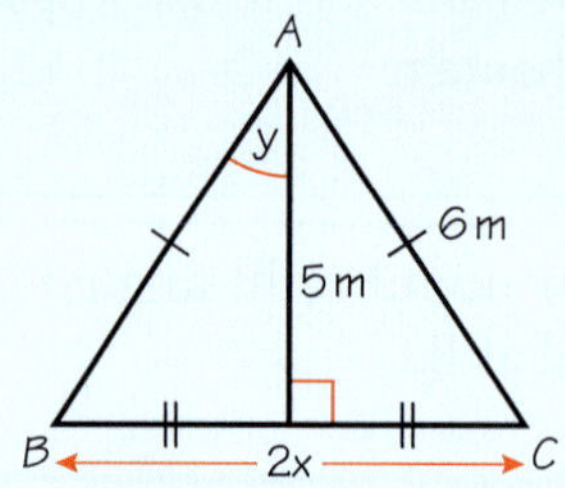

a. Let $BC = 2x$

$x^2 + 5^2 = 6^2$ Using Pythagoras' theorem.

$x^2 = 6^2 - 5^2 = 36 - 25 = 11$

$x = \sqrt{11} = 3.32$

$BC = 6.64 \approx 7$

7 m is not safe on one side of the path.

b. Let y be $\frac{1}{2}$ of $\angle A$

$\cos y = \frac{5}{6} = 0.833$

$y = 33.6°$

$\angle A = 2 \times 33.6 = 67.2°$

The dog turns about 67°.

Exercise

1 The slant height of the ramp in the diagram is 3 m. The angle between the slant height and the vertical height is 50°. Calculate the vertical height.

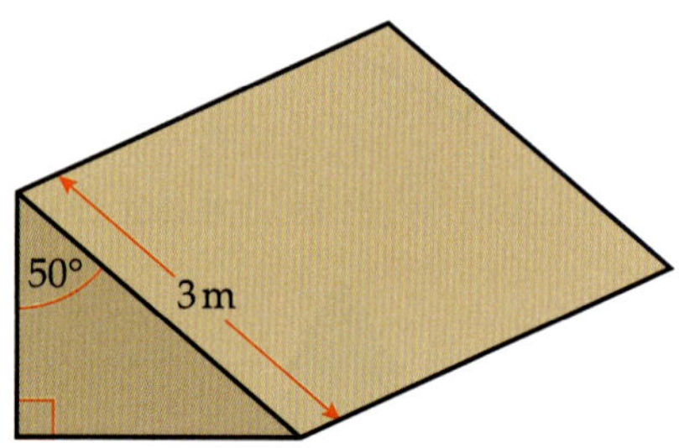

2 The angle of elevation is taken 12 m from a tower and is 65°. What is the height of tower?

3 The top of a ladder is resting 10 m up a wall. The angle between the wall and the ladder is 25°. How far from the wall is the foot of the ladder?

4 A man on a hill sights an object about 40 m out at sea. From the foot of the hill the angle of depression he used was 15°.

a) What was the vertical height of the man from sea level?

b) How far was he from the object?

5 Pat lives 2 km north of Mark's house.
Rupert's house is 1.5 km east of Mark's house.

a) Calculate the distance between Pat's house and Rupert's house.

b) What is the bearing of Rupert's house from Pat's house?

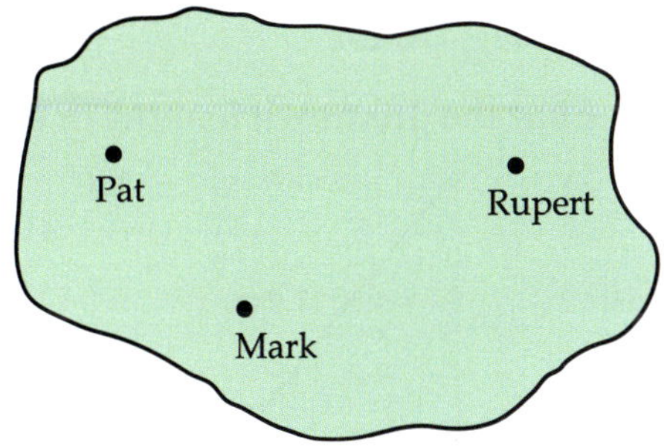

6 A rope 16 m long is tied to the top of a vertical pole and anchored on the ground. The angle of elevation from the point that it is anchored to the top of the pole is 30°.

a) What is the height of the pole?

b) How far from the foot of the pole is the rope anchored?

7 Sissy and Joan are on the same bank of a river that runs north to south. Sissy is 24 m north of Joan and sees Mr Lindsay's house is east of her. The house is on a bearing of 72° from Joan. How far is the house from each girl?

8 Calculate the angles and the sides of the kite *ABCD* if the line of symmetry is 10 cm and the shorter diagonal is 6 cm. *X* is the intersection of the diagonals.

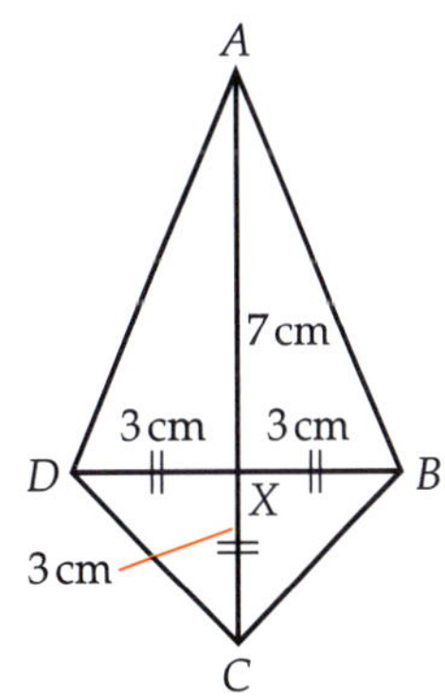

Investigation

9 Make a list of five careers in which trigonometry is used in different ways, describing briefly how it is used in each case.

Looking back

1 Find the unknown lengths in each of these right-angled triangles.

a)

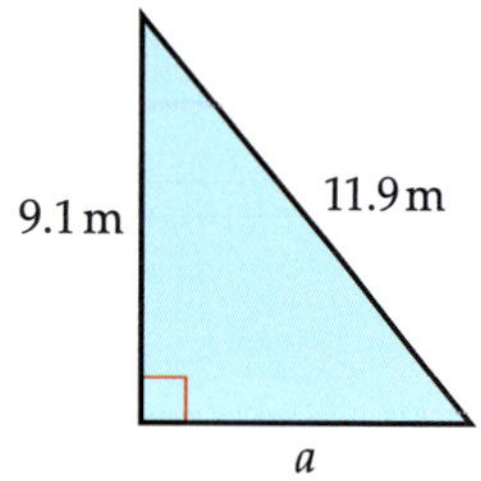

b)

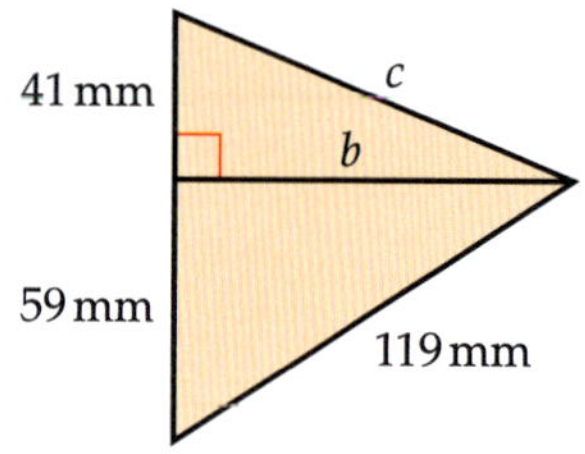

c)

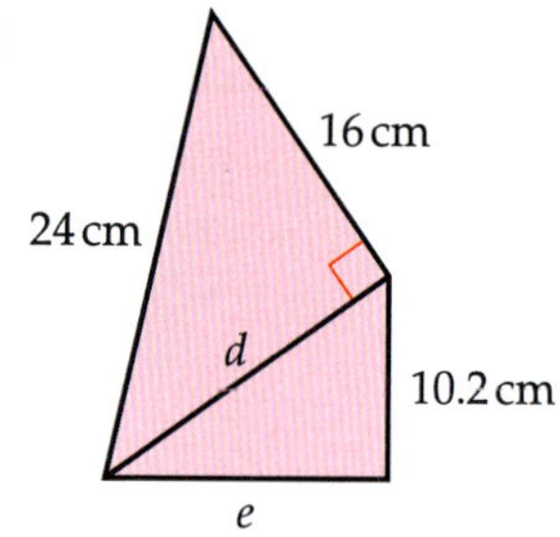

2 A square tin of biscuits has sides of length 34 cm. It is decorated with a ribbon across one diagonal. What is the length of the ribbon?

3 For the triangle shown write down in terms of a, b and c each of the following.

a) $\sin B$ **b)** $\cos B$ **c)** $\tan B$

d) $\sin C$ **e)** $\cos C$ **f)** $\tan C$

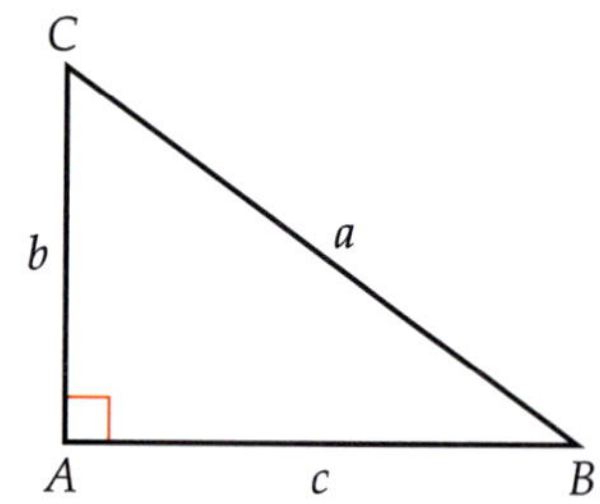

4 For the triangle shown, calculate

a) BC

b) AB.

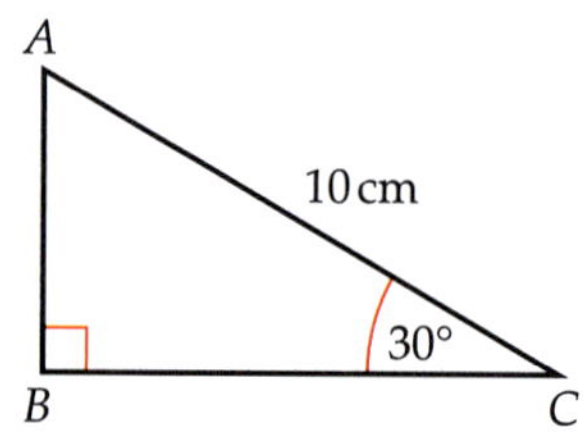

5 In a triangle PQR, $\angle Q = 90°$, $PR = 60$ mm and $PQ = 30$ mm. Calculate

a) $\angle R$ **b)** $\angle P$.

6 In a triangle ABC, $AB = 84$ cm, $\angle C = 84°$ and $\angle B = 90°$. Calculate the length of AC.

7 Jessica is standing on a grandstand 15 m high. She is looking at the top of a shed roof which has height of 8 m. She works out that the angle of depression of the top of the roof is 8°. The grandstand and the shed are both on level ground.

a) Draw a diagram to show this information.

b) Work out the distance from the grandstand to the roof of the shed.

Key words
Convex
Polygon
Pentagon
Regular polygon
Hexagon
Octagon

Polygons

Chapter 8

This chapter covers regular and irregular polygons and shows you how to calculate and use interior and exterior angles of polygons. You will also revise the construction of triangles and quadrilaterals.

Unit 8.1 Regular and irregular polygons

In this unit you will:

- learn the difference between a regular and an irregular polygon
- name a polygon according to the number of sides.

A **polygon** is a region of a plane bounded by straight lines.

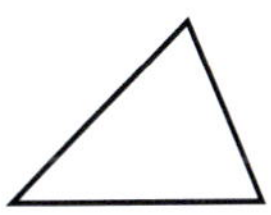

A polygon with 3 sides is called a triangle.

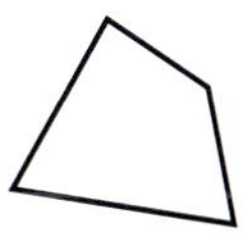

A polygon with 4 sides is called a quadrilateral.

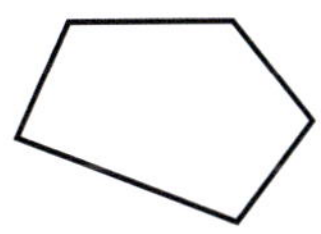

A polygon with 5 sides is called a **pentagon**.

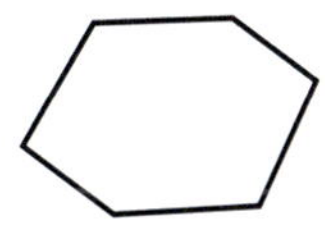

A polygon with 6 sides is called a **hexagon**.

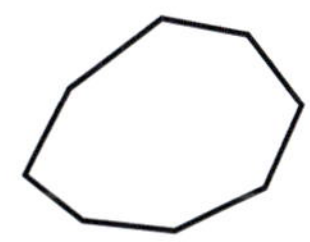

A polygon with 8 sides is called an **octagon**.

A polygon with n sides is called a n-gon.
When all the vertices point outwards, the polygon is **convex**.
All the polygons above are convex.
This polygon is concave:

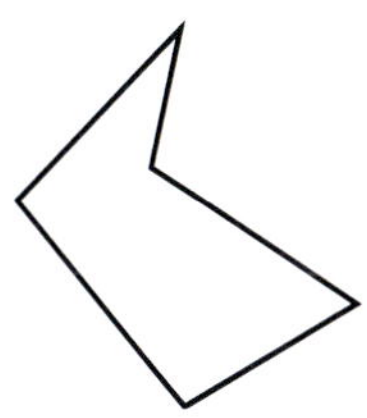

A **regular polygon** has all its sides the same length and all its angles equal.
All other polygons are irregular.

These polygons are regular.

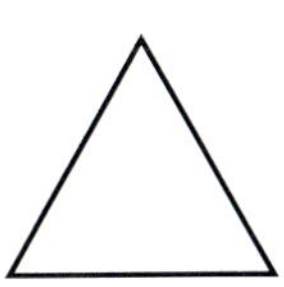

Equilateral triangle

Square

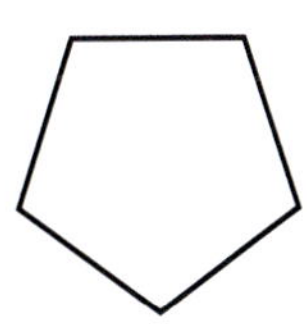

Regular pentagon

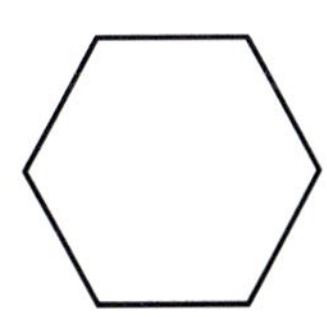

Regular hexagon

Worked example

In this polygon, AB = BC = CD = DE = EA.

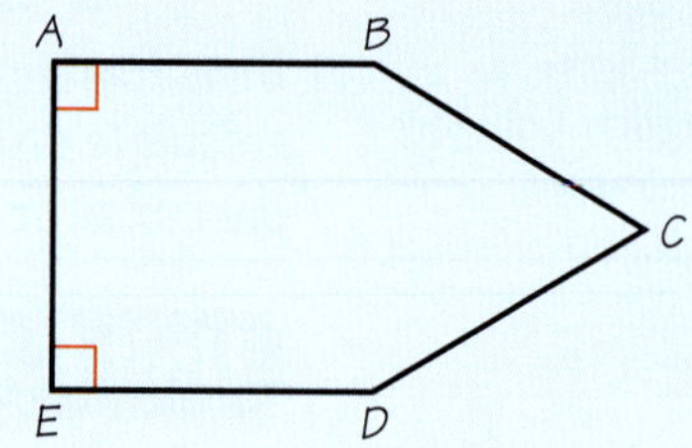

a. Name the polygon.

b. Is this a regular polygon? Give a reason for your answer.

c. By joining two vertices, show how this polygon can be divided into two regular polygons and name them.

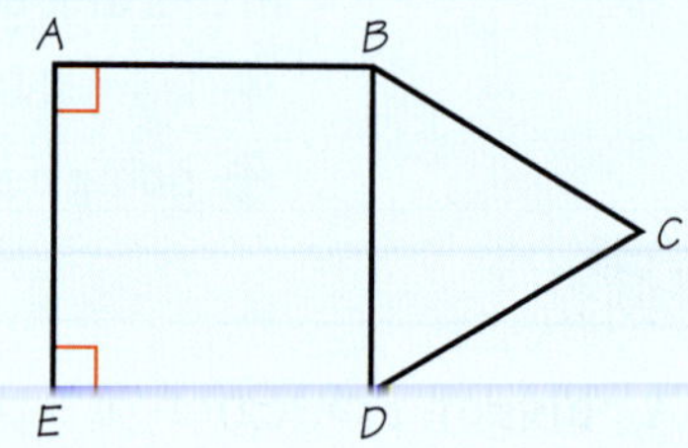

a. A pentagon.

b. No, because the angles are not all equal.

c. Join BD.
This gives a square ABCD and an equilateral triangle BDC.

Exercise

1 Name each of these polygons.

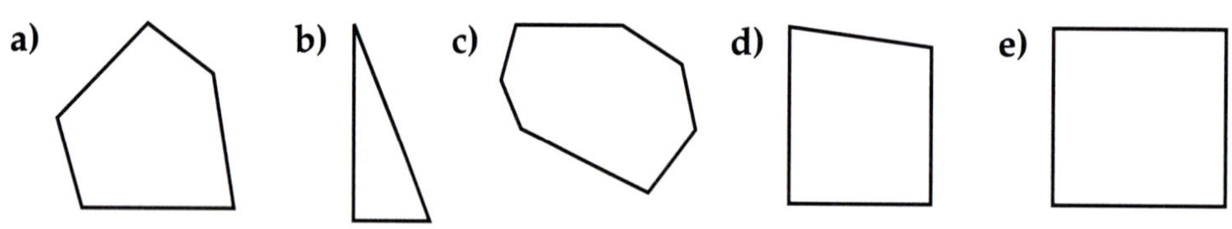

2 The polygons below are not drawn accurately. Use the information given about each one to determine whether it is regular. Give a reason for your answer.

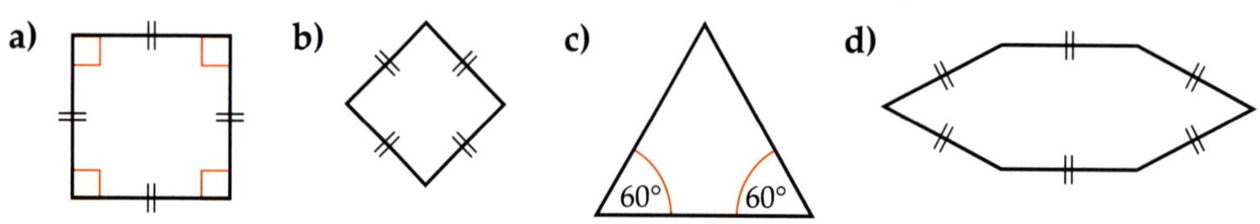

3 This is an equilateral triangle.

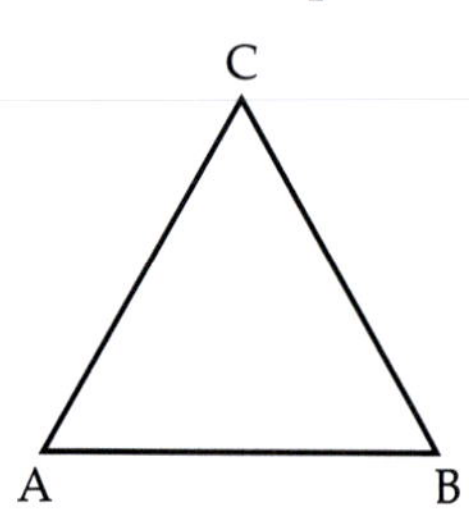

a) Copy the diagram and draw the reflection of triangle ABC in the line BC.
Label the third vertex of the reflection D.

b) On the same diagram draw the reflection of triangle ABC in AC.
Label the third vertex of the reflection E.

c) Describe EDC.

d) Reflect triangle ABC and the two images in EDC.
Describe the resulting figure.

4 In this diagram BA = AD = DC = BE = EC.

a) Name this shape

b) By adding two lines to the diagram, describe how this polygon can be divided into three regular polygons and name them.

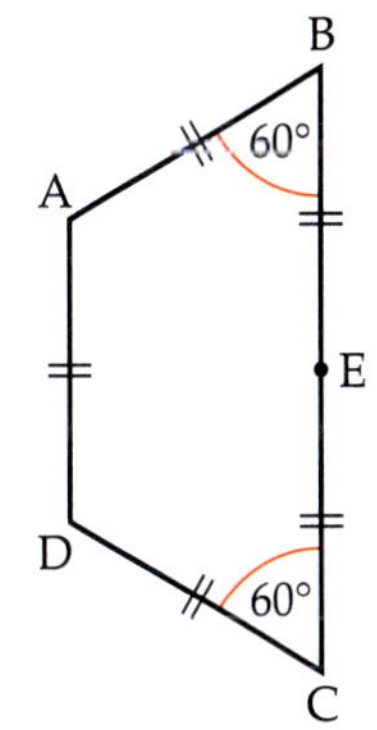

Investigation

5 Find out which regular polygons will tessellate.

6 Find a way of predicting whether a plane shape will tessellate.

Unit 8.2 Angle sums of polygons

In this unit you will:

- calculate sums of interior and exterior angles of irregular quadrilaterals, pentagons and hexagons.

Key words
Interior angle
Angle sum
Pentagon
Hexagon

In order to calculate the sum of the **interior angles** of a quadrilateral, we can split it into triangles.

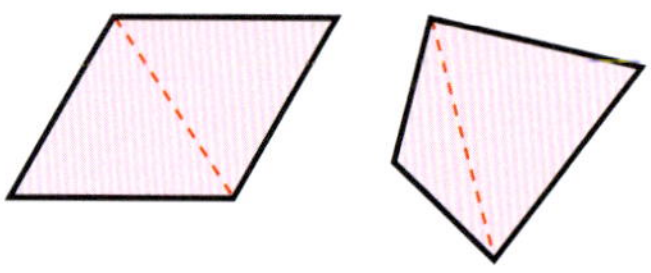

Each triangle has an **angle sum** of 180°. There are two triangles, so the total angle sum is $2 \times 180° = 360°$.

This is true of any quadrilateral, so the interior angle sum of a quadrilateral is 360°.

In order to calculate the sum of the **interior angles** of a **pentagon**, we can split it into triangles:

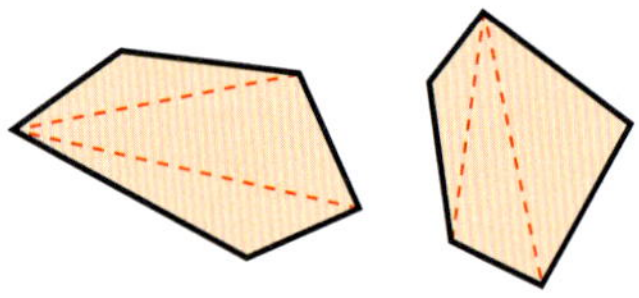

Each triangle has an angle sum of 180°. There are three triangles, so the total angle sum is $3 \times 180° = 540°$.

This is true of any pentagon, so the interior angle sum of a pentagon is 540°.

In order to calculate the sum of the **interior angles** of a **hexagon**, we can split it into triangles:

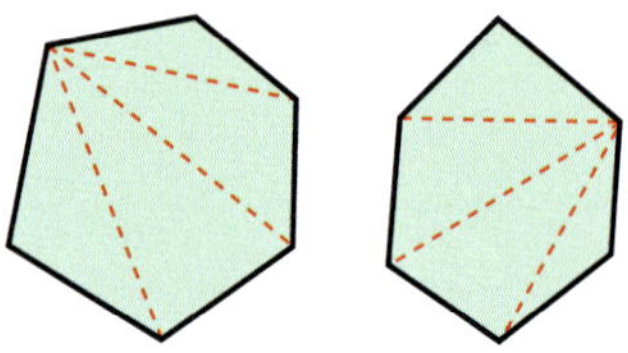

Each triangle has an angle sum of 180°. There are four triangles, so the total **angle sum** is $4 \times 180° = 720°$.

This is true of any hexagon, so the interior angle sum of a hexagon is 720°.

Worked example

Calculate the size of the lettered angles, giving reasons for your working.

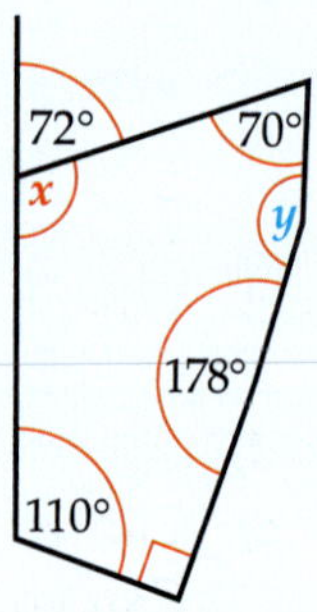

a) $x = 180 - 72$ (Angles on a straight line add up to 180°.)

$= 108°$

b) $y = 720° - 108° - 70° - 178° - 90° - 110°$

$= 164°$ (The sum of the angles of a hexagon is 720°.)

Exercise

1 Calculate the size of the lettered angles. Give reasons for your working.

a)

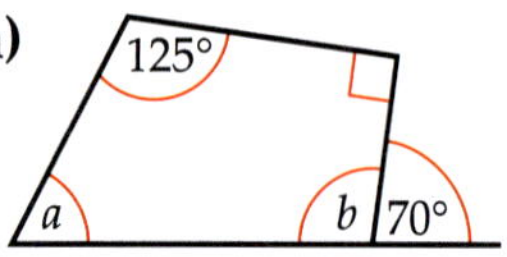

b)

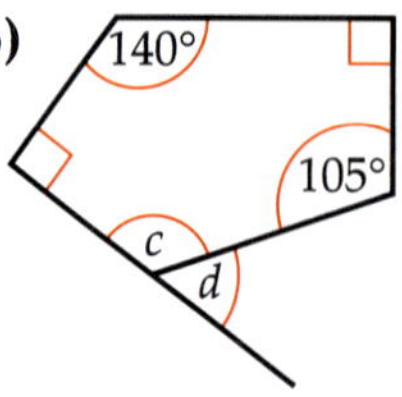

c)

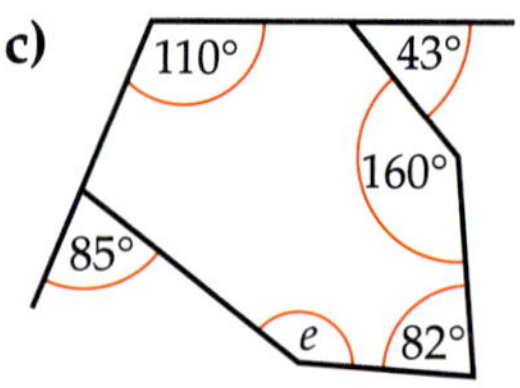

2 Calculate the sizes of the following angles, stating clearly the angle facts that you use.

a) ∠AED **b)** ∠ADE **c)** ∠ADC

d) ∠BCD **e)** ∠ABC.

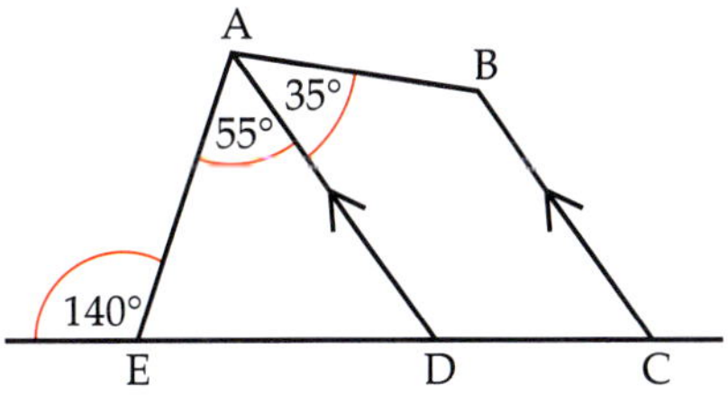

3 Calculate the value of x, stating clearly any angle facts you use.

a)

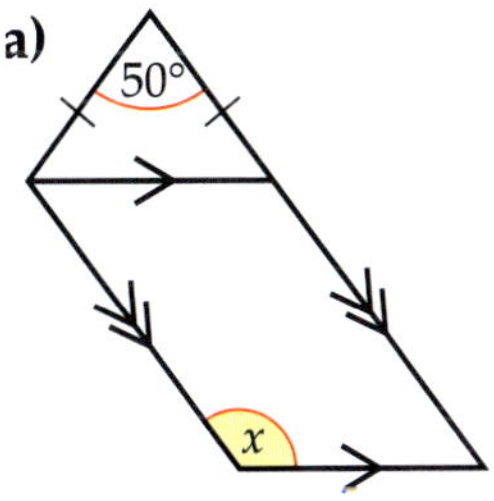

b)

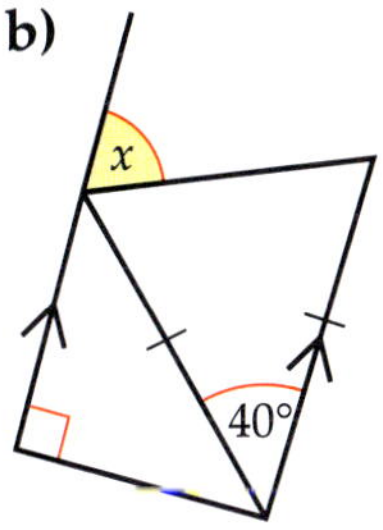

4 Calculate the value of x, stating clearly any angle facts you use.

a)

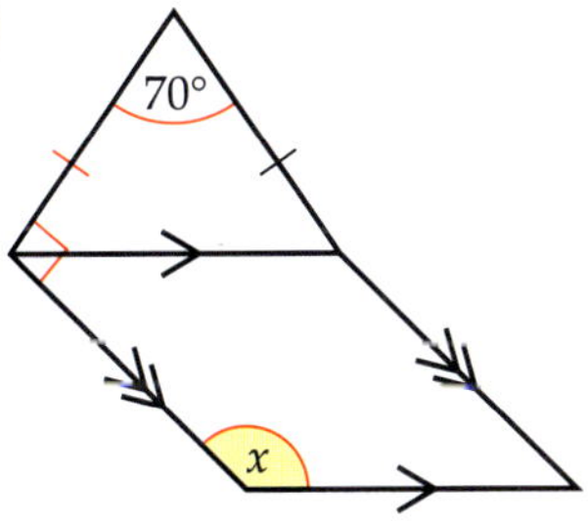

b)

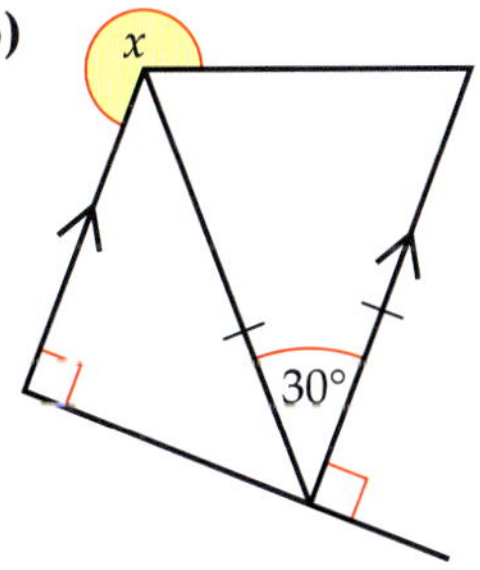

5 **a)** Draw any pentagon and measure the size of the interior angles. Mark them on your diagram. Calculate the size of each of the exterior angles and add them together. Write down your total.

b) Draw any hexagon and measure the size of the interior angles. Mark them on your diagram. Calculate the size of each of the exterior angles and add them together. Write down your total.

c) What do you notice about the total sum of the exterior angles of the pentagon and hexagon?

d) Is the same true for other polygons?

e) Can you explain why this happens? Can you give a proof?

Try triangles and quadrilaterals first.

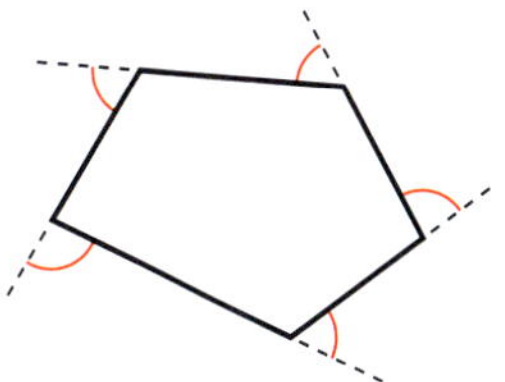

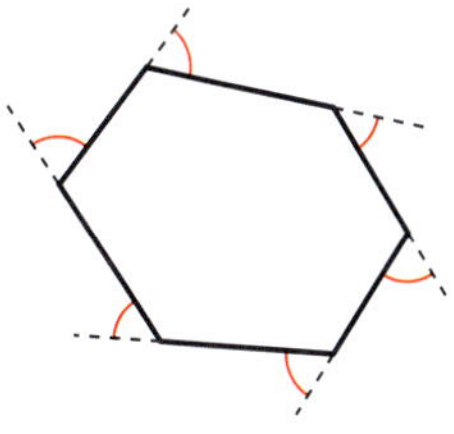

Investigation

6 To find the sum of the angles of a polygon, it can be split into triangles.
A pentagon can be split into three triangles, and a hexagon can be split into four triangles.
Investigate the number of sides of a polygon and the number of triangles it can be split into.
What do you notice?

Unit 8.3 Regular polygons

In this unit you will:

- find, calculate and use the interior and exterior angles of regular polygons.

Key words
Polygon
Interior
Regular
Exterior
Vertex

A **polygon** with n sides can be split into $n - 2$ triangles. Each triangle has an angle sum of 180°, so the **interior** angle sum of any polygon is $(n - 2) \times 180°$.

There are always two fewer triangles than the number of sides of the polygon.

A **regular** polygon has all sides of equal length and all angles equal.

The sum of the interior angles of a pentagon is
$(n - 2) \times 180° = 3 \times 180° = 540°$
Each interior angle of a regular pentagon is $540° \div 5 = 108°$.

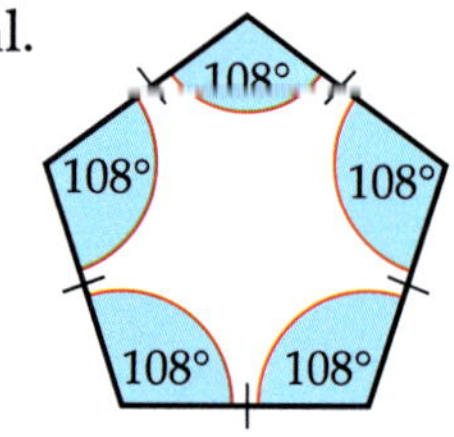

The sum of the **exterior** angles of any polygon is 360°.
Imagine this pentagon as a path. At each **vertex** there is a change of direction.
By walking around the complete polygon so that you finish by facing in the same direction as you started, a whole turn of 360° will be made.
The exterior angles sum to 360°.

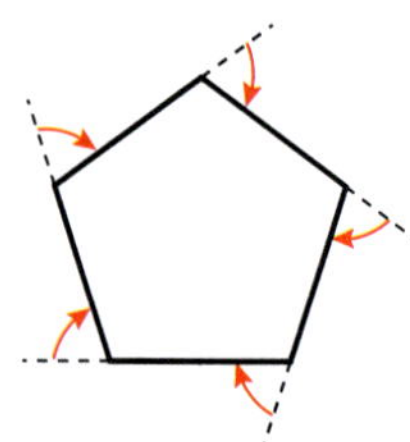

If the exterior angle sum is 360°, then each exterior angle of a regular polygon is $\frac{360°}{n}$.

For a regular pentagon, this is:
$\frac{360°}{n} = \frac{360°}{5} = 72°$

Since the interior and exterior angles sum to 180°, each interior angle of a regular polygon is $180° - \frac{360°}{n}$.

For a regular pentagon, this is:
$180° - \frac{360°}{n} = 180° - \frac{360°}{5} = 108°$

Worked examples

1. Calculate the sizes of the exterior and interior angles of a regular 10-sided polygon.

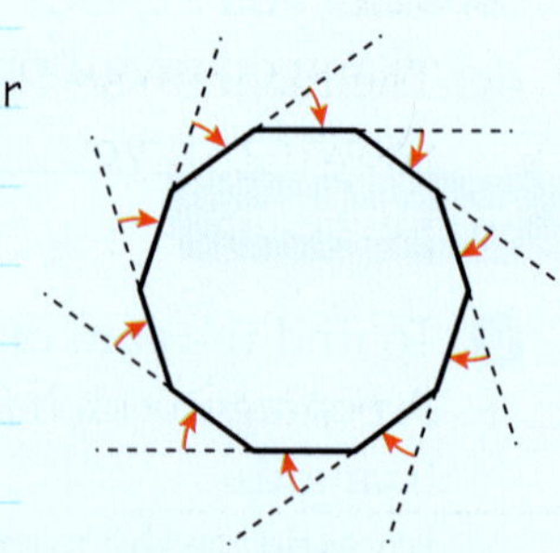

1. Each exterior angle of a regular polygon $= \frac{360°}{n} = \frac{360°}{10}$
$= 36°$

Each interior angle $= 180° - 36° = 144°$.

The interior and exterior angles sum to 180°.

Substitute **10** into the formula.

2. A regular polygon has interior angles of 150°. How many sides does it have?

2. The exterior angle $= 180° - 150°$

$= 30°$

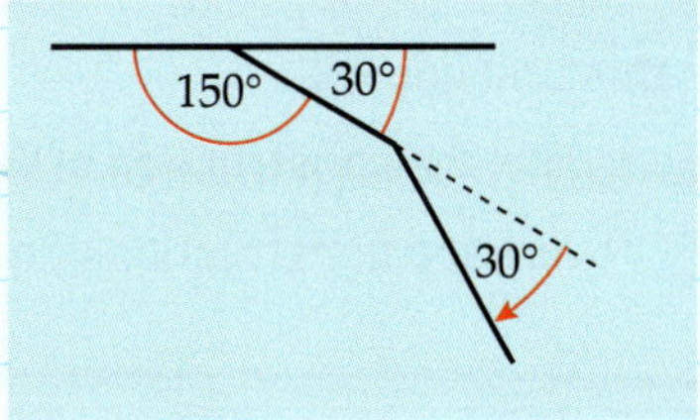

Each exterior angle of a regular polygon $= \frac{360°}{n}$, where n is the number of sides.

If $\frac{360°}{n} = 30°$, then $30° \times n = 360°$ and $n = 360° \div 30° = 12$.

The polygon has 12 sides.

Exercise

1. Calculate the size of each interior angle in a regular octagon.

2. Calculate the size of the interior and exterior angles of a regular 18-sided shape.

3. Calculate the size of the interior and exterior angles of regular polygons with the following number of sides:
   a) 8 **b)** 15 **c)** 24 **d)** 60.

4. The interior angle of a regular polygon is 162°. How many sides does it have?

5. A regular polygon has exterior angle x and interior angle 160°. How many sides has the polygon?

6. In a regular polygon, each interior angle is 120° greater than each exterior angle. Calculate the number of sides that the polygon has.

Investigation

7. The diagram shows a regular octagon. Work out angle x, showing all your working.

   Investigate other triangles formed by joining adjacent vertices to point P.

   Is it possible to calculate all the angles at P?

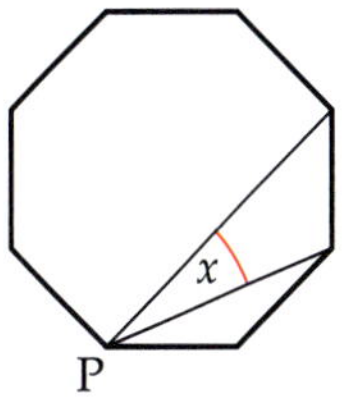

Unit 8.4 Constructing triangles and quadrilaterals

Key words
Hypotenuse

In this unit you will:

- review the construction of triangles
- review the construction of quadrilaterals.

Last year you learnt how to construct unique triangles given:

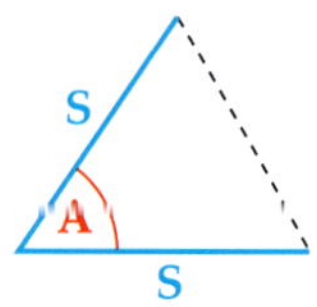

two sides and the included angle (SAS)

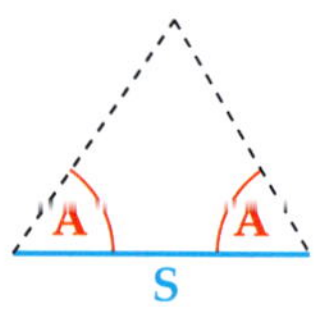

two angles and the side between them (ASA)

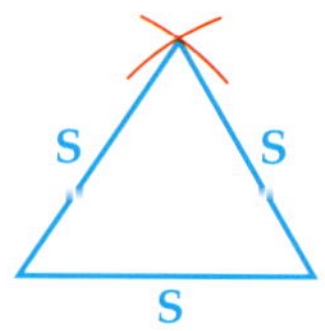

three sides (SSS)

If we are given this information, we can be sure that the triangle we are drawing is the only one of its kind.

The **hypotenuse** is the longest side of a right-angled triangle.

In Year 2 we practised constructing the perpendicular from a point on a line. Draw two arcs equidistant from P. From each arc, draw two more arcs with the same radius that intersect. Join P to the intersection, to give the perpendicular.

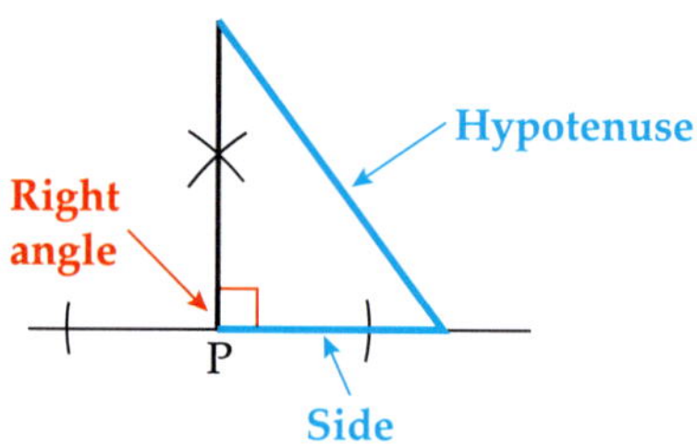

If we know the lengths of the **hypotenuse** and one of the sides, we can construct a right-angled triangle (RHS).

Last year you also learnt how to construct a quadrilateral using the properties of that quadrilateral.

Worked examples

1. Using a ruler and a pair of compasses only, construct a right-angled triangle ABC with A = 90°, BC = 5 cm and AC = 3 cm. Measure and write down the length of AB and the sizes of ∠B and ∠C.

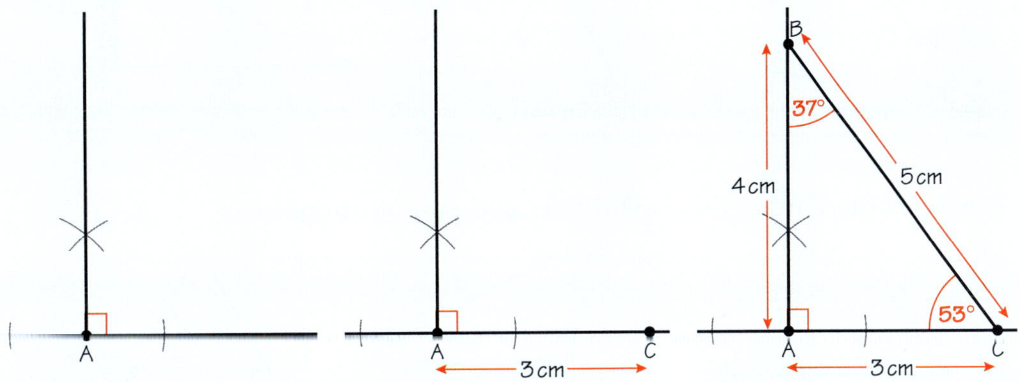

First, draw a line more than 3 cm in length. Mark a point A. Construct the perpendicular from A and draw in the right angle.

Measure a horizontal distance 3 cm from A and mark in the point C.

Open your compasses to 5 cm and place the point on C. Draw an arc that intersects with the line perpendicular to AC. Mark this point B.

AB = 4 cm.
∠ABC = 37° and ∠BCA = 53°.

Finally, measure the length of AB and the size of the angles ∠ABC and ∠BCA.

2. Construct this polygon using a ruler and protractor.

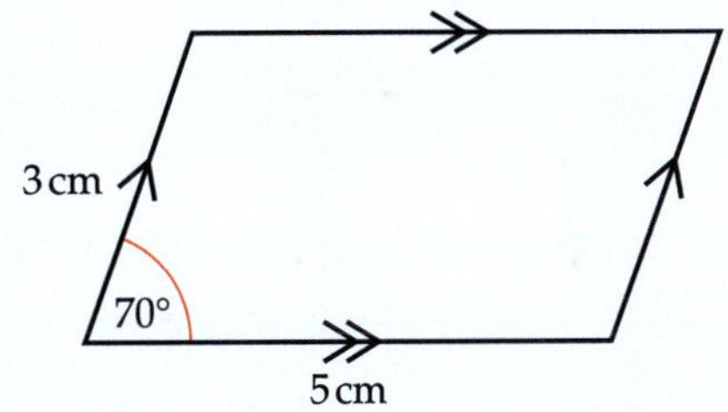

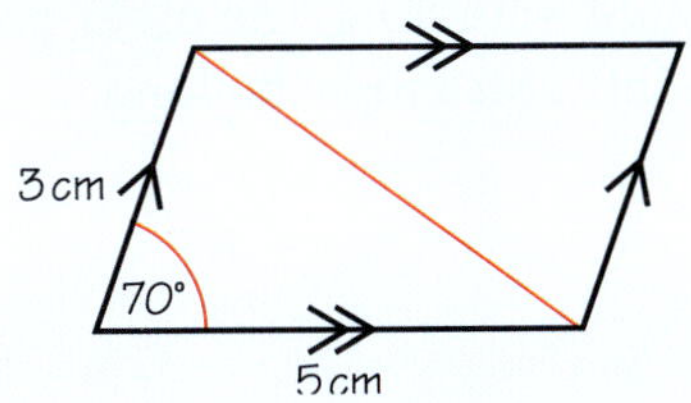

Draw a sketch, marking all the measurements needed. We can use the fact that the opposite sides of a parallelogram are the same length and that a diagonal divides the parallelogram into two identical triangles.

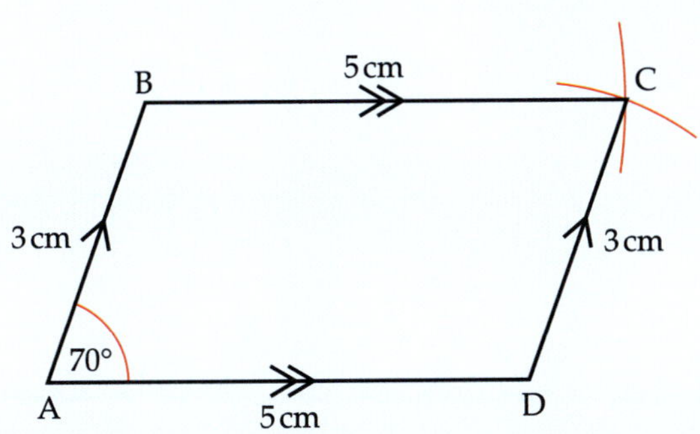

Draw AD first, then ∠A, then AB. Use compasses to draw arcs from D (radius 3 cm) and from B (radius 5 cm) to intersect at C. Join DC and BC.

Exercise

1. Construct these triangles using a ruler and protractor. Measure and write down the sizes of the angles and sides not given.

a)

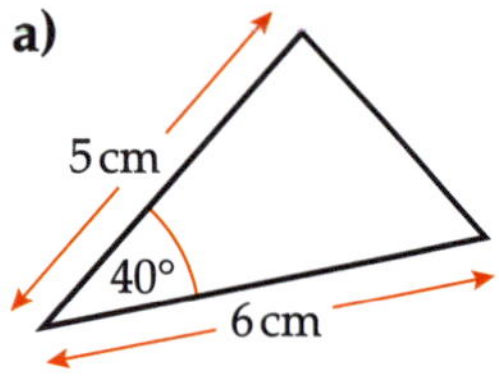

b)

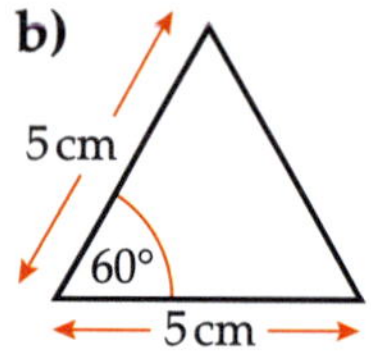

c)

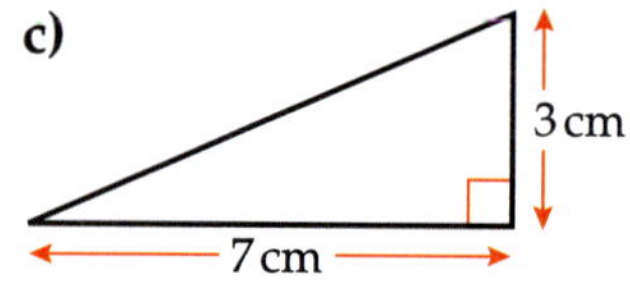

2. Construct these triangles using a ruler and pair of compasses.

a)

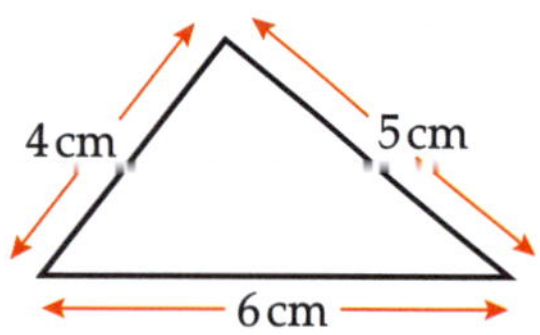

b)

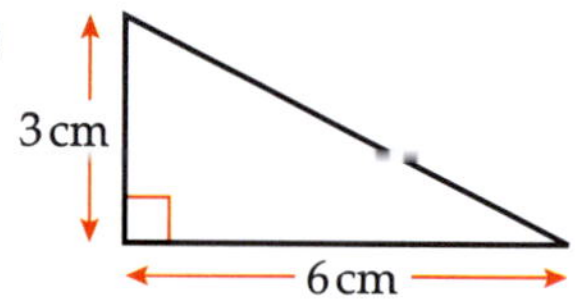

c)

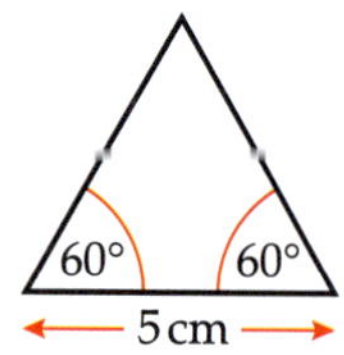

3. Construct these quadrilaterals using a ruler, protractor and a pair of compasses.

a) Measure BD.

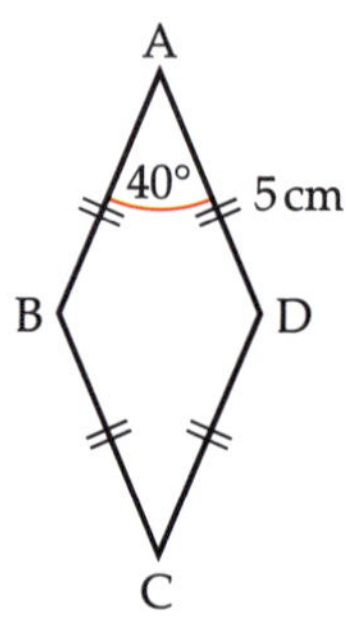

b) Measure AD and ∠D.

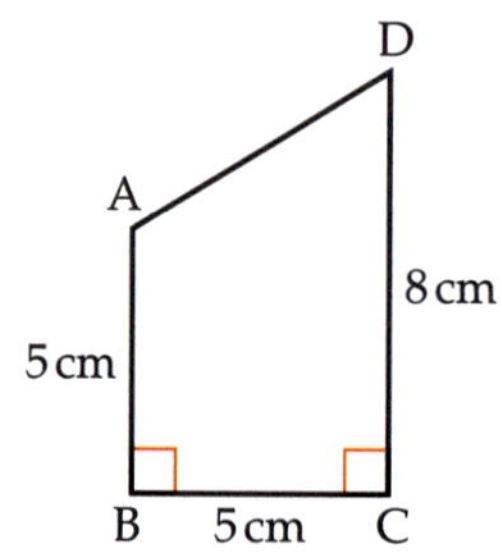

c) Measure DC.

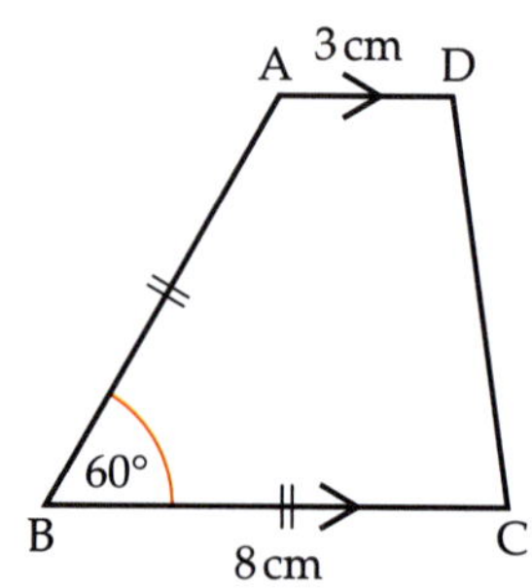

4. Construct the parallelogram ABCD in which AB = 5 cm, BC = 8 cm and ∠B = 45°. Measure the diagonal AC.

5. Construct the kite in which one pair of adjacent sides are 5 cm long, the other pair of adjacent sides are 9 cm long and the equal angles are 150°. Measure the long diagonal.

Key words
Polygon
Regular
Inscribe

Unit 8.5 Constructing regular polygons

In this unit you will:

- construct polygons using the properties of compound shapes
- construct regular polygons using a circle and the central angle.

We can construct a **polygon** when we are given enough information about it by dividing the shape into triangles and quadrilaterals.

This pentagon

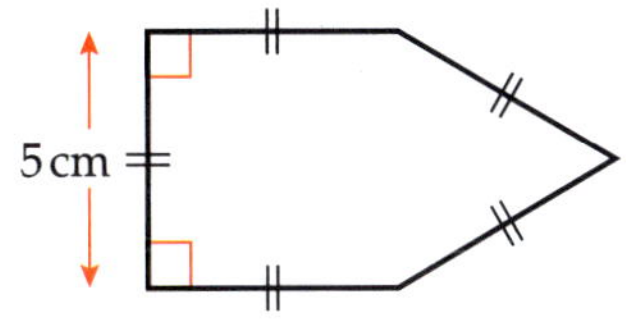

can be divided into a square and an equilateral triangle:

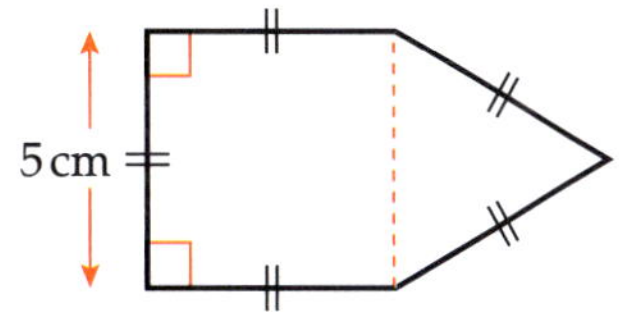

We can construct a **regular** polygon by **inscribing** it in a circle. We do this by dividing the 360° angle at the centre of the circle by radii into the same number of equal parts as the number of sides of the polygon.

This gives equal sectors of the circle. We then draw the chords joining adjacent radii.

This shows a regular octagon inscribed in a circle.

There are 8 identical sectors, each with a central angle of 45° because 360° ÷ 4 = 45°.

Inscribing a figure in a circle means that every vertex touches the circle.

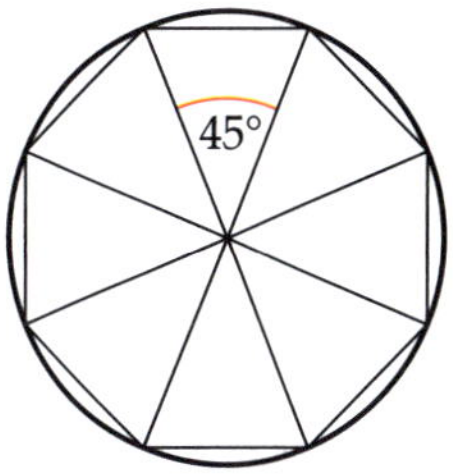

Worked example

Draw a circle of radius 5 cm. Use the circle to construct a regular pentagon.

The central angle for each sector = 360° ÷ 5 = 72°.

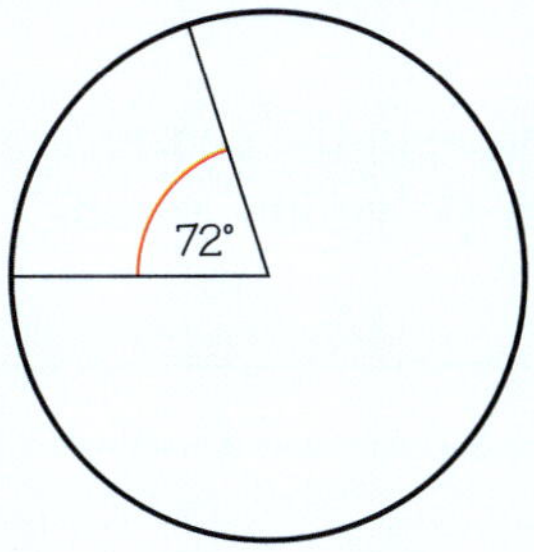

Draw the circle. Then draw one radius and use a protractor to draw another radius at 72° to the first radius.

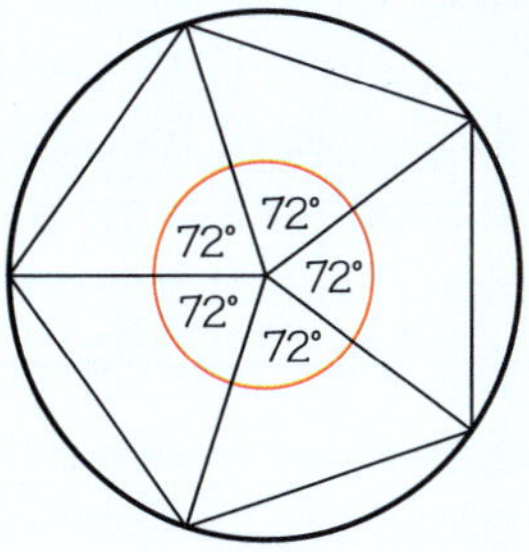

Use a protractor to draw the other radii so that the central angles are all 72°. Join the ends of the radii.

Exercise

1. Draw a circle of radius 6 cm. Use this circle to construct a regular pentagon. Measure and write down the length of a side of the pentagon.

2. Draw a circle of radius 7 cm. Use this circle to construct a regular 9-sided polygon. Measure and write down the length of a side of the polygon.

3. Draw a circle of radius 7 cm. Use this circle to construct a regular 7-sided polygon. Measure and write down the length of a side of the polygon.

4. **a)** Draw a circle of radius 6 cm. Use this circle to construct a regular hexagon.
   b) Measure and write down the length of the side of the hexagon.
   c) Hence describe a method for drawing the same regular hexagon without drawing the central angles.

5. **a)** This diagram shows a regular hexagon inscribed in a circle with centre O.

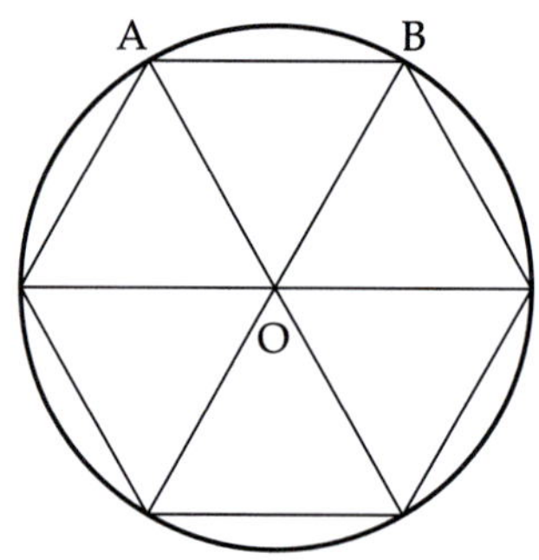

   Describe triangle AOB.
   b) Construct a regular hexagon with sides 4 cm long using only a ruler and pair of compasses.

6. Draw a circle of radius 6 cm. Use this circle to construct a regular octagon. Measure and write down the length of a side of the octagon.

Investigations

7. It is possible to construct a regular octagon without drawing a circle or bisecting an angle and using a ruler and pair of compasses only. Investigate how this can be done.

8. **a)** Construct a regular pentagon inscribed in a circle of radius 8 cm.
   b) Draw the diagonals of the pentagon. What is the smaller shape created in the centre?
   c) Draw the diagonals of this shape. What is the smaller shape in the centre?
   d) Try going through the same procedure with other regular polygons, joining each vertex to the next-but-one vertex.

Looking back

1 Find the perimeter of each of the following regular polygons. Give your answers correct to one decimal place.

a)

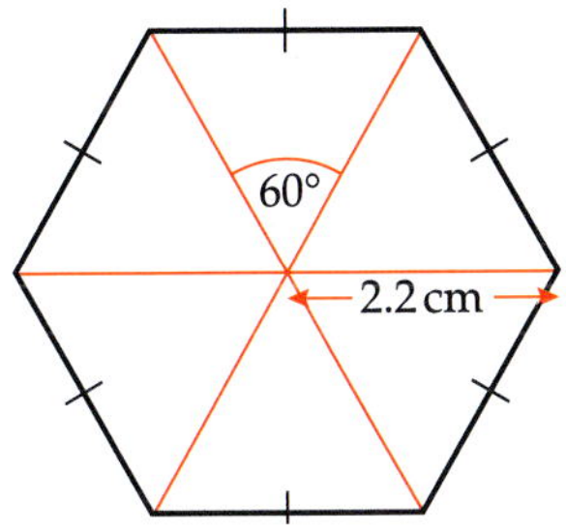

b)

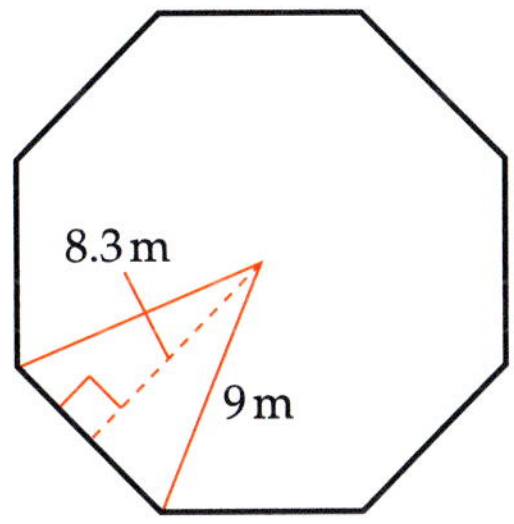

2 Using ruler, protractor and pair of compasses, construct the rhombus ABCD with sides of length 6 cm and one internal angle of 72°. Measure the lengths of the long and short diagonals of your rhombus.

3 By drawing constructions and measuring, copy and complete this table showing the perimeter of n-sided regular polygons inscribed inside a circle of radius 5 cm.

Number of sides (n)	Radius of circle (cm)	Perimeter of regular polygon (cm)
3	5	
4	5	
5	5	
6	5	
8	5	

Investigate for different-sized circles and different values of n.

Case study - Quilt maker

Susan Ross-Morrison

I am a retired teacher. I started making quilts as a hobby because I've always loved sewing and I hate to sit around doing nothing.

Anyone with an imagination can design a quilt. Whether you think of a pattern of strictly repeating designs, or whether you let your design take its own course – it doesn't matter, you should let your imagination run. What matters is that you use basic techniques and get your measurements right.

People normally use 100% cotton, but you can also make a work of art by combining different textures and thicknesses. You should try to use the correct equipment if possible, but to experiment and try out this line of art, you can start with what is available to you, for example, rulers, cutting mats and scissors.

Lay out the whole quilt on the floor or a big board to see the completed effect before you sew it. Once you are happy, stitch the first row of squares together (6 mm from the edge). Do the same with row 2 and then join row 1 to row 2 and so on.

Stitch a border on to the four edges of your quilt using mitred corners.

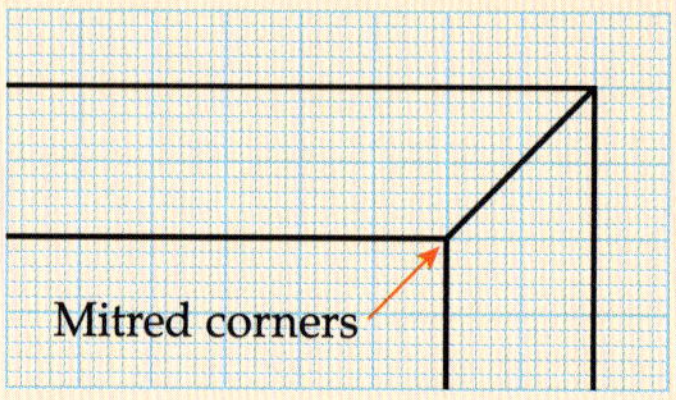

Cut the batting 4 cm bigger than the quilt on all sides and tack it onto the whole back of the quilt. For the batting to stay in place, either

a) stitch through the quilt and the batting along a few strips on the right side of the quilt, following the line between the strips

b) stitch beads or buttons onto the quilt in a design of your own by hand.

Turn the quilt over and now work on the back. Fold the border over the sides of the batting. Place a backing on the quilt. I use a sheet of the same width as the quilt so that the quilt does not slide off the bed. Plain flannelette or other fabric is also suitable. Tack this in place and then stitch right around it.

Remember to autograph your quilt. For example:

To Lulu, with lots of love Susan, 25 December 2005

I do mine in embroidery on the backing sheet.

Key words
Centre of rotation

Chapter 9

Rotation

In this chapter you will learn about rotation and rotational symmetry of plane shapes, and practise writing coordinates for an image after a rotation.

Unit 9.1 Rotation in the Cartesian plane

In this unit you will:

- rotate an object in the Cartesian plane
- write the coordinates for the image after a rotation
- observe patterns in the coordinates of the image and object given the centre and angle of rotation.

In a rotation all the points move in the same direction and through the same angle.
A rotation is done in an anti-clockwise direction, unless stated otherwise.
The distance from the **centre of rotation** to the object is equal to the distance from the centre to the equivalent point on the image.

The centre of rotation is the invariant point.

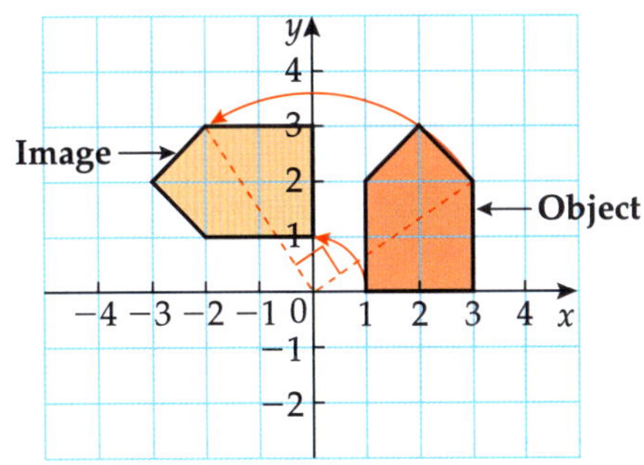

All the points are rotated 90° about the centre (0, 0).

The origin.

Note the coordinates:

Object	Image
(1, 0)	(0, 1)
(3, 0)	(0, 3)
(3, 2)	(−2, 3)
(2, 3)	(−3, 2)
(1, 2)	(−2, 1)

The coordinates of the object and the image, for a rotation of 90° about the origin, form the pattern $(x, y) \rightarrow (-y, x)$.

Worked example

Rotate the figure A about the point (1, 1) by 270°.
Note the coordinates of the image and the object.

Object	Image
(1, 1)	(1, 1)
(3, 2)	(2, −1)
(2, 2)	(2, 0)
(1, 5)	(5, 1)

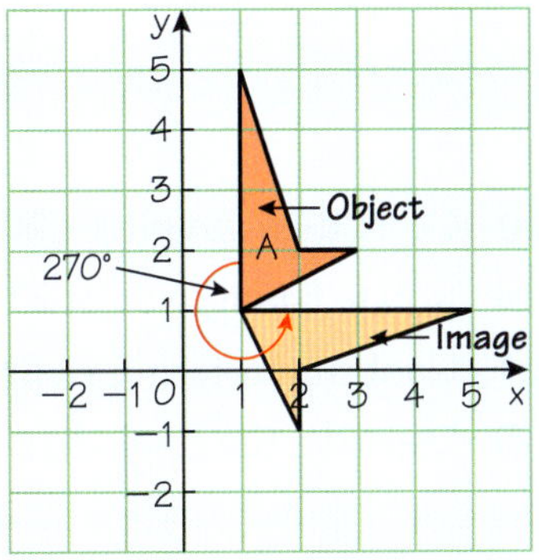

The centre (1, 1) is invariant.

Note that 270° anti-clockwise is 90° clockwise.

Exercise

1 Copy the figure E.

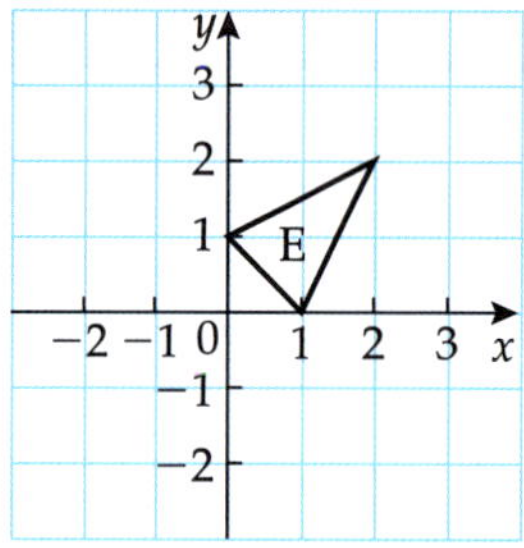

a) Use the origin as the centre and rotate E by 90° to F.

b) Rotate E by 180° to G.

c) Rotate E by 270° to H.

d) What is the angle of rotation from

i) F to G **ii)** F to H **iii)** G to H?

Remember: join the vertices to the centre of rotation. Measure the angle anti-clockwise. Distance from centre to object = distance from centre to image.

2 Copy this diagram.

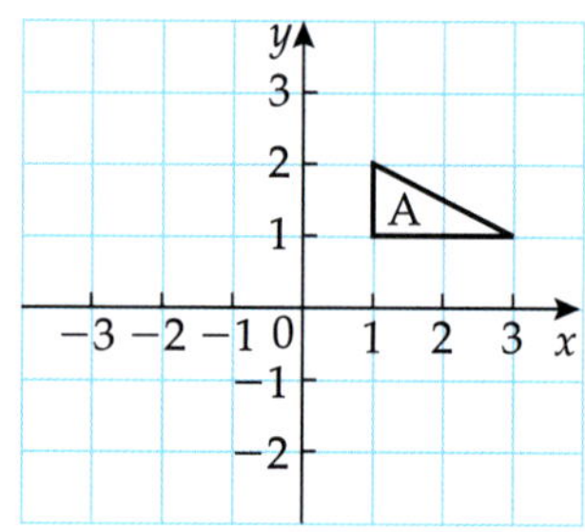

a) Draw the image of A after a rotation of 90° clockwise about (1, 1). Label the image B.

b) Draw the image of A after a rotation of 90° anti-clockwise about (0, 0). Label the image C.

c) Write the coordinates of the object and the two images parts in **a)** and **b)**. Is there any pattern between the object and any of the two images?

3

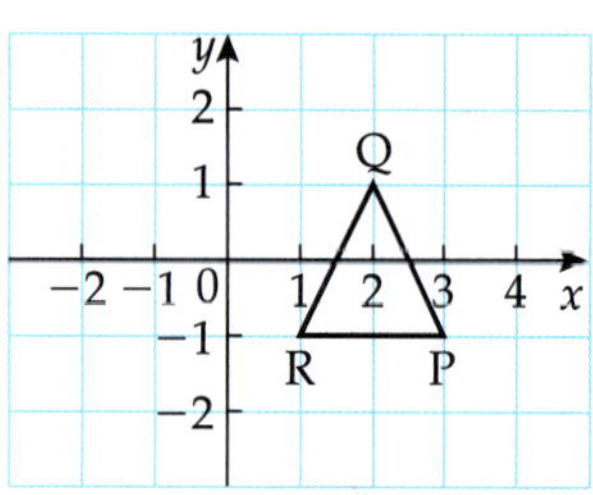

a) Copy the triangle PQR.

b) Rotate triangle PQR 90°, using the middle of PR as the centre of rotation.

c) Continue rotating each image 90° until the image overlaps the original object.

d) Describe the final shape obtained through all the rotations.

Key words
- Rotational symmetry
- Order

Unit 9.2 Describing rotations

In this unit you will:
- use perpendicular bisectors to find the centre of rotation
- find the angle of rotation
- find the rotational symmetry of a shape.

If you are given an object and an image you can find the centre and angle of rotation.

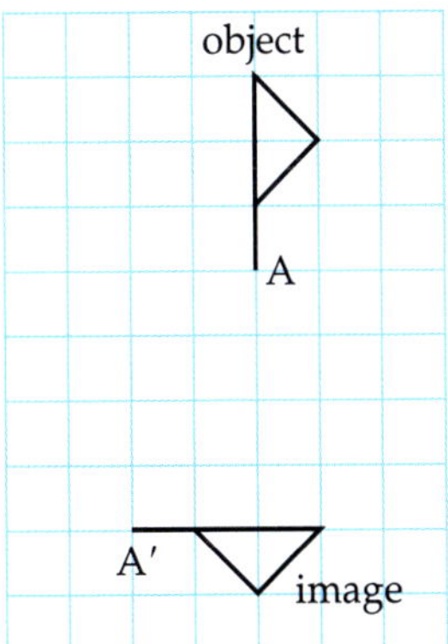

Choose a point on the object and its corresponding point on the image.

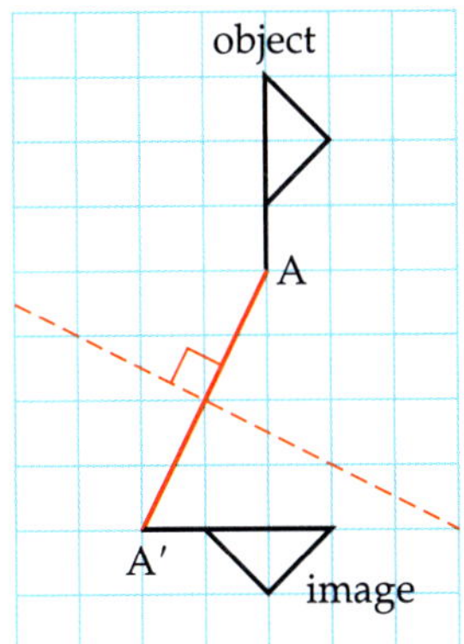

Join them with a straight line and construct the perpendicular bisector to that line.

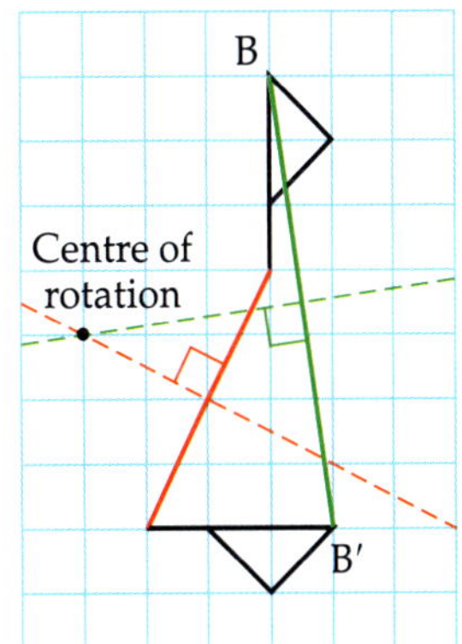

Join a second pair of points and construct the perpendicular bisector to that line. The point where the lines cross is the centre of rotation.

You can find the angle of rotation using a protractor.

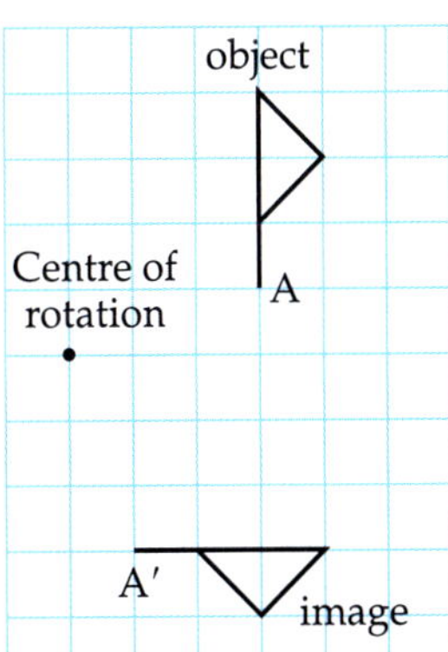

Choose a point on the object and its corresponding point on the image.

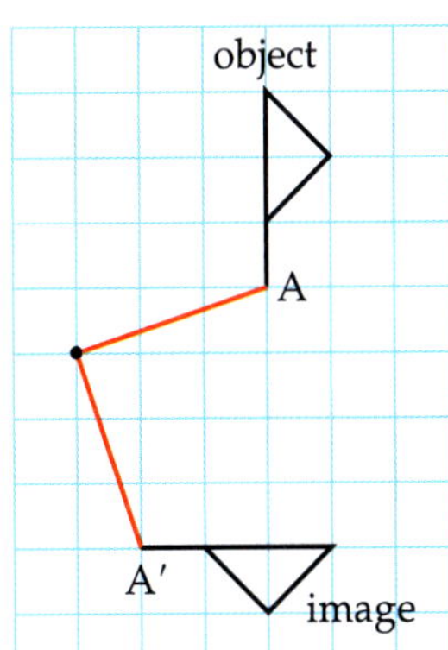

Join each point to the centre of rotation.

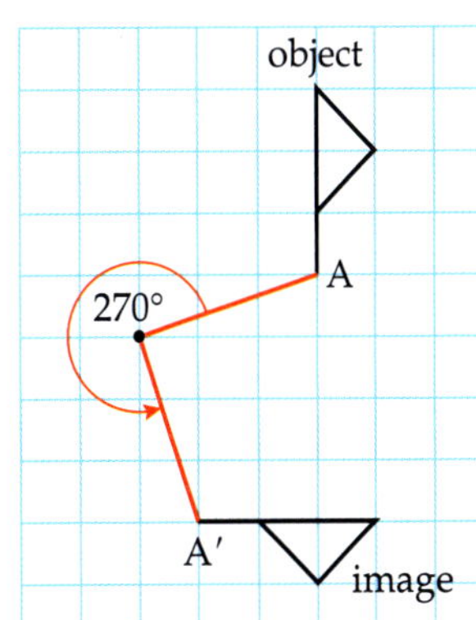

The angle between the lines is the angle of rotation. Remember that the standard direction for a rotation is anti-clockwise.

The **order of rotational symmetry** is the number of ways a shape fits onto itself in a complete 360° turn.

Order of rotation symmetry 2

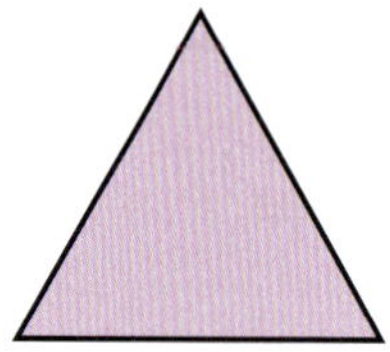

Order of rotation symmetry 2

Order of rotation symmetry 5

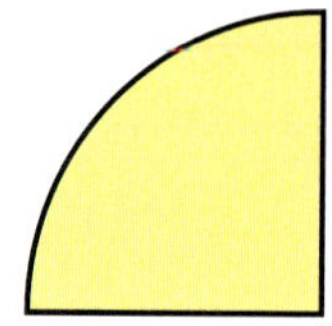

Order of rotation symmetry 1

Any shape looks identical after a full 360° turn, so all shapes have rotational symmetry of at least order 1. Regular polygons have the same order of rotational symmetry as their number of sides.

Worked examples

1. Fully describe this rotation:

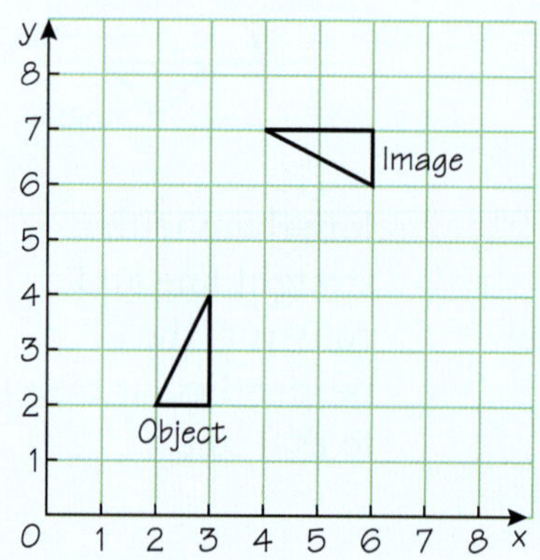

2. What is the order of rotational symmetry of a parallelogram?

1.

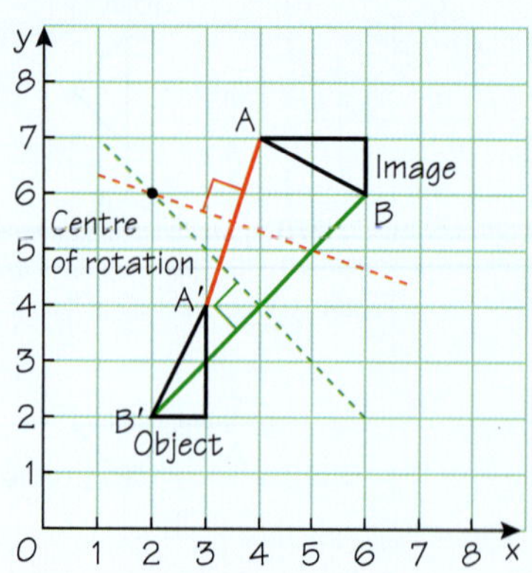

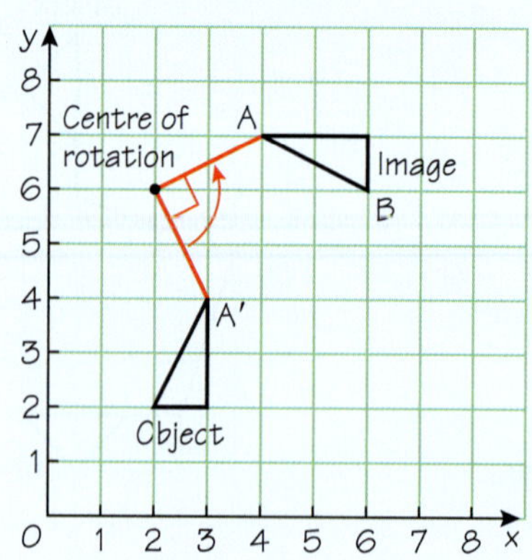

This is a rotation of 90° about the point (2, 6).

2. The order of rotational symmetry is 2.

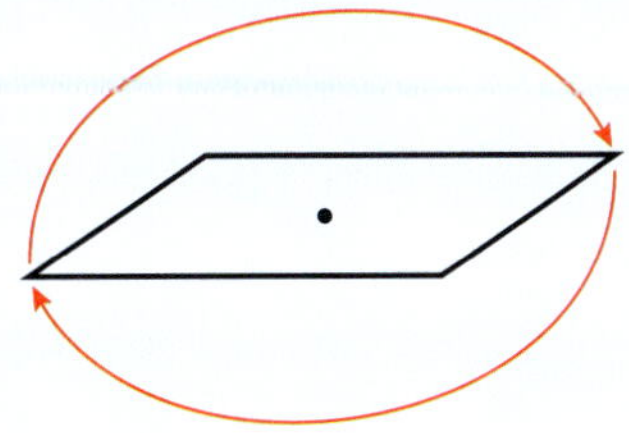

Exercise

1. Write down the order of rotational symmetry of each shape:

a) 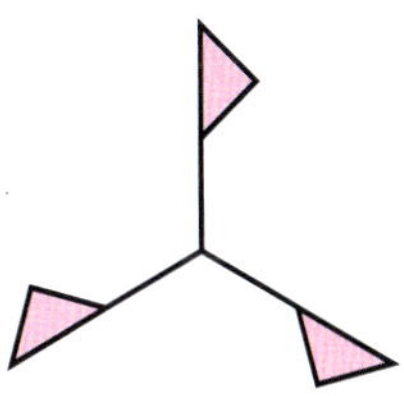

3

b) 

2

c) 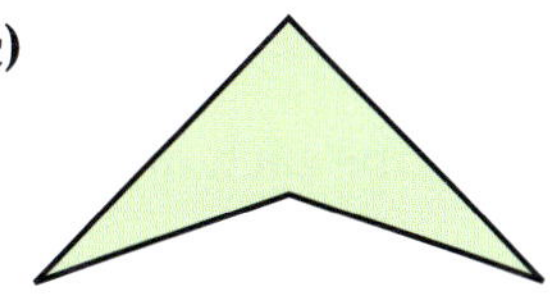

1

d) 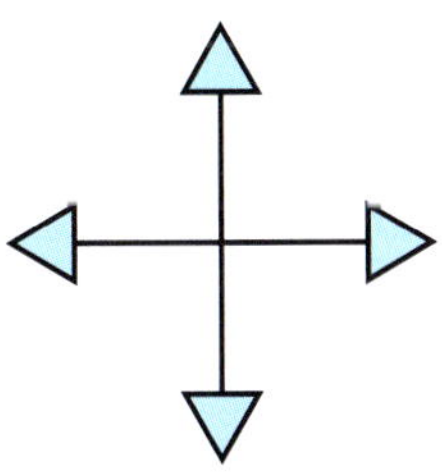

4

e) 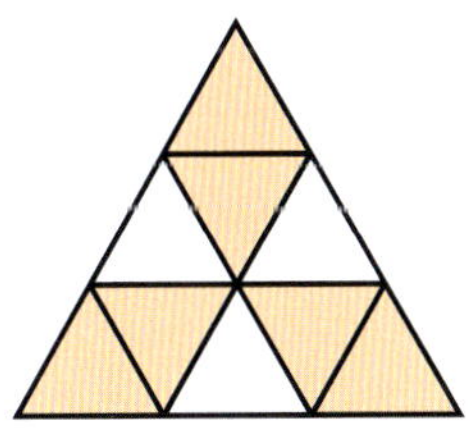

3

f) 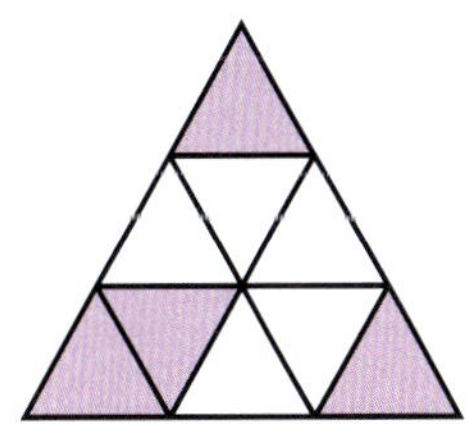

1

2. What is the order of rotational symmetry of each of the following shapes:

a) Square **b)** Rectangle **c)** Trapezium
d) Equilateral triangle **e)** Semicircle **f)** Regular hexagon?

3. 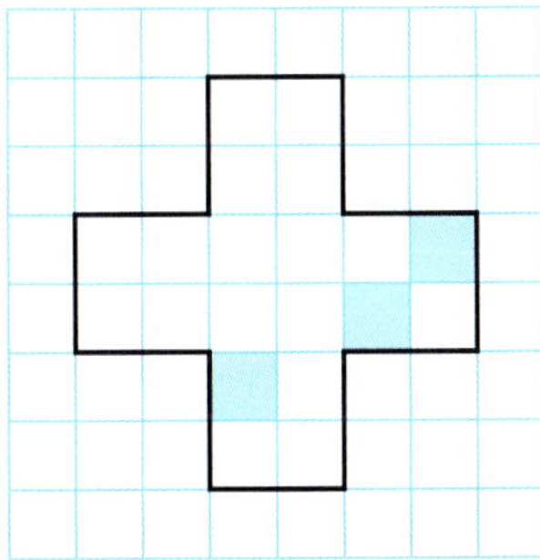

a) What is the order of rotational symmetry of this shape?
b) On a copy of the shape, shade three more small squares so the shape has order of rotational symmetry 2.
c) On another copy of the shape, shade five more small squares so the shape has order of rotational symmetry 4.

4. Trace each pair of shapes and use perpendicular bisectors to find the centre of rotation.

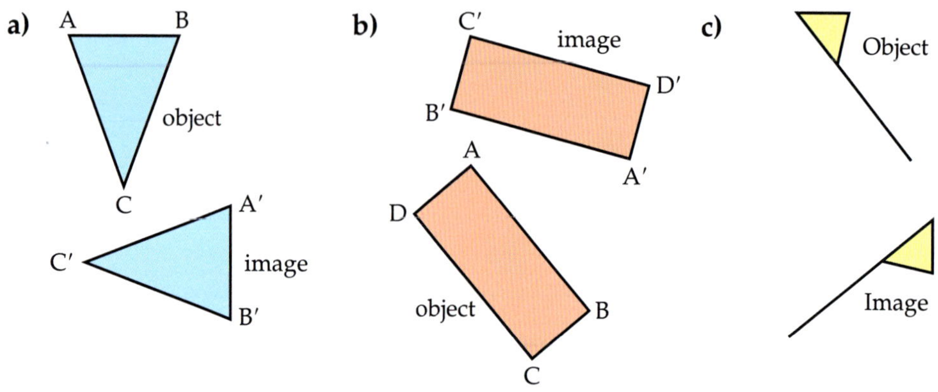

5. Fully describe each of the following rotations.

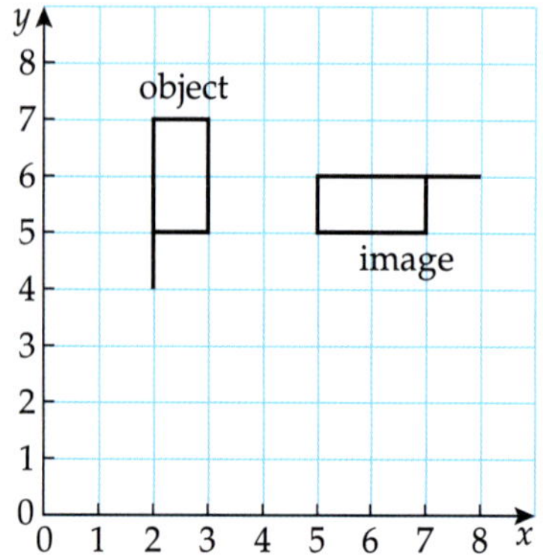

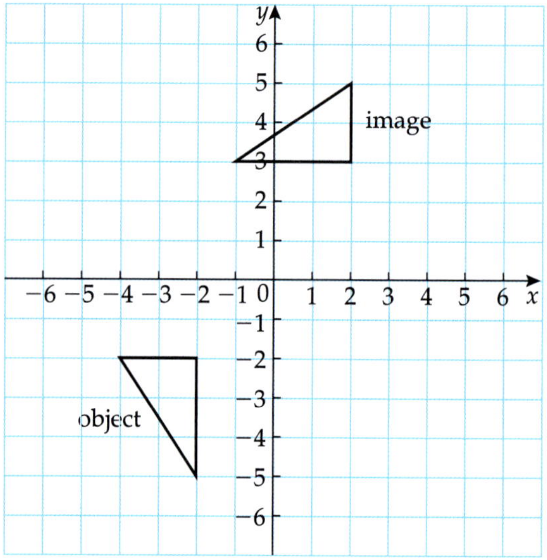

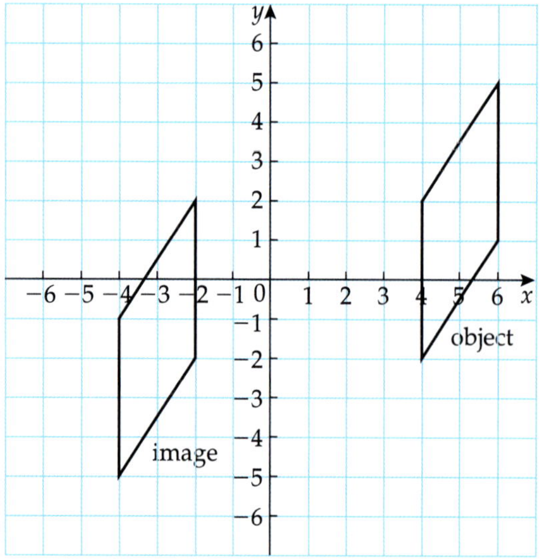

Investigations

6 Investigate the rotational symmetry of different roadsigns. Can you find a roadsign with order of rotation 2?

7 Try to find patterns with rotational symmetry in tiles and rugs. What is the highest order of rotational symmetry you can find?

Looking back

1 What is the order of rotational symmetry of each of the following shapes?

a) 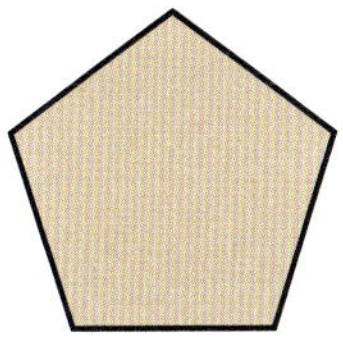

b) 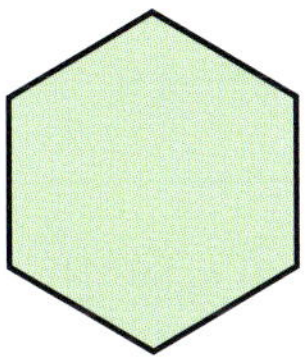

c) 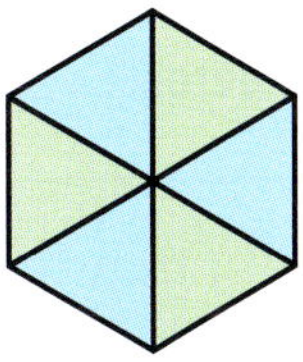

2 Fully describe the following rotation.

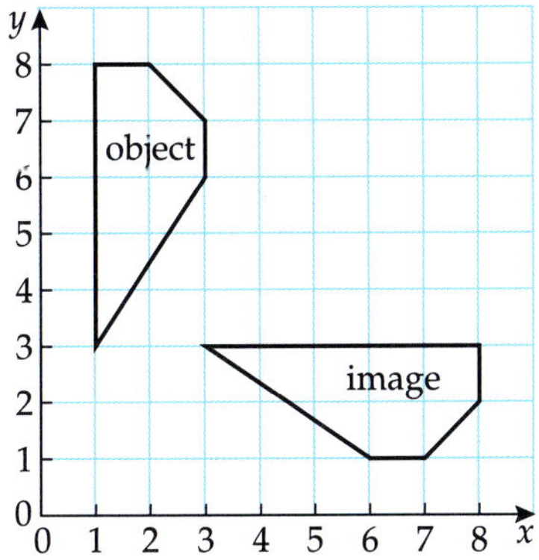

3 Copy the triangles on squared paper. Put in the x- and y-axes.

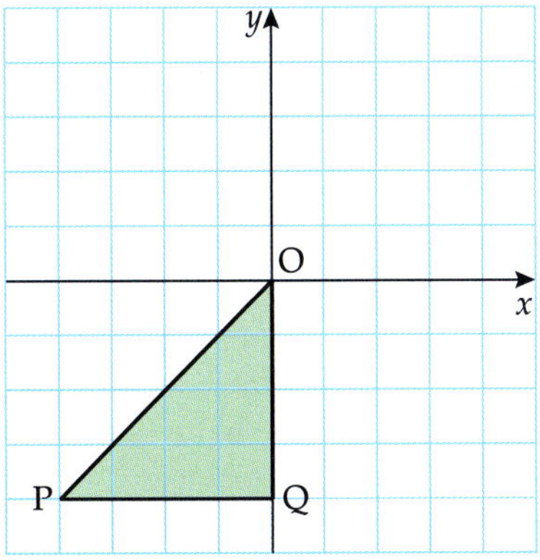

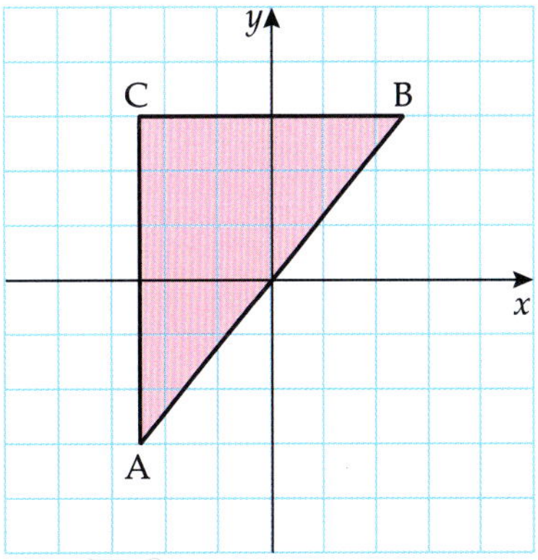

a) Rotate OPQ by 180° about the origin, to ONM. Join PM and QN. Give the special name for the shape PQNM.

b) Rotate ABC by 180° about the origin, to DBC. Give the special name for the shape ABCD.

Chapter 10

Enlargement and similar triangles

Last year you learned about transformations of shapes by reflection and translation, and Chapter 9 dealt with rotation. Shapes can also be transformed by enlargement, and this chapter looks at scale factors, ratio and proportion, and the specific situation of similar triangles.

Key words
Enlargement
Scale factor
Centre of enlargement
Ratio

Unit 10.1 Enlargement

In this unit you will:

- understand and use the language and notation associated with enlargement
- enlarge 2-D shapes given a centre of enlargement and a positive whole number scale factor.

Enlargement is a type of transformation. When a shape is enlarged, the lengths of all the sides are multiplied by a **scale factor**. The angles stay the same during enlargement.

Every enlargement has a **centre of enlargement**. Lines joining equivalent points on the object and image meet at the centre of enlargement.

When a shape is enlarged, the **ratios** of corresponding sides on the image and the object are equal to the scale factor. For any enlargement, if the scale factor is n, then the ratio of the corresponding sides on the object and the image is $1 : n$.

The diagram shows a triangle ABC that has been enlarged by scale factor 3 with centre of enlargement at O.

The lengths of the sides of triangle A′B′C′ are three times the lengths of the corresponding sides of triangle ABC. For example, $A'B' = 3 \times AB$.

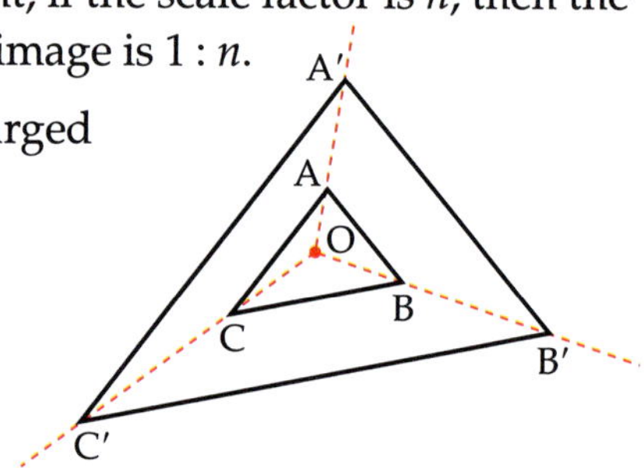

Worked example

Describe this enlargement.

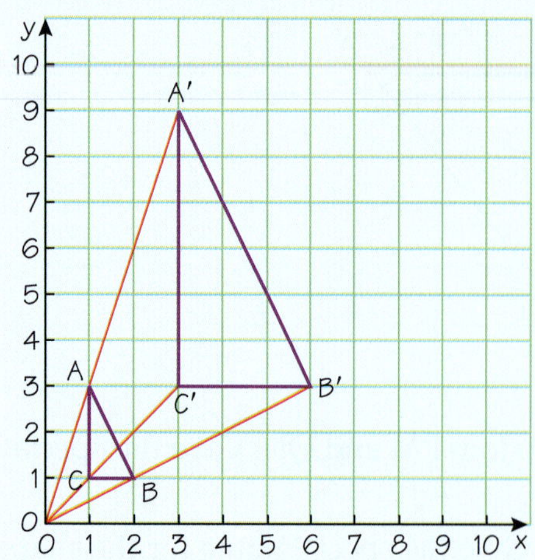

Scale factor = 3

Centre of enlargement = (0, 0)

The side lengths of the image A′B′C′ are three times the size of the side lengths of the object ABC, so the scale factor is 3.

Lines joining equivalent points on the object and image meet at (0, 0), so this is the centre of enlargement.

Exercise

1 Trace each shape and enlarge it, using the centre of enlargement O, by the given scale factor.

a) 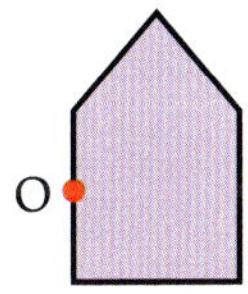

scale factor 3

b) 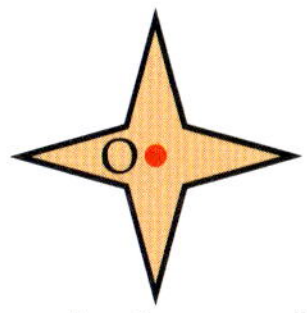

scale factor 2

c) 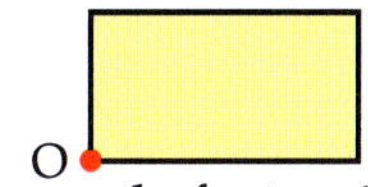

scale factor 4

2 Describe this enlargement:
ABCD → A′B′C′D′

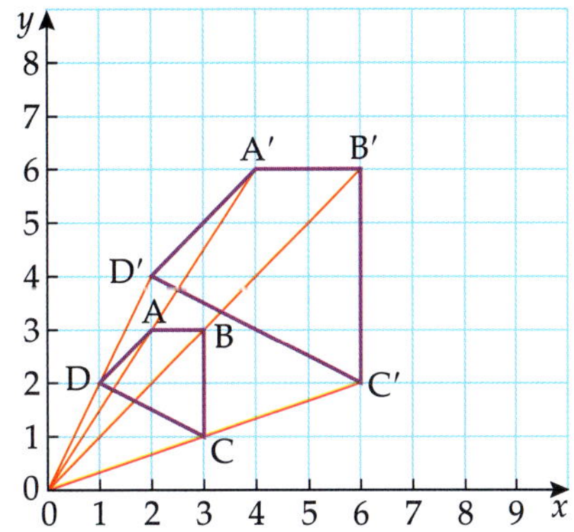

3 Describe this enlargement.

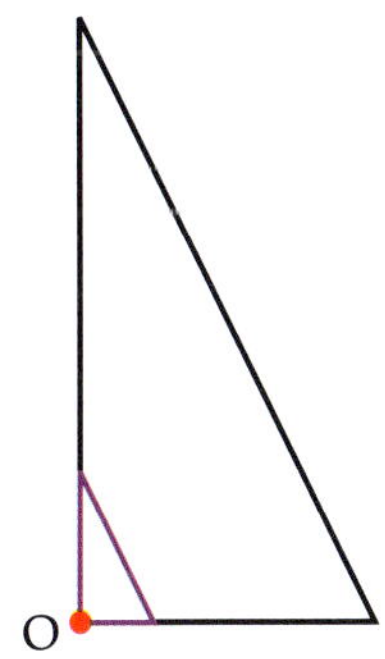

4 A_2 and A_3 are enlargements of A_1.

a) Trace the diagram and find the centre of enlargement that maps
 i) A_1 onto A_2
 ii) A_2 onto A_3
 iii) A_1 onto A_3.

b) What do you notice about the three centres of enlargement?

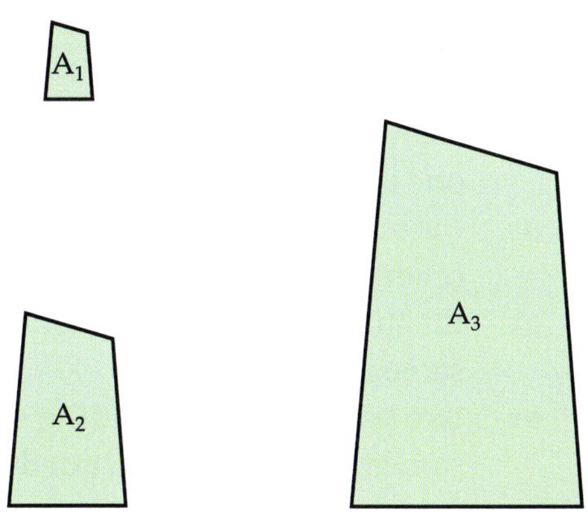

5 a) What centre of enlargement maps triangle ABP onto triangle ACK? What is the scale factor of this enlargement?

b) What centre of enlargement maps triangle JKL onto triangle JAD? What is the scale factor of this enlargement?

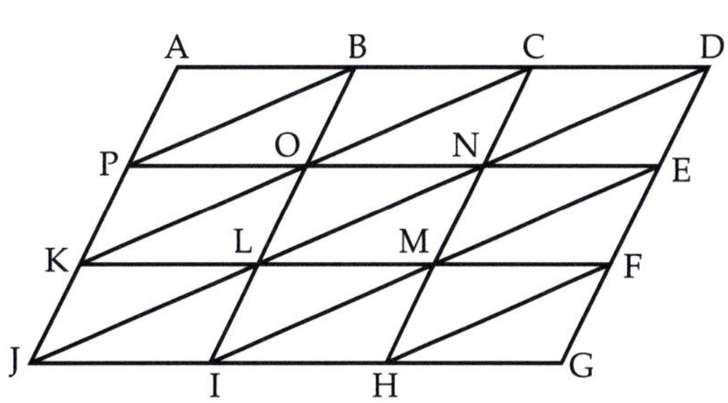

Unit 10.2 Similar shapes, ratio and proportion

Key words
Similar
Ratio
Simplest form
Proportion
Included angle

In this unit you will:

- identify similar shapes and their properties
- understand and use ratio related to enlargement of 2-D shapes
- identify the scale factor of an enlargement as a ratio of any two corresponding line segments
- reduce a ratio to its simplest form.

If a shape is enlarged, the image is mathematically **similar** to the object. This means that all the angles are the same and that the **ratios** of corresponding sides on the image and the object are equal to the **scale factor**.

Rectangle A′B′C′D′ is an enlargement of rectangle ABCD by scale factor 3.
The ratio of the sides AB : A′B′ is 1 : 3.
The ratio of the sides CD : C′D′ is 2 : 6, which cancels to give 1 : 3 in its **simplest form**.

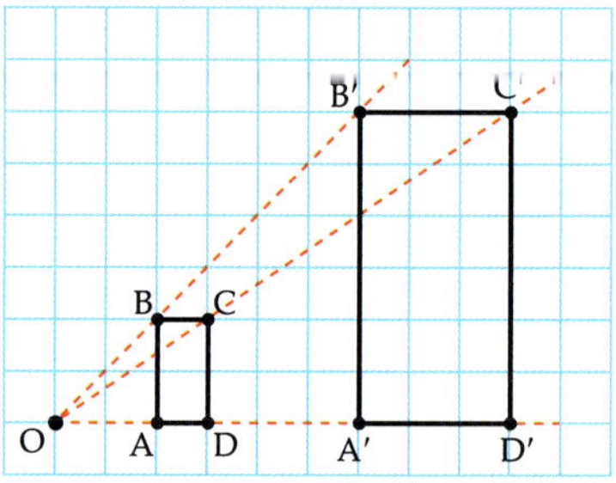

Two quantities are in **proportion** if their ratio stays the same.
Since ABCD was enlarged by scale factor 3, we know that the dimensions of A′B′C′D′ are three times the dimensions of ABCD. This means that the sides of rectangle ABCD and rectangle A′B′C′D′ are in proportion.

Since ABCD was enlarged by scale factor 3, each point on the image A′B′C′D′ is three times as far away from the centre of enlargement as the equivalent point on ABCD. For example, OA′ is three times the distance OA.

These two triangles are similar.
Their corresponding angles are equal, and the lengths of their sides are in proportion.

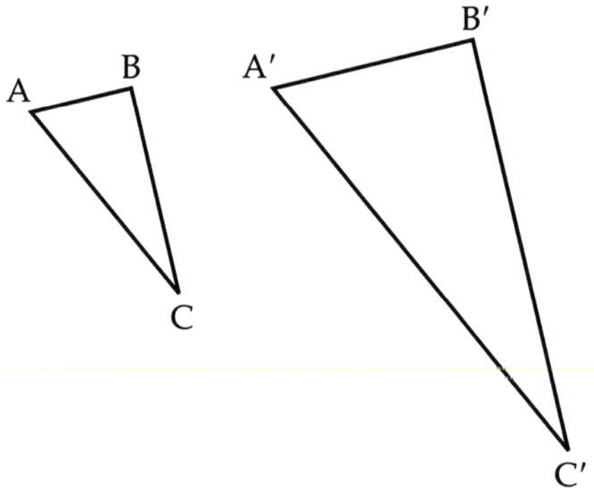

Two triangles are similar if:

one is an enlargement of the other
or the angles of one triangle are equal to the other triangle
or the three corresponding sides are in the same ratio
or two pairs of corresponding sides are in the same ratio and the **included angles** are equal.

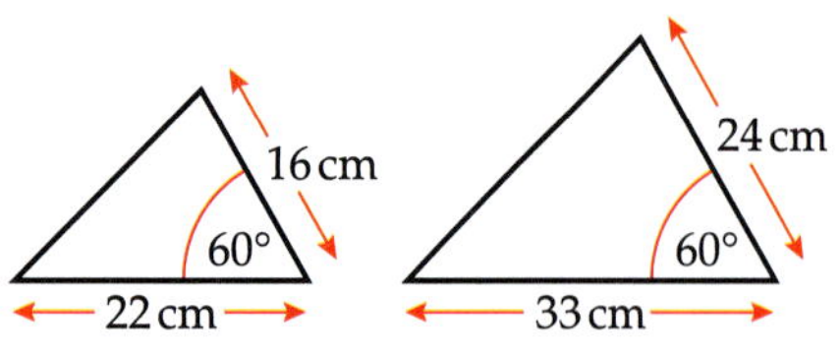

Similar triangles: two pairs of corresponding sides in the same ratio and included angles equal.

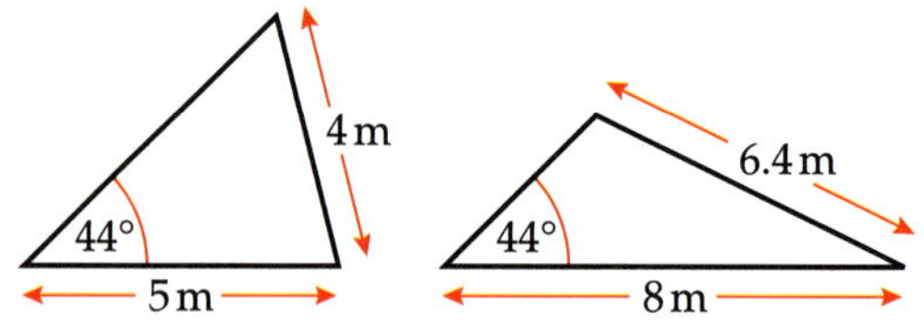

Non-similar triangles: two pairs of corresponding sides in the same ratio but the equal angles are not the included angles.

Worked examples

1. Triangle A′B′C′ is an enlargement of triangle ABC. The side lengths of A′B′C′ are in proportion to the side lengths of ABC.
   a. Write the ratio of the corresponding sides AB and A′B′ in its simplest form. Repeat for BC and B′C′.
   b. Write the side lengths of ABC as a fraction of the side lengths of A′B′C′.
   c. Find the scale factor of the enlargement.
   d. Find the size of the missing length A′C′.

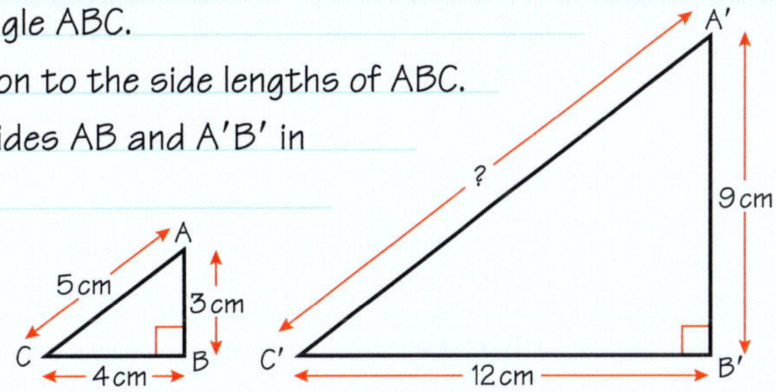

2. Which of the following pairs of triangles are similar?

a.

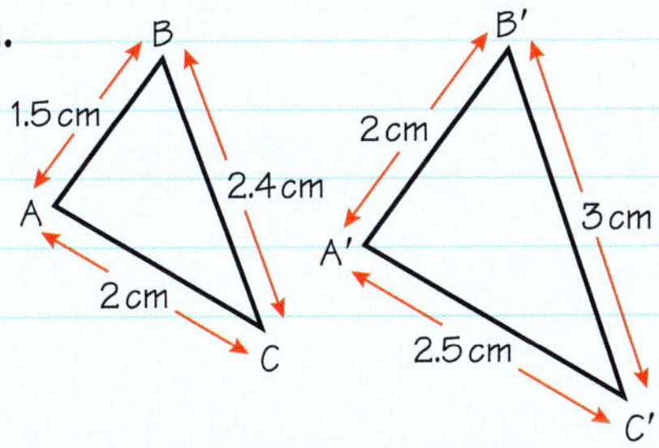

b.

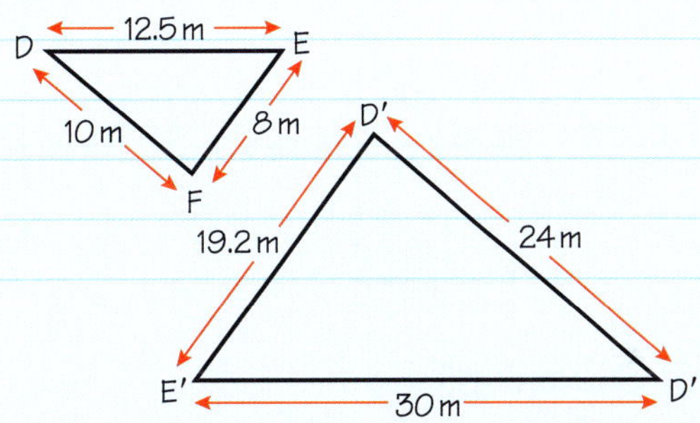

1. **a.** The ratio of AB to A′B′ is $3:9 = 1:3$ and the ratio of BC to B′C′ is $4:12 = 1:3$.
   b. Side lengths of ABC are $\frac{1}{3}$ of A′B′C′.
   c. Scale factor is $\frac{9}{3} = 3$.
   d. The length AC = 5 cm so $A'C' = 5 \times 3 = 15$ cm.

> The ratio of corresponding lengths in the object and the image is equal to the scale factor.

> Since the scale factor is 3, the side lengths of A′B′C′ are 3 times the lengths of the corresponding sides on ABC.

2. **a.** $AB: A'B' = 1.5:2 = 3:4$
   $BC: B'C' = 2.4:3 = 4:5$
   $CA: C'A' = 2:2.5 = 4:5$
   The corresponding are not in the same ratio so the triangles are not similar.

> You can simplify ratios with decimals by finding an equivalent ratio with whole number values then cancelling.
>
> $2.4 : 3$ (×5) → $12 : 15$ (÷3) → $4 : 5$

   b. $DE: D'E' = 12.5:30 = 5:12$
   $EF: E'F' = 8:19.2 = 5:12$
   $FD: F'D' = 10:24 = 5:12$
   The corresponding sides are in the same ratio so the triangles are similar.

> $12.5 : 30$ (×2) → $25 : 60$ (÷5) → $5 : 12$

Exercise

1. ABCD is to be enlarged by scale factor 2, where D is the centre of enlargement.

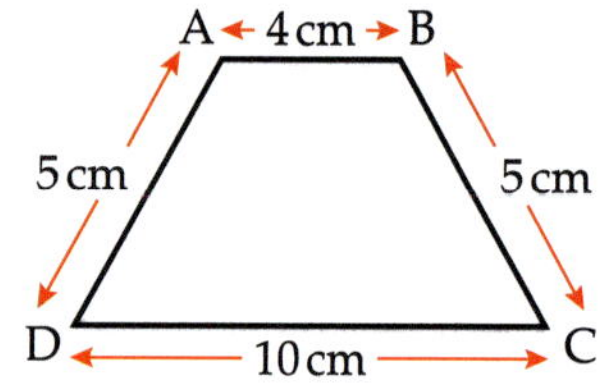

   a) Calculate the lengths of A′B′, B′C′, C′D′ and D′A′ on the enlarged shape.

   b) Check your answers to part **a)** by drawing ABCD accurately and enlarging it by scale factor 2, using D as the centre of enlargement.

2. In each of these diagrams, the side lengths of shape A are in proportion to the side lengths of shape B.

i)

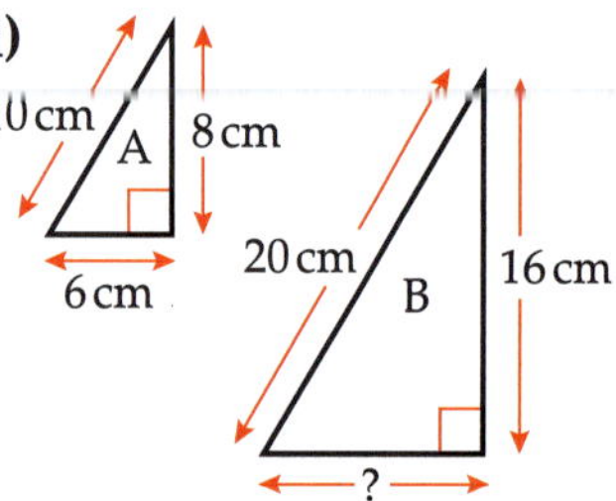

ii)

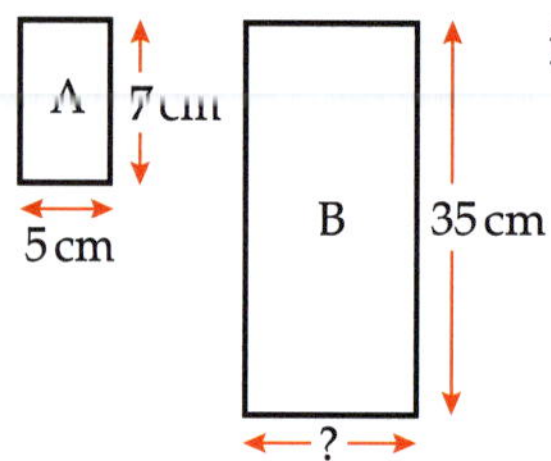

iii)

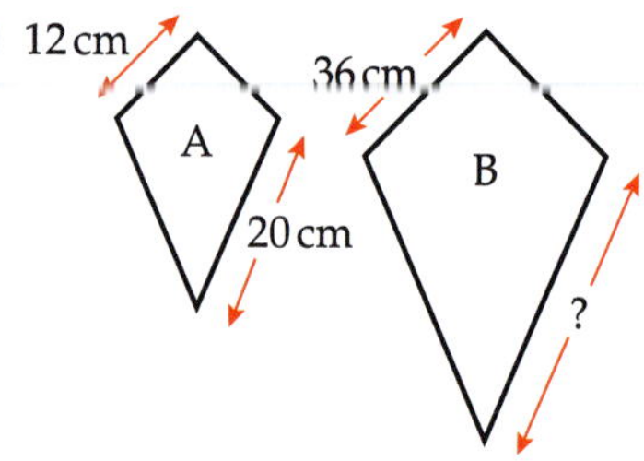

For each enlargement:

a) write the ratios of the corresponding sides of A and B in their simplest form.

b) write the lengths of shape A as a fraction of those of shape B.

c) find the scale factor of the enlargement.

d) find the size of the missing length.

3. In each of these diagrams, shape X and shape Y are similar. Calculate the missing length for each pair of shapes.

i)

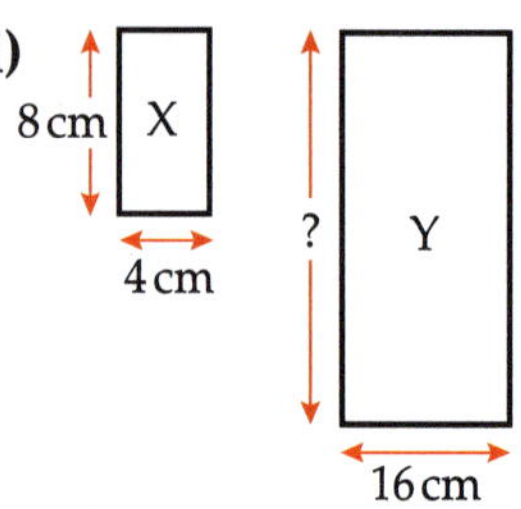

ii)

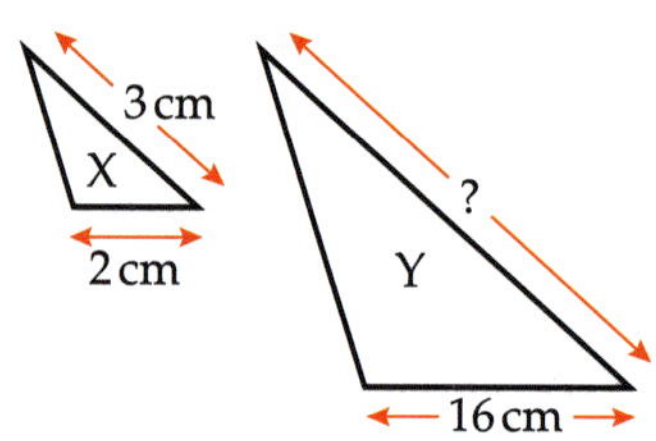

iii)

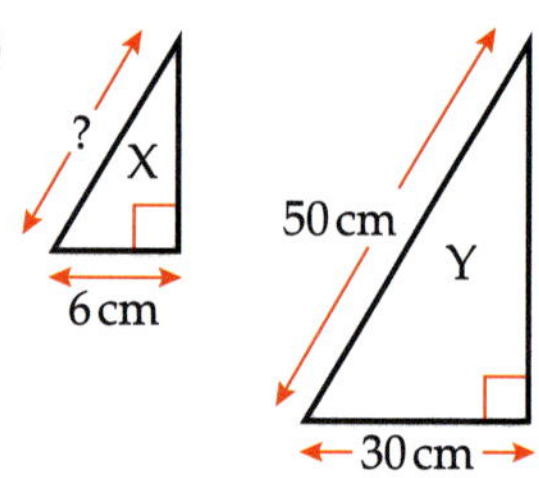

4 Are the side lengths of these triangles in proportion? Explain your answer.

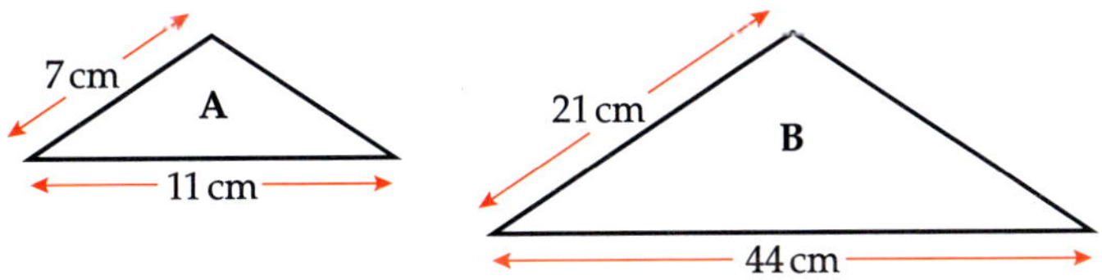

5 Which of the following pairs of triangles are similar?

a)

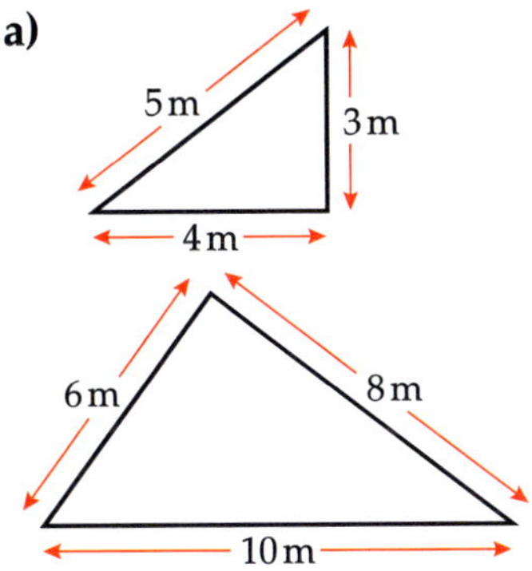

b)

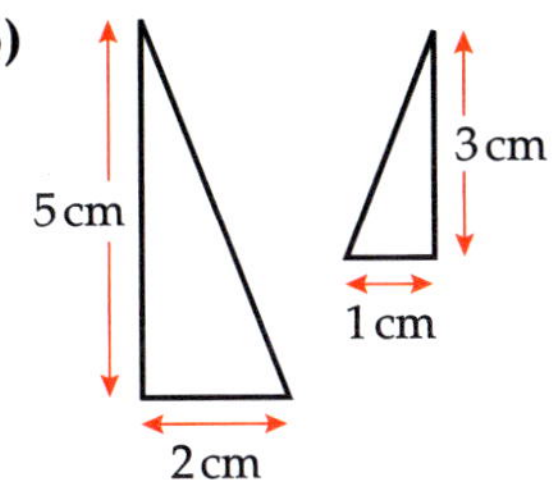

c)

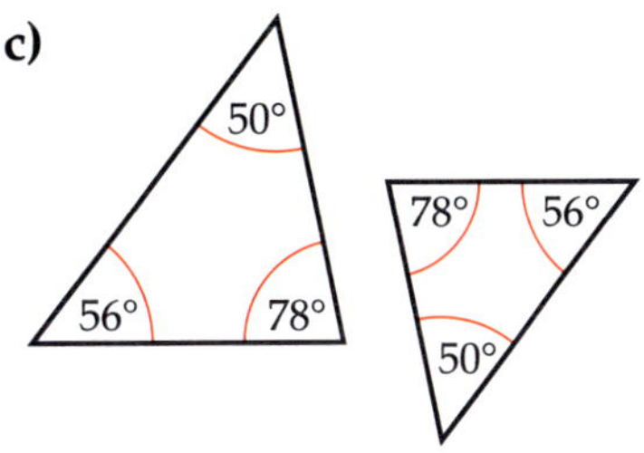

d)

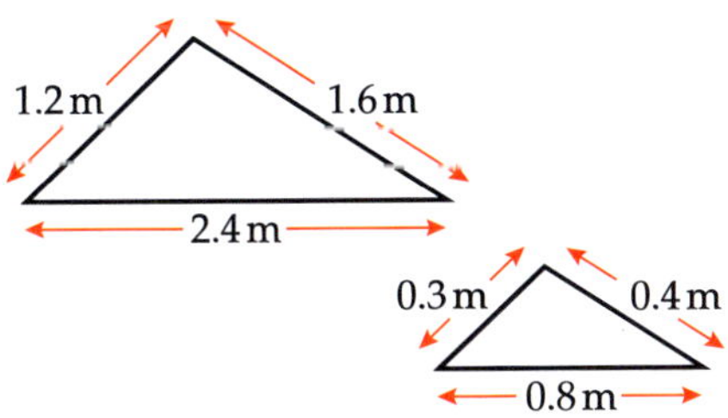

e)

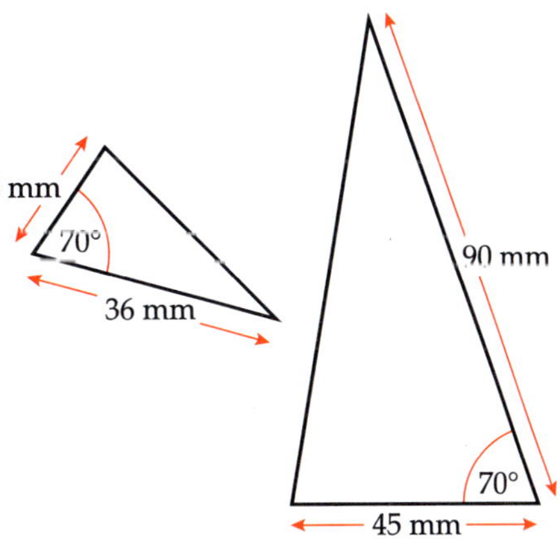

6 Which of the following shapes are always similar? Explain your answers.

a) Right-angled triangles
b) Squares
c) Equilateral triangles
d) Rhombuses

Investigation

7 Investigate the proportions of metric paper sizes.

a) Can you divide up a sheet of A3 paper to make sheets of

i) A4 paper ii) A5 paper iii) A6 paper?

b) Is an A3 piece of paper an enlargement of an A4 piece of paper? How can you prove this?

Unit 10.3 Fractional and negative scale factors

In this unit you will:

- enlarge 2-D shapes given a centre of enlargement and a fractional scale factor
- enlarge 2-D shapes given a centre of enlargement and a negative scale factor
- describe enlargements with different scale factors and explore their properties.

Key words
Object
Image
Fractional scale factor
Negative scale factor

You can calculate the scale factor of an enlargement by measuring the distance of points on the **object** and **image** from the centre of enlargement, O:

$$\text{Scale factor} = \frac{\text{Distance from O to point on image}}{\text{Distance from O to corresponding point on object}}$$

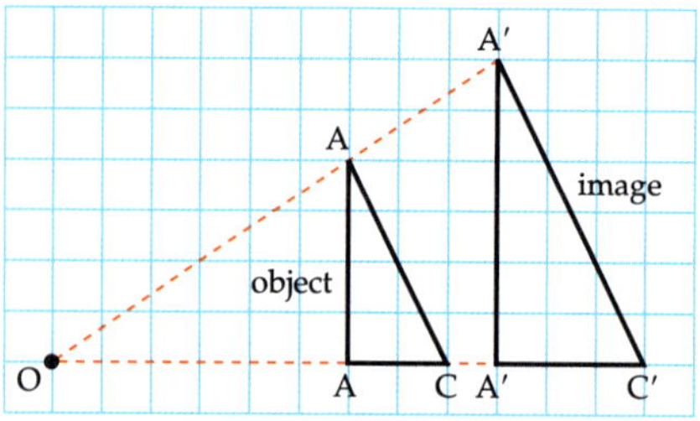

The length OA′ is 9 units and the length OA is 6 units.

Scale factor $\frac{OA'}{OA} = \frac{9}{6} = 1.5$

The scale factor of the enlargement is 1.5. The lengths of the sides on triangle A′B′C′ are 1.5 times the lengths of the corresponding sides on triangle ABC. OB′ = 3 × OB and OC′ = 3 × OC.

Scale factors can be whole numbers or fractions. If the scale factor of an enlargement is less than 1, the image will be smaller than the object. This diagram shows an enlargement of a rectangle by scale factor $\frac{1}{3}$.

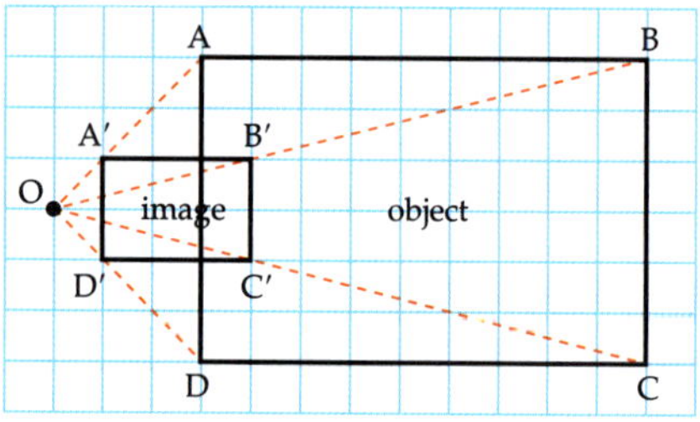

ABCD is the object and A′B′C′D′ is the image. OA′ = $\frac{1}{3}$OA and OB′ = $\frac{1}{3}$OB. The lengths of the sides on A′B′C′D′ are $\frac{1}{3}$ the lengths of the corresponding sides on ABCD.

Scale factors can also have negative values. This diagram shows an enlargement of a trapezium by scale factor −2.

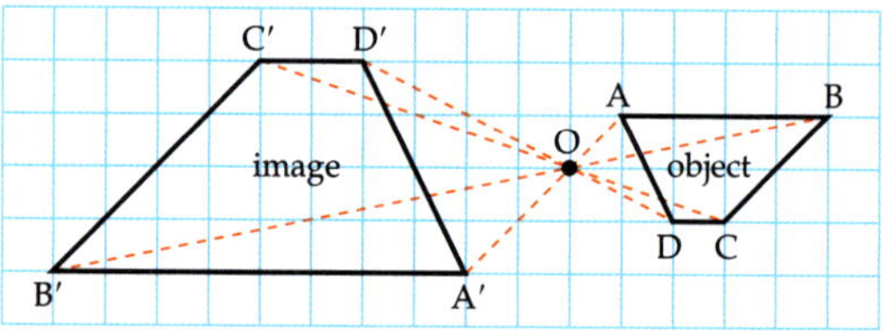

OA′ = −2 × OA. The minus sign means that the line OA′ is drawn in the opposite direction to the line OA. The lengths of the sides on A′B′C′D′ are 2 times the lengths of the corresponding sides on ABCD, but the image has been rotated through 180°.

Worked example

Enlarge this shape by scale factor $-\frac{1}{2}$ using centre of enlargement (4, 3).

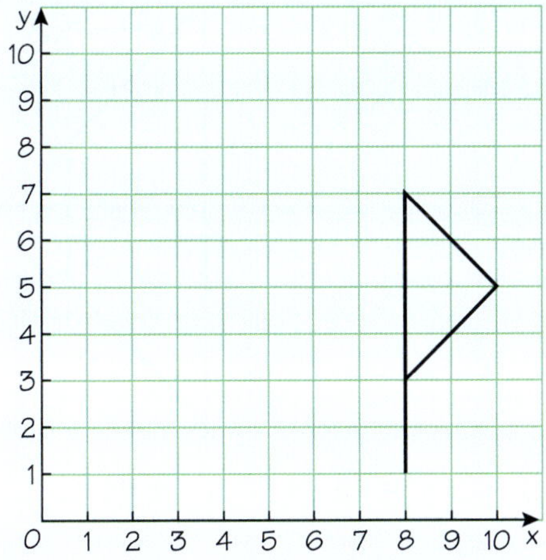

$OA' = -\frac{1}{2}OA$. So the length OA' is $\frac{1}{2}$ the length OA in the opposite direction.

$OB' = -\frac{1}{2}OB$. So the length OB' is $\frac{1}{2}$ the length OB in the opposite direction.

etc...

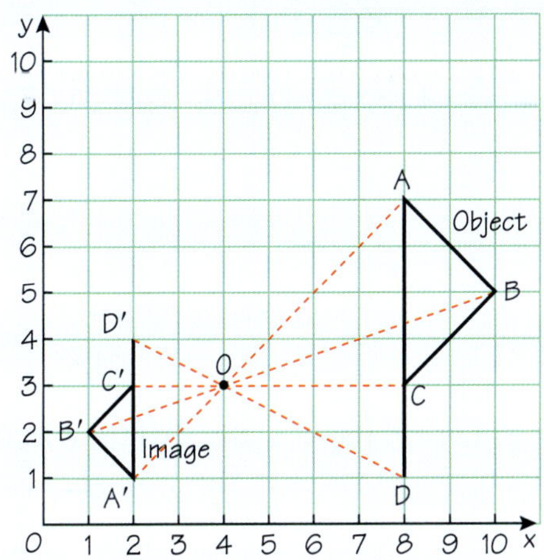

Exercise

1 Calculate the scale factor for each of the following enlargements.

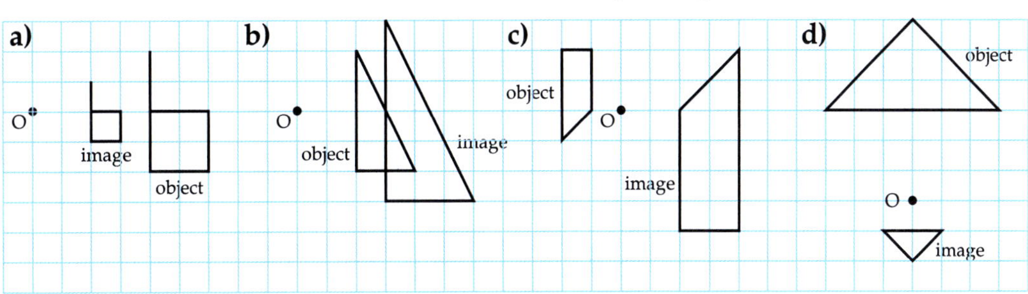

2 On a copy of this diagram enlarge this shape by scale factor $\frac{1}{3}$, using centre of enlargement (2, 1).

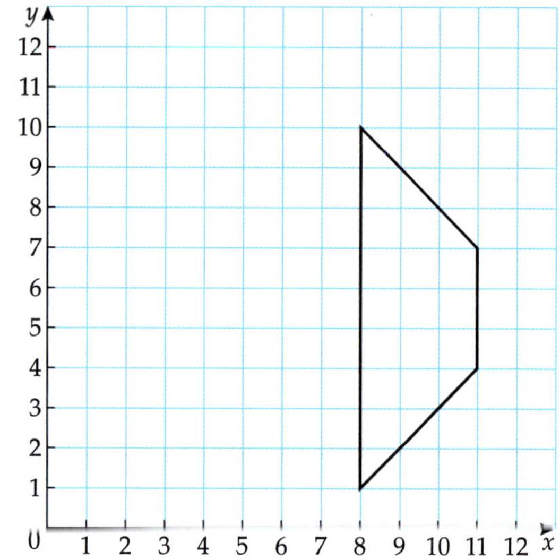

3 On a copy of this diagram enlarge this shape by scale factor −2, using centre of enlargement (−1, 0).

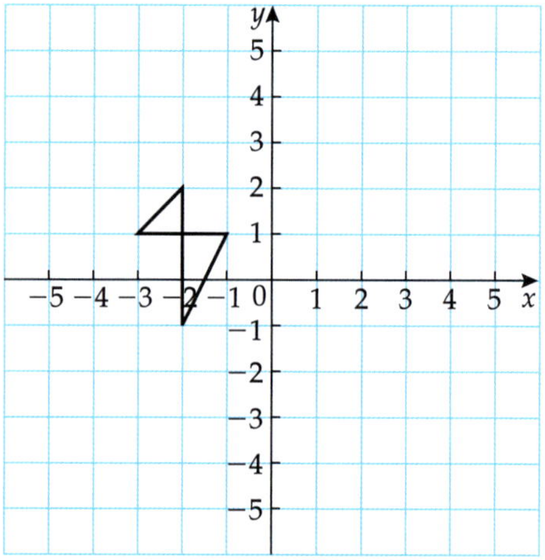

4 A triangle with sides of length 20 cm, 10 cm and 16 cm is enlarged using different scale factors. Copy and complete this table showing the lengths of sides of the triangle after each enlargement:

Scale factor	Side 1	Side 2	Side 3	Image rotated?
1	20 cm	10 cm	16 cm	No
−2	40 cm			Yes
$\frac{1}{4}$		2.5 cm		
		15 cm		No
−3.5	70 cm			
			8 cm	Yes

5 Lora has a photograph of herself enlarged with scale factor 3.5. On the original photograph, she was 4 cm tall. How tall is she on the enlarged photograph?

6

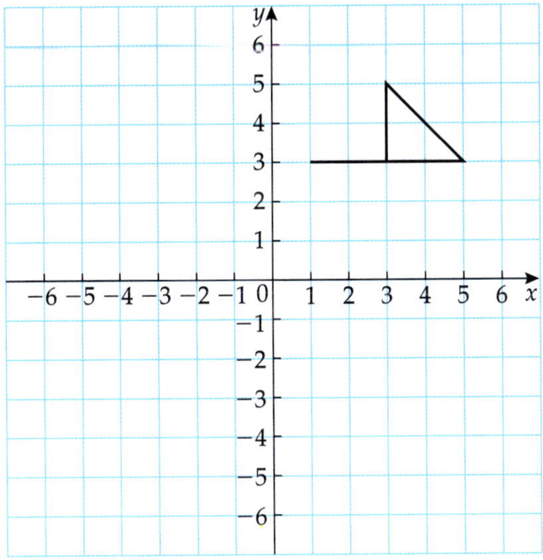

a) On a copy of this diagram enlarge this shape by scale factor $-\frac{1}{2}$, using centre of enlargement (1, −1). Label this image X.

b) On the same diagram enlarge the original shape by scale factor $\frac{1}{2}$, using the same centre of enlargement. Label this image Y.

c) Describe fully the transformation that maps X onto Y.

7 Copy and complete this sentence:
A 180° rotation about a point O is the same as an enlargement with scale factor ☐ and centre of enlargement O.

8 Shape A is mapped onto shape B by an enlargement with scale factor 4. What is the scale factor of the enlargement that maps shape B onto shape A?

Investigation

9 Draw a rectangle on squared paper and write down its area.
Enlarge your rectangle by scale factor 2. What is the area of your new rectangle?
Try enlarging your original rectangle by different scale factors. Record your results in a table like this one:

Scale factor	Length of rectangle	Height of rectangle	Area

What do you notice?

Unit 10.4 Problems involving enlargement

Key words
Similar
Ratio
Proportion

In this unit you will:

- solve problems involving ratio and proportion.

When a shape is enlarged, the image is mathematically **similar** to the object. The angles on the object and the image are the same, and the **ratios** of corresponding sides on the image and the object are equal to the scale factor. An object and its enlarged image are in **proportion**.

We can use this information to solve problems to do with enlargement.

Worked example

Ali has a photograph that measures 10.5 cm by 15.5 cm. He takes the photo to be enlarged so that its sides are four times their original size. Allowing for a 2 cm border all round, what size frame should he buy?

The photograph and its enlargement are proportional.

If the original photograph measures 10.5 cm by 15.5 cm and is enlarged to four times the size, the ratios of the sides will be 1 : 4.

The enlargement will therefore measure 42 cm by 62 cm, so the frame needs to measure 46 cm × 66 cm.

Exercise

1 The dimensions of two pictures, A and B, are in proportion. Picture B is an enlargement of picture A.

a) Calculate the width of picture B.

b) Calculate the perimeter of picture A and of picture B. Write the ratio of the perimeters in its simplest form.

c) Explain why the ratio of the perimeters is the same as the ratio of the lengths.

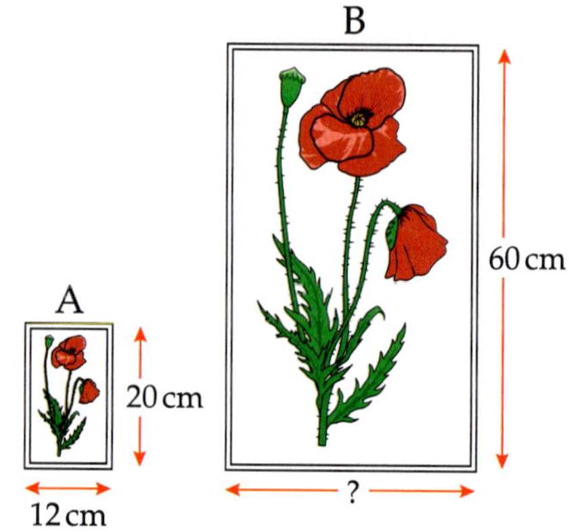

2 a) Plot the following points on a coordinate grid.
A(2, 1) B(3, 1) C(3, 3) and join them up to make a right-angled triangle.

b) Measure and write down the size of the angles.

c) Enlarge the triangle by scale factor 2, centre (0, 0). Label the image A′B′C′. Measure the angles of the enlarged triangle.

d) What can you say about the angles of ABC and A′B′C′?

e) Explain why angles remain unchanged when a shape is enlarged.

Investigations

1. During their holiday in London, Alf and Sue take one of their photos to be enlarged. The original photo measures 5 cm by 8 cm and costs 40p.
The enlargement measures 10 cm by 16 cm and the chemist wants to charge £1.60 for it. Alf thinks the enlargement should cost 80p. How do Alf and the chemist calculate the cost of the enlargement? How much do you think the enlargement should cost?

2. We use microscopes to give us enlarged images of very tiny objects.
Investigate other pieces of equipment that are used to enlarge images.

3. 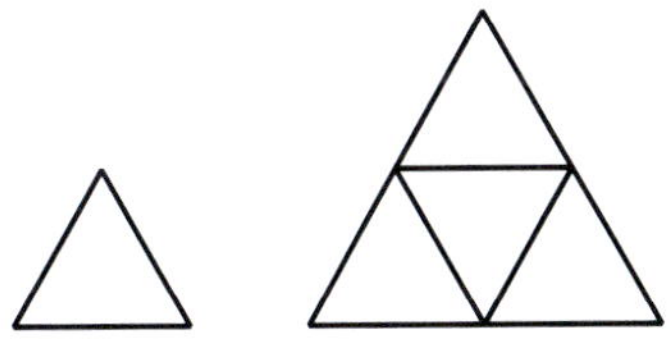

You will need isometric paper.
All the triangles in this diagram are equilateral. The small triangles are the same size.

a) Write down the ratio of the lengths of the sides of shape A to shape B.

b) Write down the ratio of the perimeter of shape A to the perimeter of shape B in its simplest form.

c) Compare the areas of shapes A and B.

d) Investigate the ratios of the lengths, perimeters and areas for other enlargements made of small triangles. What do you notice?

Looking back

1. Which of the following pairs of triangles are similar? Give reasons for your answers.

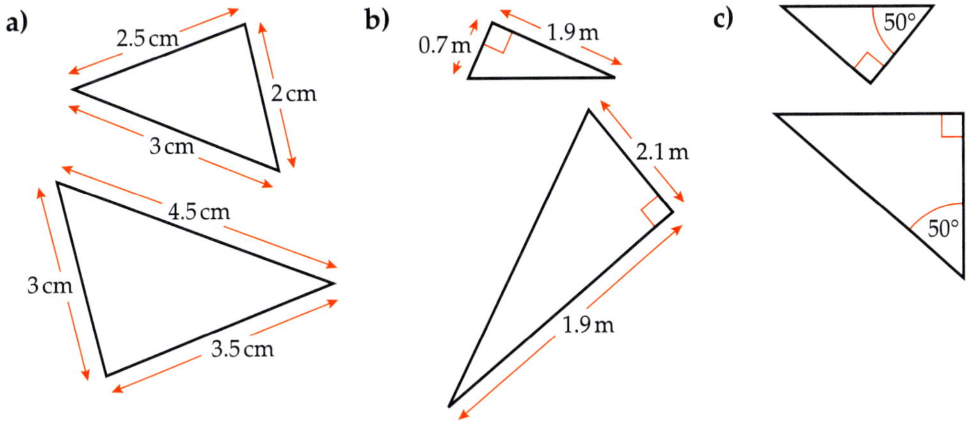

2 Copy these shapes on squared paper.

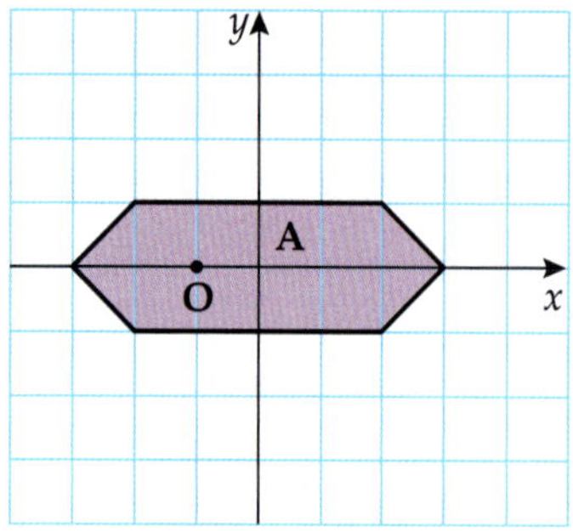

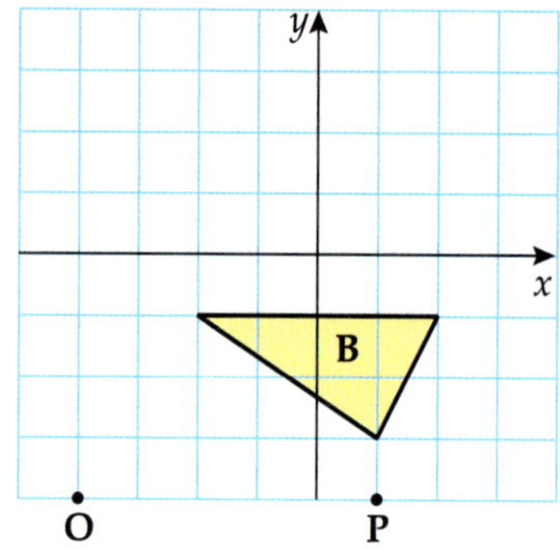

a) Enlarge shape A by $1\frac{1}{2}$, centre of enlargement O.

b) Enlarge shape B by

i) $\frac{1}{2}$, centre O **ii)** 2, centre P.

3 State whether each of the following pairs of triangles are similar or not. Give reasons.

a)

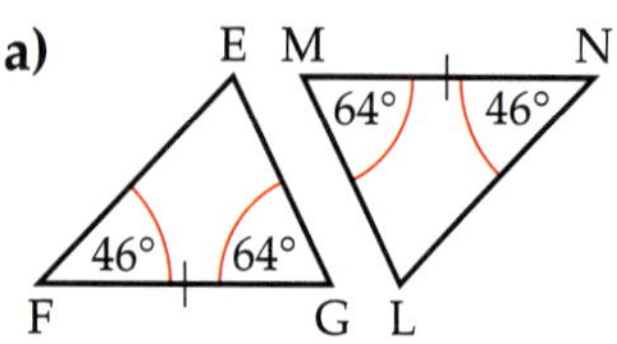

b)

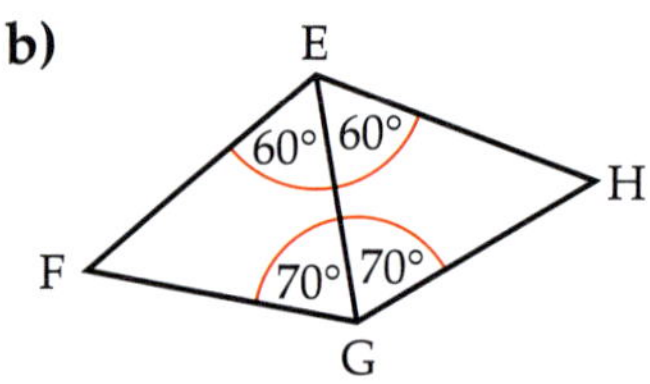

c)

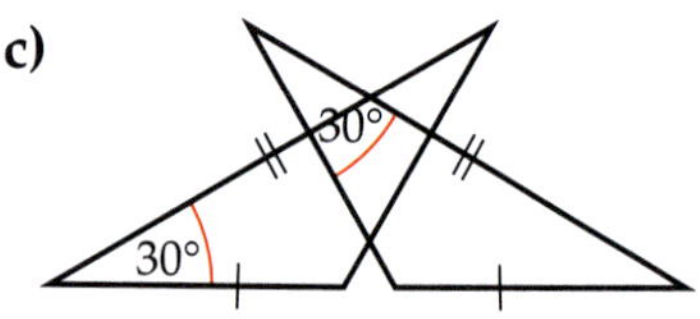

d)

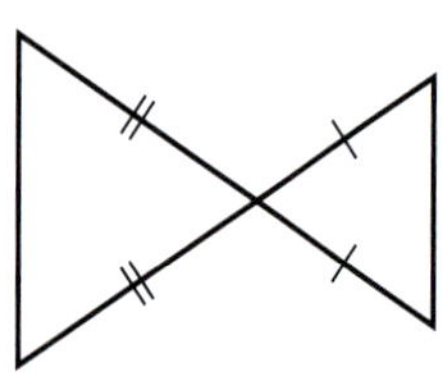

e)

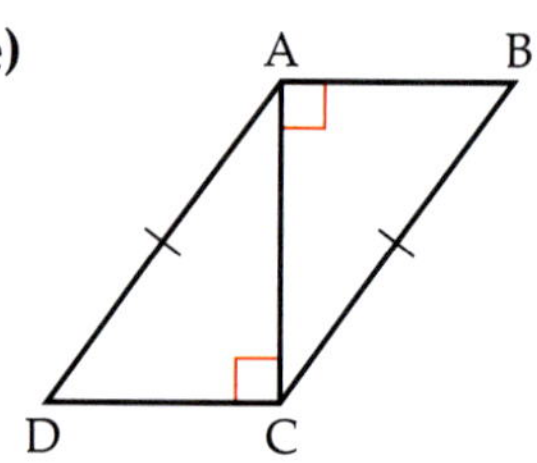

f)

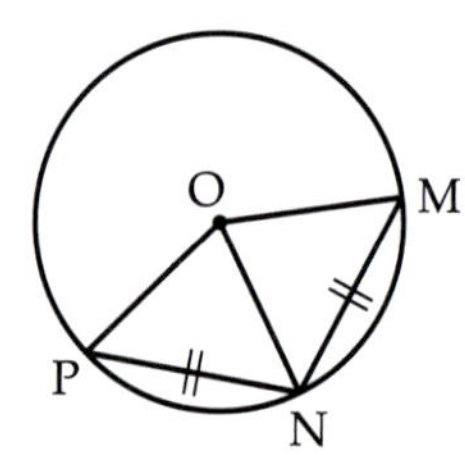

4 Copy these shapes and the given centre of enlargement, O. Enlarge them by the given scale factor in each case.

a)

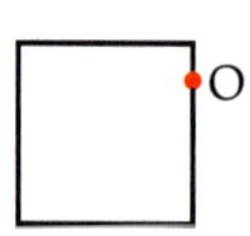

Scale factor 2

b)

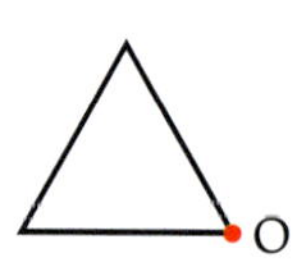

Scale factor 3

c)

Scale factor 2

d)

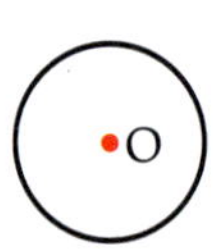

Scale factor 2.5

e)

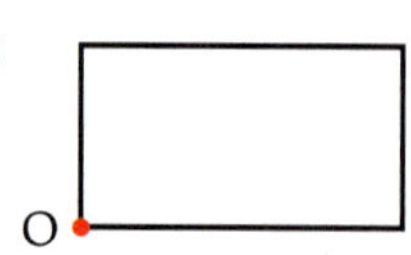

Scale factor 3.5

5 On a copy of this diagram enlarge this shape by scale factor $-\frac{1}{3}$, using the origin as the centre of enlargement.

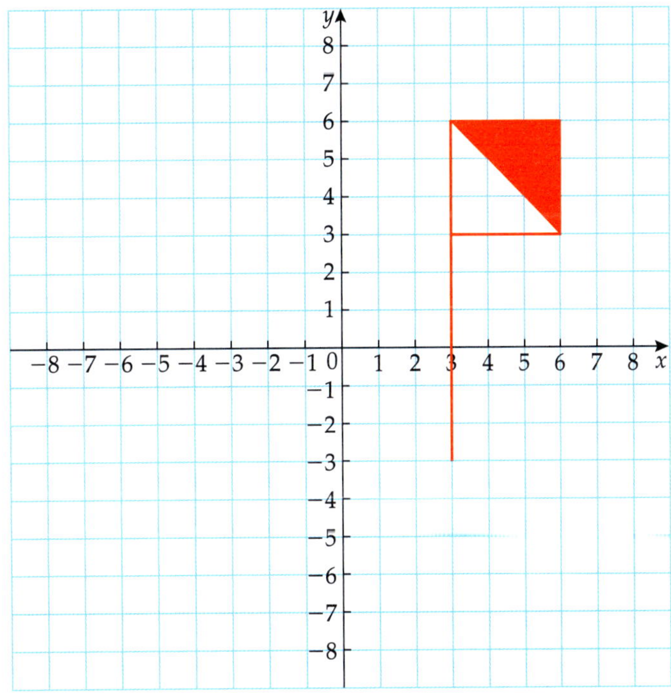

Project: Tiles

Mr De Sousa is tiling his bathroom.
This is the pattern he makes:

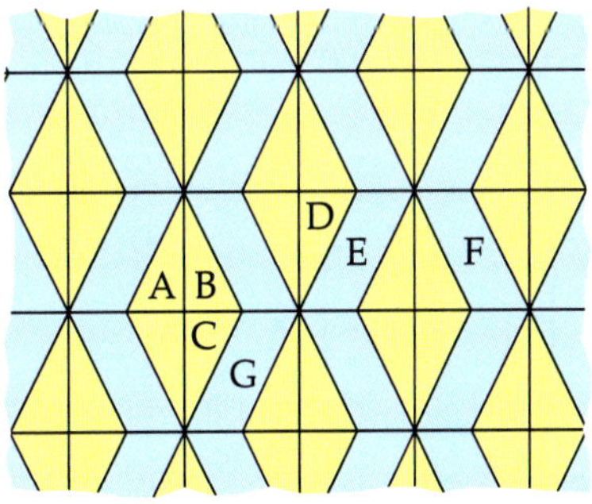

What shapes of tile is Mr De Sousa using?
What transformation maps each of the following tiles:

- Tile A onto tile B
- Tile A onto tile C
- Tile C onto tile D
- Tile E onto tile F?

What two transformations could both map tile E onto tile G?

Mr De Sousa is also tiling his kitchen.
He uses equilateral triangles of different sizes.
What is the scale factor of the enlargement that maps tile H onto tile I?
What is the scale factor of the enlargement that maps tile J onto tile K?

Use combinations of shapes to make your own tile patterns. What transformations do you need to use? You could use some of the following shapes:

- Trapezium
- Isosceles triangle
- Regular hexagon
- Rhombus
- Right-angled triangle
- Square

Can you create a tile pattern that uses enlargement?

Make a table like this showing the properties of different transformations:

Type of transformation	How you describe it	Preserves	Does not preserve	Other properties
Rotation	Centre of rotation, angle and direction of turn	Angles and lengths of sides	Orientation	Object and image are congruent

Tile patterns that fit together exactly are called **tessellations**. Some shapes, such as a regular hexagon, tessellate on their own. Others such as a regular octagon need other shapes to make a tessellation.

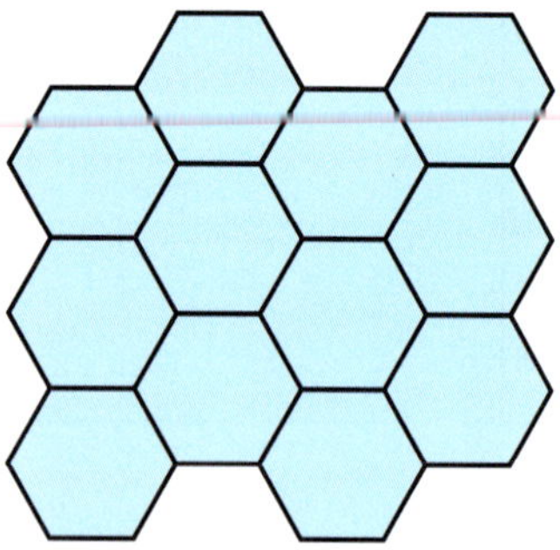
Translation using regular lengths

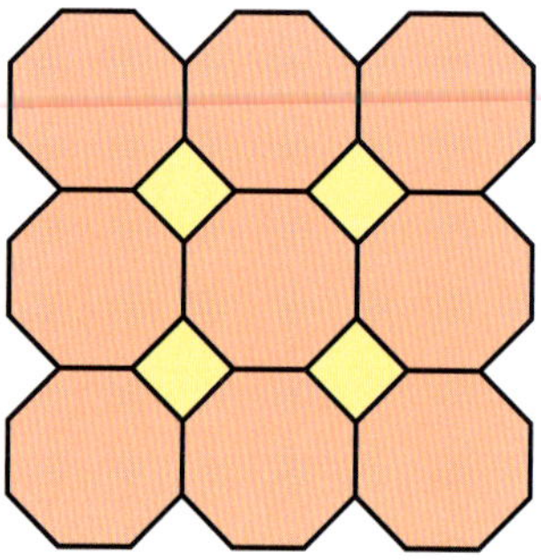
Translation using regular octagons and squares

You can make your own tessellating tile like this:

Start with a square.

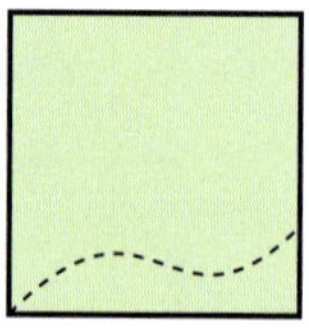
Draw a shape to cut out of one side.

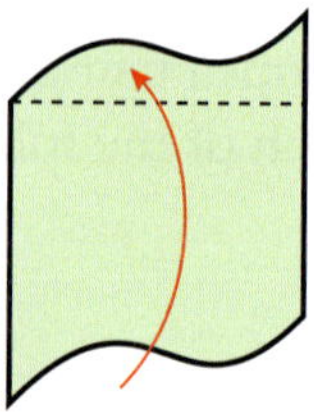
Place the cut-out section on the opposite side.

Your new shape will tessellate like this.

What transformations of the finished tile do you need to use to create the tessellation? Try placing the cut-out section on the side of the square:

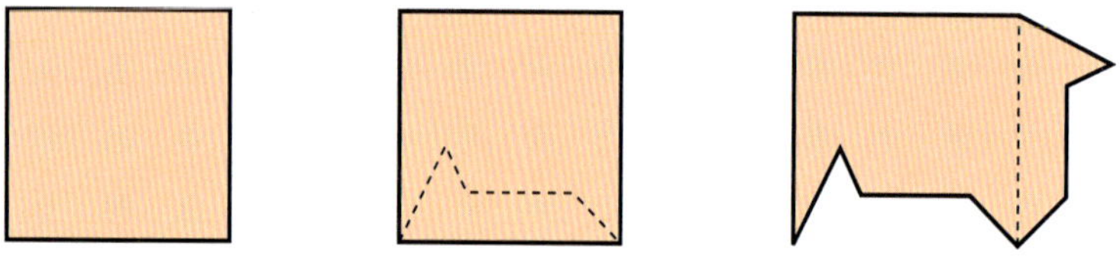

What transformations do you need to use to create a tessellation using this tile?

Investigate how you can make tessellating tiles by starting with other shapes, such as a triangle or a parallelogram.

Revise and consolidate 2

1 **a)** Construct a triangle ABC with a right angle at A, a hypotenuse BC of length 5 cm and a side AC of length 4 cm.

b) Measure the length AB.

c) Measure the angles.

2 If £10 = \$120.

a) how many dollars would I get for £3.50?

b) how many pounds would I get for \$188?

3 What would be the cost of a car, initially \$18 000, after a 15% increase?

4

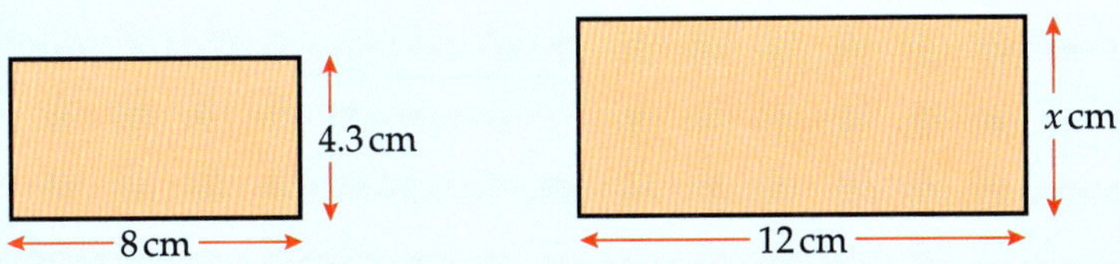

a) In these similar rectangles, what is the ratio of the sides?

b) Write the ratio in the form 1 : □

c) Find the length marked x.

5 Solve the following equations.

a) $x^2 + 2 = 27$ **b)** $a^2 - 7 = 42$ **c)** $3r^2 = 48$ **d)** $30.25 = \frac{b^2}{4}$

6

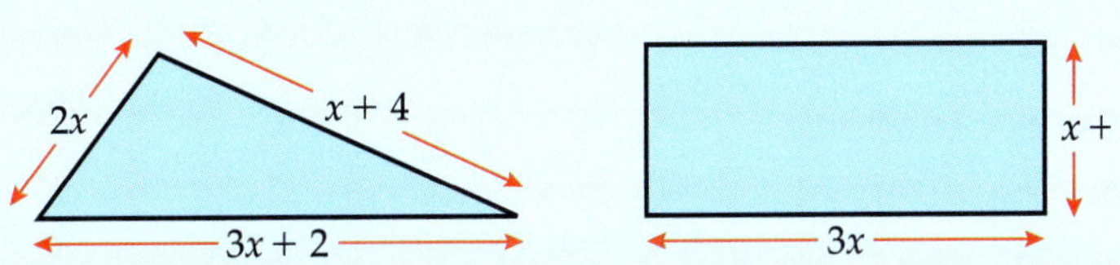

a) Write an equation to show that the perimeters are equal.

b) Solve your equation to find x.

c) The height of the triangle is 7 cm. Write a formula for its area.

d) Using your answers, find the area of both shapes.

7 On a full set of axes marked from $^{-}5$ to $^{+}5$, plot the points A(2, 2), B(4, 2) and C(2, 4).

a) Reflect ABC in the y-axis and label this P.

b) Reflect P through the x-axis and label this Q.

c) What single transformation would map ABC onto Q?

d) Translate ABC 2 left and 2 down and label this R.

e) Rotate R 90° clockwise about the point (0, 0) and label this S.

f) What other single equivalent transformation would map R to S?

g) What combination of transformations would map S onto P?

h) Enlarge R by a scale factor 2 with (0, 0) as the centre of the enlargement. Write down the coordinates of the vertices of the triangle.

8

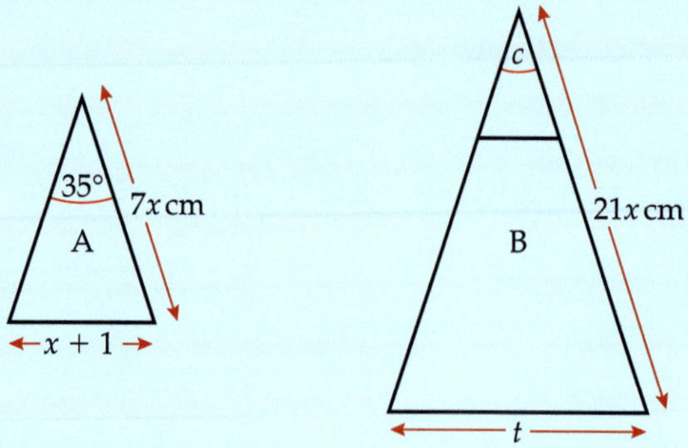

Triangle B is an enlargement of triangle A

a) What is the ratio of the corresponding sides?

b) What is the scale factor of the enlargement?

c) What is the length marked t?

d) What is the size of angle c?
Explain your answer.

e) If $x = (3.2)^2$ find the lengths of the sides.

9 Bicycles have increased in price this year by 20%. If I paid \$438 this year, what would the bicycle have cost me last year?

10 As part of a history project, Sian needs to know the height of a tower. The tower casts a shadow of 8.75 m and close by a tree 1.3 m high casts a shadow of 35 cm.

a) Using the shadows to help, find the height of the tower.

b) From the top of the tower you can see 5 km in every direction.
What is the area you can see from the tower?

11 Solve the following simultaneous equations:

a) $3x - y = 4$ and $y = 2x$

b) $2x + 3y = 12$ and $2x + 7y = 20$

c) $t - 2s = -7$ and $3t + s = 7$

12 **a)** Find the length of the missing side in each of these right-angled triangles.

i)

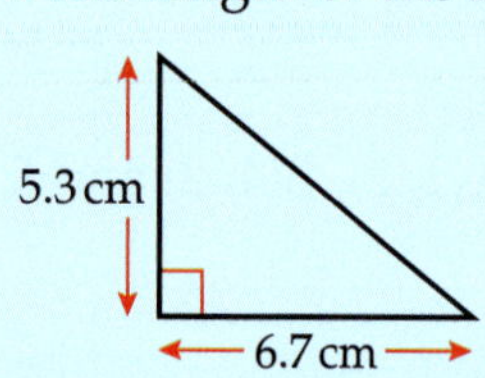

ii)

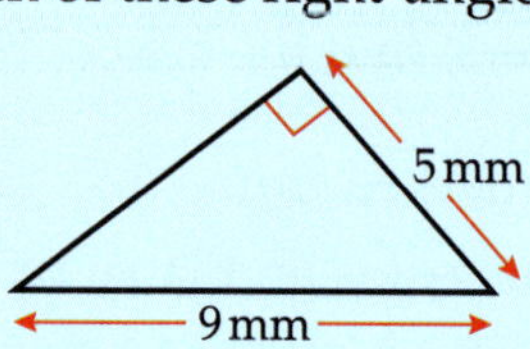

b) Is a triangle with sides 4 cm, 5 cm and 7 cm a right-angled triangle?

13 **a)** A bicycle originally priced at \$1250 was in the sale for \$900.
What percentage decrease is this?

b) David bought a jacket in a "20% off" sale. He saved \$15.
How much was the original price of the jacket?

14 Jane's grandfather gives her a choice of present for her 18th birthday. She can either have:

Option 1: \$1000 to invest at a compound interest rate of 4% per month, for one year or

Option 2: £2 on her birthday, doubling to \$4 the next month, doubling to \$8 the next month, doubling each month for a year.

Derive a formula for the amount of money Jane receives each month for each choice and calculate which is the best option.

15 **a)** Find the perimeter of the shape.

b) Find the area of the shape.

c) Given that $a = 2b$ and the perimeter is 20 cm, find values for a and b.

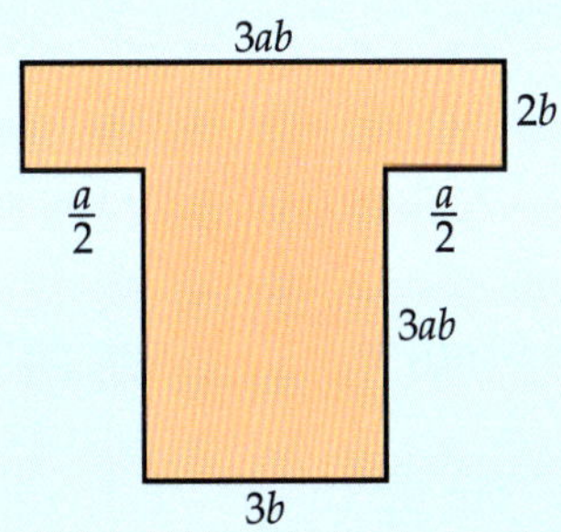

16 **a)** On a coordinate grid plot the points A $(^{-}3, ^{-}4)$, B $(^{-}3, ^{-}2)$ and C $(0, ^{-}2)$. Enlarge ABC by a scale factor of 3 from the point $(^{-}5, ^{-}6)$.
Label this A′B′C′. Now enlarge A′B′C′ by a scale factor $\frac{1}{2}$ from the point $(^{-}5, ^{-}6)$ and label this A″B″C″.

b) Measure the length of the side AB.

c) Explain how you can calculate the length of A″B″ from the length of AB using the scale factors. Measure your diagram to check your calculations.

d) What is the ratio of corresponding sides for ABC and A″B″C″.

e) What is the ratio of the perimeters for ABC and A″B″C″?

f) What is the ratio of the areas for ABC and A″B″C″?

g) Join BC′ and find the length of this line without measuring.

17 A balloon is fixed to a string 8 m long. It makes a 60° angle with the horizontal. How high is the balloon off the ground?

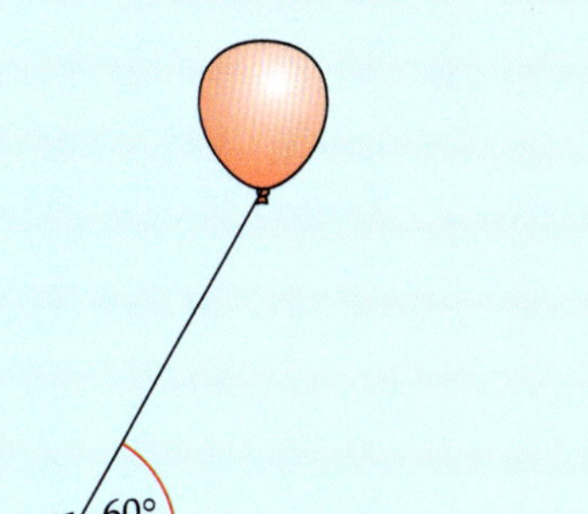

⓲ A ladder 1.5 m from the foot of a wall reaches 4 m up the wall.
To be safe to work on, the angle between the floor and the ladder must be more than 40° and less than 75°. Is the ladder safe to use?

⓳

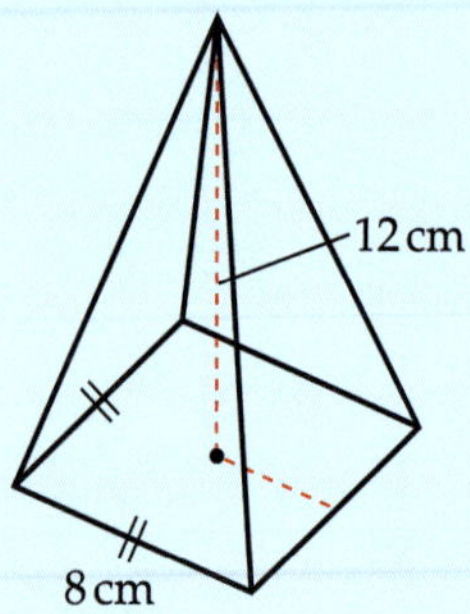

Find **a)** the volume **b)** the surface area of this square-based pyramid.

⓴ **a)** Mark the following points on squared paper or using ICT.
A(1, 1) B(3, 1) C(3, 3) and D(1, 2).
Join up the points to make a quadrilateral.

b) Enlarge ABCD by scale factor 3 using centre (0, 0).

c) On the same diagram, enlarge ABCD by scale factor 4 using centre (0, 0).

d) What is the relation between the coordinates of the vertices of ABCD and each of its enlarged images?

Key words
Union
Intersection

Chapter 11

Set theory

In this chapter you will review and extend your knowledge of sets and Venn diagrams which you met last year. You will learn to use Venn diagrams to model real situations.

Unit 11.1 The union and intersection of three sets

In this unit you will:

- find the union and intersection of three sets.

You already know that:

$A \cup B$ is the set of elements belonging to A or B.
$A \cap B$ is the set of elements belonging to both A and B.

We can expand this to three sets:

$A \cup B \cup C$ is the set of elements belonging to A or B or C.
$A \cap B \cap C$ is the set of elements belonging to A and B and C.

Worked examples

1. A = {O, V, E, N}   B = {N, O, V, E, R}   C = {V, O, T, E}

   Find **a.** $A \cap B \cap C$

   b. $A \cup B \cup C$

2. U = {whole numbers from 1 to 10}

   U is the universal set.

   A = {odd numbers}

   B = {factors of 15}

   C = {factors of 21}

   Find **a.** $A \cap B \cap C$

   b. $A \cup B \cup C$.

1. **a.** $A \cap B \cap C$ = {O, V, E}

   b. $A \cup B \cup C$ = {O, V, E, N, R, T}

2. A = {1, 3, 5, 7, 9}   B = {1, 3, 5}   C = {1, 3, 7}

   a. $A \cap B \cap C$ = {1, 3}

   b. $A \cup B \cup C$ = {1, 3, 5, 7, 9}

Exercise

1. Find A ∩ B ∩ C and A ∪ B ∪ C.
   - **a)** A = {chair, table, stool}, B = {bed, stool}, C = {dresser, bed, chair}
   - **b)** A = {B, O, T, H, E, R}, B = {O, L, D, E, R}, C = {Y, O, U, N, G, E, R}
   - **c)** A = {D, E, L, I, A}, B = {S, H, E, I, L, A}, C = {O, P, H, E, L, I, A}
   - **d)** A = {2, 4, 6, 8, 10, 12}, B = {1, 2, 3, 4, 5, 6}, C = {6, 7, 8, 9, 10}
   - **e)** A = {Donna, Sarah, Sharon, David},
     B = {Jonathan, Sandra, Sharon, David},
     C = {Donna, Sarah, Jonathan, Ricky}

2. The following table shows the universal set and sets A, B and C. In each case find
   i) A ∪ B **ii)** B ∪ C **iii)** A ∩ B ∩ C **iv)** A ∪ B ∪ C.

	U	A	B	C
a)	{letters in the alphabet}	{letters in BIOLOGY}	{letters in HISTORY}	{letters in CHEMISTRY}
b)	{letters in the alphabet}	{letters in GRAPE}	{letters in PEAR}	{letters in APPLE}
c)	{whole numbers from 0 to 10}	{even numbers}	{multiples of 3}	{factors of 12}
d)	{whole numbers from 1 to 25}	{multiples of 5}	{factors of 50}	{square numbers}
e)	{integers from −5 to 7}	{natural numbers}	{positive odd numbers}	{prime numbers}
f)	{natural numbers less than 20}	{factors of 35}	{factors of 49}	{factors of 42}
g)	{multiples of 5 between 1 and 51}	{factors of 100}	{multiples of 10}	{factors of 50}
h)	{whole numbers less than 20}	{triangular numbers}	{factors of 36}	{factors of 24}
i)	{whole numbers less than 14}	{prime numbers}	{prime factors of 60}	{odd numbers less than 4}
j)	{composite numbers between 5 and 15}	{multiples of 2}	{multiples of 6}	{multiples of 4}

Investigation

3. Review the definitions of **natural**, **whole**, **integer**, **irrational** and **real** numbers. Represent them as sets and describe the relationship between the sets.

Key words

Venn diagram

Shading

Unit 11.2 Venn diagrams

In this unit you will:

- use Venn diagrams to represent relationships between sets
- draw Venn diagrams illustrating the intersection of three sets.

Before we move onto representing three sets in a **Venn diagram**, make sure you remember how to represent subsets, the intersection and union of sets, and the complement of a set using Venn diagrams.

A ⊂ B

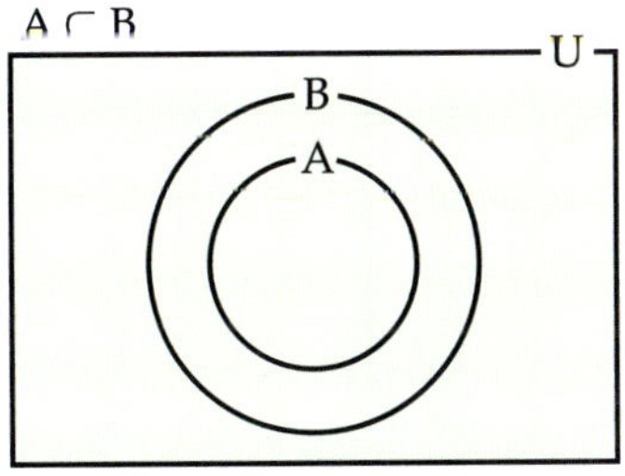

A ∩ B

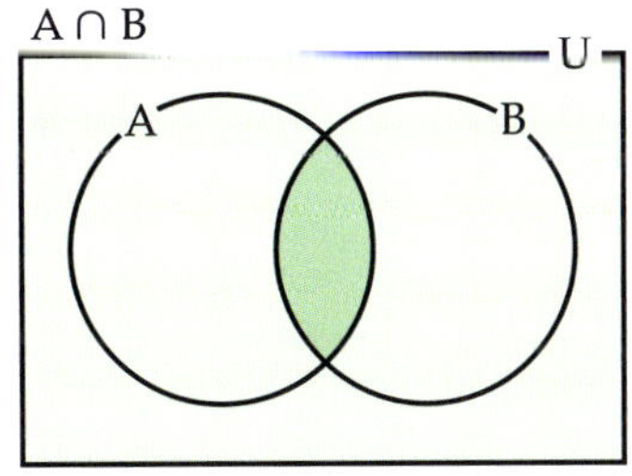

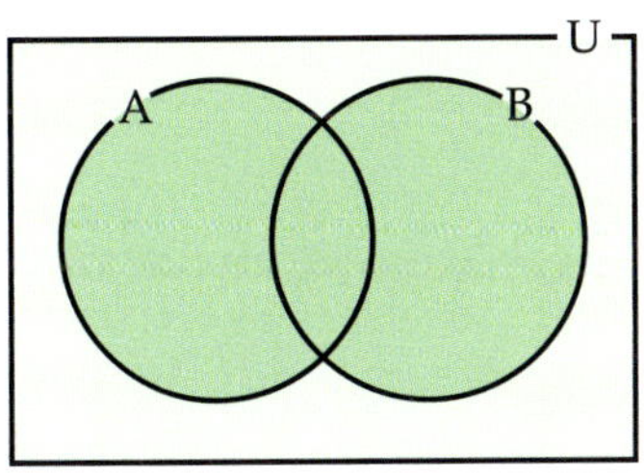

A ∪ B

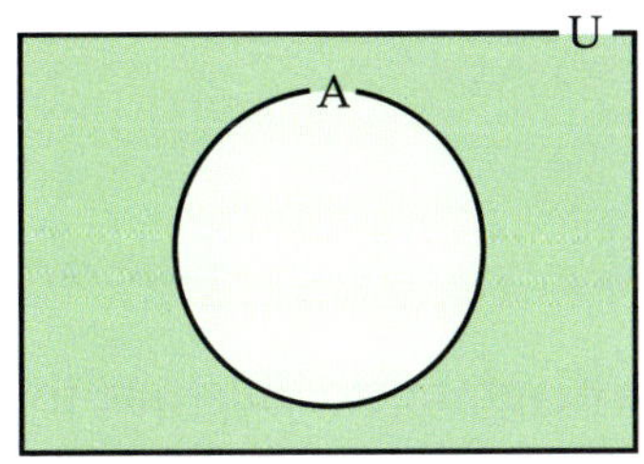

A′ (the shaded portion)

To draw a Venn diagram illustrating the intersection of three sets A, B and C:

- List the members of A, B and C.
- List the members of A ∩ B ∩ C, A ∩ B, A ∩ C, B ∩ C.

Draw a circle or rectangle to represent the universal set U, and three intersecting circles to represent A, B and C.

Place elements in each portion of the Venn diagram as follows:

- elements in A ∩ B ∩ C
- elements in A ∩ B only
- elements in B ∩ C only
- elements in A ∩ C only
- elements in A only
- elements in B only
- elements in C only
- elements in U but not in A, B or C

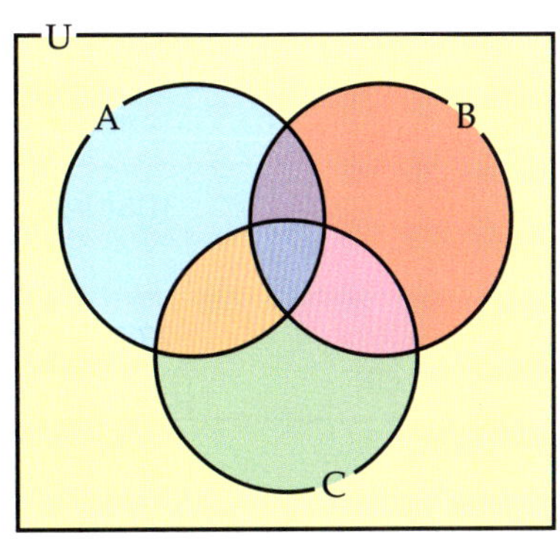

Worked example

Draw a Venn diagram illustrating A ∩ B ∩ C.

U = {whole numbers from 1 to 10}

A = {prime numbers}   B = {factors of 12}   C = {even numbers}

A = {2, 3, 5, 7}

B = {1, 2, 3, 4, 6}

C = {2, 4, 6, 8, 10}

A ∩ B ∩ C = {2}

A ∩ B = {2, 3}

A ∩ C = {2}

B ∩ C = {2, 4, 6}

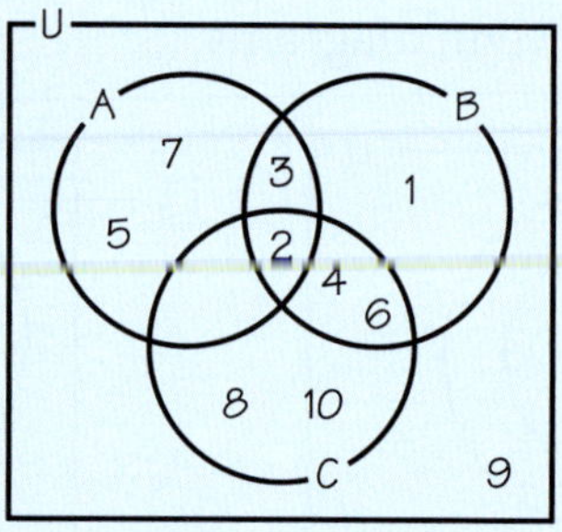

Exercise

1 Draw Venn diagrams to illustrate the following sets.

	U	A	B	C
a)	{whole numbers from 1 to 10}	{odd numbers}	{prime numbers}	{factors of 14}
b)	{M, A, C, H, I, N, E}	{N, A, M, E}	{M, I, N, E}	{C, H, A, I, N}
c)	{H,U,N,G,R,Y,O,W,E}	{G, R, O, W}	{R, O, G, U, E}	{H, O, G}
d)	{days of the week}	{days beginning with T}	{days with 8 letters}	{days between Monday and Friday}
e)	{whole numbers from 10 to 20}	{composite numbers}	{multiples of 3}	{multiples of 4}
f)	{integers from −2 to 10}	{natural numbers}	{square numbers}	{factors of 18}
g)	{odd numbers less than 20}	{multiples of 5}	{factors of 100}	{prime numbers}

2 For each Venn diagram above, list the members of the following sets.

i) A ∪ B ∪ C   **ii)** (A ∪ B ∪ C)′

3 Draw a Venn diagram to show A ∩ B ∩ C, where A = months beginning with J, B = months with 31 days and C = months with 30 days. U = months of the year.

a) Which month is A ∩ C?   **b)** Which month is (A ∪ B ∪ C)′?

Looking back

1. Find $A \cap B \cap C$ and $A \cup B \cup C$
   A = {apple, orange, grape, melon}
   B = {grape, banana, orange}
   C = {orange, lemon}

2. Find $A \cap B \cap C$ and $A \cup B \cup C$ for U = {positive integers less than 10}
   a) A = {prime numbers}
   B = {odd numbers}
   C = {square numbers}
   b) A = {factors of 24}
   B = {multiples of 4}
   C = {square numbers}

3. Show your answers to questions 1 and 2 as Venn diagrams.

Chapter 12

Relations and functions

This chapter shows you how to distinguish between relations and functions, and how to use functional notation. You will use functional notation to obtain ordered pairs of linear and quadratic functions, and represent functions graphically.

Key words
- Relation
- Mapping
- Function
- One-to-one
- Many-to-one

Unit 12.1 Relations and functions

In this unit you will:

- review mapping diagrams and number lines
- identify one-to-one and many-to-one mappings
- identify functions
- use function notation to describe functions
- evaluate values of functions given functional notation.

Numbers that are related to each other by a rule form **relations**.

For example: S is the number of people who walk into a classroom. T is the total number of people in the classroom. This relation can be displayed in a **mapping** diagram.

$S \rightarrow T$
$1 \rightarrow 26$
$2 \rightarrow 27$
$10 \rightarrow 35$
$S \rightarrow 25 + S$

This is a one-to-one mapping.

S 'maps to' $25 + S$.

Mappings can also be displayed using number lines.

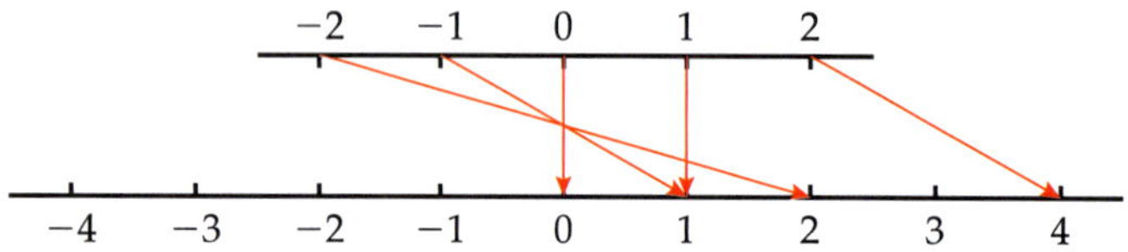

This is a many-to-one mapping.

When one number is mapped to one, and only one, other number, then the relation is said to be one : one. When a number is mapped only once but the number it is mapped to is mapped many times the relation is many : one. Both these types of relations are called **functions**.

Functions are special relations where a number in the first row or column of the mapping only ever appears once.

Functions are described using the notation

$f : x \rightarrow ax + b$

or $f(x) = ax^2 + bx + c$.

The function f maps x to $ax + b$.

The function f of x is equal to $ax^2 + bx + c$.

$f(x)$ does not mean $f \times x$.

Worked examples

1. a. Draw a mapping diagram for $x \rightarrow 3x + 2$, for $-2 \leqslant x \leqslant 3$, $x \in \{I\}$.

   b. Write an equation that describes this mapping using function notation.

2. If $f(x) = 5 + 5x + x^2$ what is $f(2)$?

1. a.

x	$\rightarrow$	y
-2	$\rightarrow$	-4
-1	$\rightarrow$	-1
0	$\rightarrow$	2
1	$\rightarrow$	5
2	$\rightarrow$	8
3	$\rightarrow$	11
x	$\rightarrow$	$3x + 2$

   b. $f(x) = 3x + 2$

2. $f(x) = 5 + 5x + x^2$

   $f(2) = 5 + (5)(2) + 2^2$ (Substitute $x = 2$ into the equation)

   $f(2) = 5 + 10 + 4$

   $f(2) = 19$

Exercise

1 Which of the following mappings are of functions?

a)

p	$\rightarrow$	q
-3	$\rightarrow$	-27
-2	$\rightarrow$	-8
-1	$\rightarrow$	-1
0	$\rightarrow$	0
1	$\rightarrow$	1
2	$\rightarrow$	8
3	$\rightarrow$	27
p	$\rightarrow$	p^3

b)

M	$\rightarrow$	N
10	$\rightarrow$	50
20	$\rightarrow$	350
30	$\rightarrow$	850
40	$\rightarrow$	1550
50	$\rightarrow$	2450
M	$\rightarrow$	$M^2 - 50$

c)

x	$\rightarrow$	y
-2	$\rightarrow$	4
-1	$\rightarrow$	1
0	$\rightarrow$	0
1	$\rightarrow$	1
2	$\rightarrow$	4
x	$\rightarrow$	$\sqrt{x}$

d)

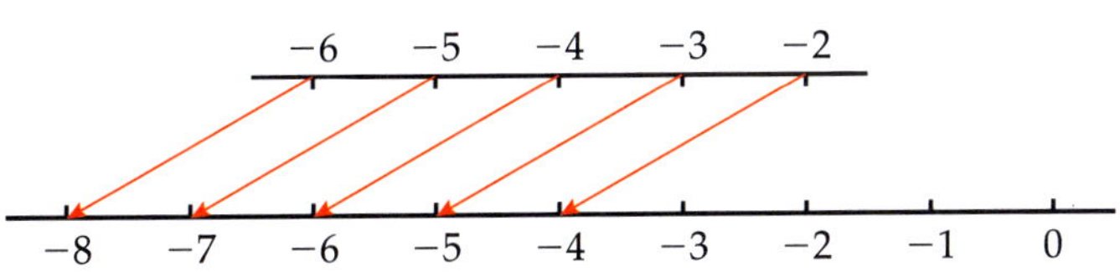

e)

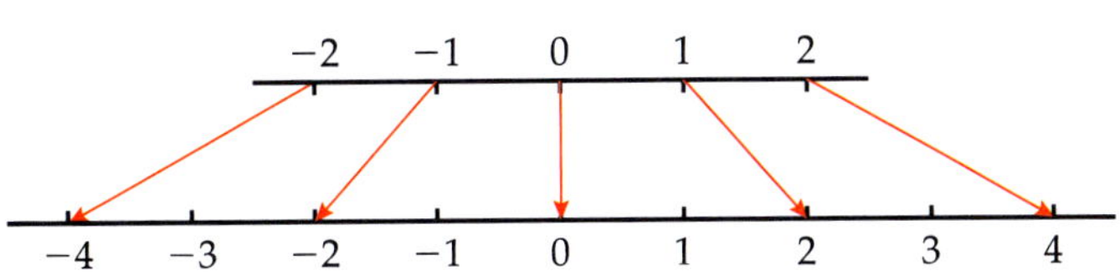

2 **a)–e)** For each of the functions above, describe what type of relation it is.

3 Write each of the following mappings as functions, using function notation.

a) $x \to x^3$ **b)** $x \to 3 + x$ **c)** $x \to x^2 + x$

d) $x \to \frac{5}{x}$ **e)** $x \to 2 - x^2$ **f)** $x \to 2x - 11$

g) $x \to \frac{2}{x+1}$

4 **a)** Draw the mapping $p \to 3p$ on a number line for $p \in \{I\}$ and $-3 \leqslant p \leqslant 3$.

b) Is this relation a function? Why?

5 If f: $x \to x^2 + 4x - 6$ evaluate

a) $f(4)$ **b)** $f(-5)$ **c)** $f(10)$ **d)** $f(100)$

6 If $f(x) = x^3 + 20$ evaluate

a) $f(-2)$ **b)** $f(1)$ **c)** $f(5)$ **d)** $f(100)$

7 **a)–g)** For each function in question **3**, evaluate $f(-1)$ and $f(-5)$.

Key words
Functions
Ordered pairs
Coordinates

Unit 12.2 Graphs of relations and functions

In this unit you will:

- create ordered pairs from mapped functions and equations of functions
- draw graphs of functions using tables and ordered pairs.

Functions can be described using mappings, equations or a set of ordered pairs or coordinates.

	A	B
Mapping	$-2 \to -5$ $-1 \to -3$ $0 \to -1$ $1 \to 1$ $2 \to 3$ $x \to 2x - 1$	$-2 \to 12$ $-1 \to 3$ $0 \to 0$ $1 \to 3$ $2 \to 12$ $x \to 3x^2$
Equation	$y = 2x - 1$	$y = 3x^2$
Ordered pairs	$(-2, -5), (-1, -3), (0, -1), (1, 1), (2, 3)$	$(-2, 12), (-1, 3), (0, 0), (1, 3), (2, 12)$

Remember, x is the independent variable.

In a set of ordered pairs for any function, no pairs will have the same first element.

Functions can also be represented using graphs. In a graph of a function you can draw any vertical line and it will only pass through the graph once. Linear and quadratic equations are both examples of functions.

'Linear' $\Rightarrow$ straight-line graphs.
'Quadratic' $\Rightarrow$ parabolic graphs, x^2 is the highest power.

Using your knowledge about linear and quadratic equations (see Chapter 5) you can plot graphs of functions.

Worked examples

1. a. What is the equation that maps the following mapping diagram?

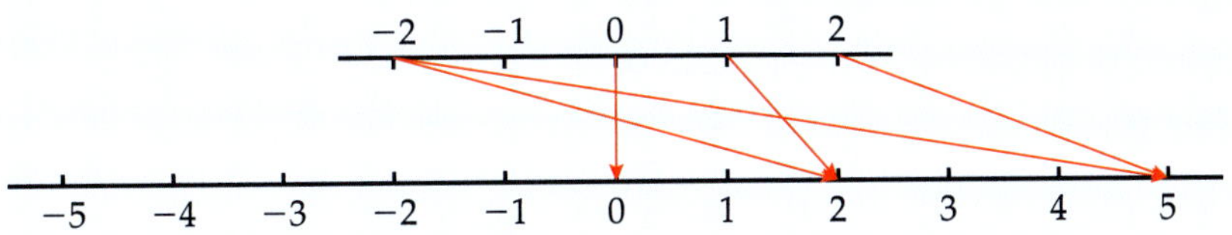

b. Use your equation to complete the table.

x	−5	5	20	−100
y				

c. Is this relation a function? Why?

2. a. Write a set of ordered pairs for the function $y = x^2 - x + 1$ for $-4 \leqslant x \leqslant 4, x \in \{I\}$

b. Use these pairs to plot a graph of $y = x^2 - x + 1$.

1. a.

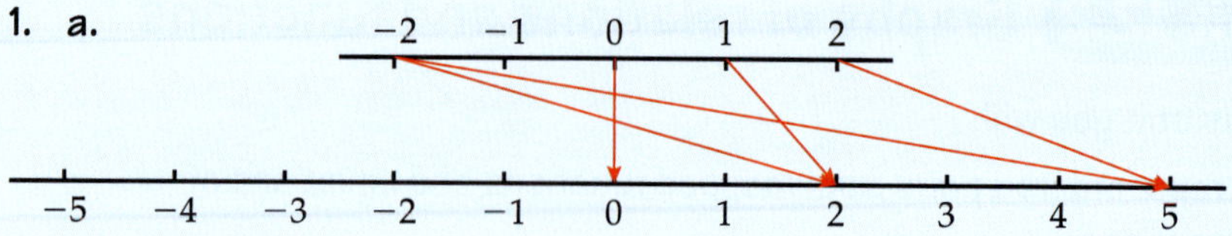

The mapping is $f: x \rightarrow x^2 + 1$, or $y = x^2 + 1$.

> The value when $x = 0$ helps to calculate constants of addition or subtraction.

b.

x	−5	5	20	−100
y	26	26	401	10 001

> Repetition of the mapped values implies a quadratic function.

c. This relation is a function because the value for x is never repeated: it is a many-to-one function as the y-value is repeated.

2. a. (−4, 21), (−3, 13), (−2, 7), (−1, 3), (0, 1), (1, 1), (2, 3), (3, 7), (4, 13)

> Put $x = -4$ into the equation and so on...

b.

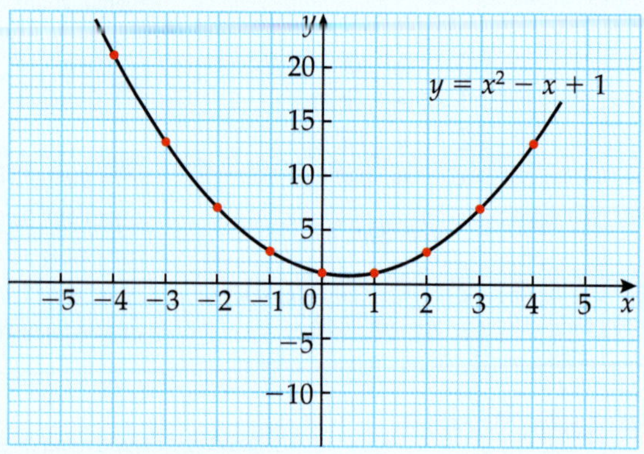

Exercise

1 Write equations for the functions represented by the following mapping diagrams.

a)

x		y
−2	→	−5
−1	→	−2
0	→	1
1	→	4
2	→	7

b)

x		y
−2	→	12
−1	→	11
0	→	10
1	→	9
2	→	8

c)

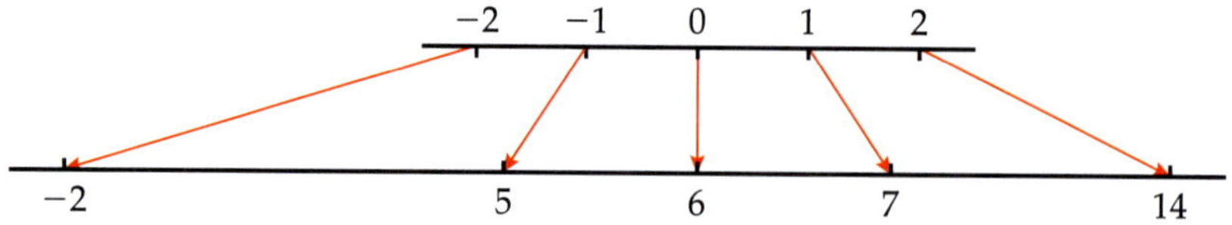

d)

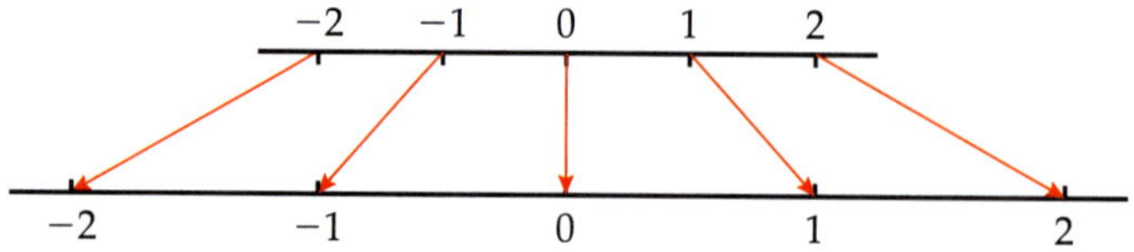

2 Write ordered pairs for the following mappings.

a) $x \rightarrow \frac{1}{x}, 1 \leqslant x \leqslant 5, x \in \{I\}$

b) $x \rightarrow 9x - x^2, -3 \leqslant x \leqslant 3, x \in \{I\}$

c) $x \rightarrow 14x - 2, -2 \leqslant x \leqslant 4, x \in \{I\}$

d) $x \rightarrow 3x^2 - 5x + 11 - x^2 - x, -2 \leqslant x \leqslant 2, x \in \{I\}$

3 Write functions for the mappings in question 2 using function notation.

4 **a)** For the function $f(x) = x^2 - 2x - 8$ write ordered pairs for $-3 \leqslant x \leqslant 6, x \in \{I\}$

b) Plot the graph of $f(x)$ for $-3 \leqslant x \leqslant 6, x \in \{R\}$.

5 **a)** Write a set of coordinates for $f(x) = 2x^2 + 3x - 1, -3 \leqslant x \leqslant 3, x \in \{I\}$

b) Draw a graph of this function.

Unit 12.3 Further graphs of functions

In this unit you will:

- draw graphs of functions using the algebraic form of the function
- determine the algebraic form of a function by analysing its graph

Key words
Algebraic form
Linear function
Quadratic function
Intercept
Root

You can plot graphs by looking at the **algebraic form** of the function (that is, the equation). You can also find the equation by looking at the graph.

Two common types of functions are **linear functions** and **quadratic functions**. The graph of a linear function is a straight line, and the graph of a quadratic function is a parabola.

<table>
<tr><th>Function</th><th>Linear</th><th>Quadratic</th></tr>
<tr><td>Algebraic form</td><td>$f(x) = mx + c$</td><td>$f(x) = ax^2 + bx + c$</td></tr>
<tr><td>Graphical form</td><td>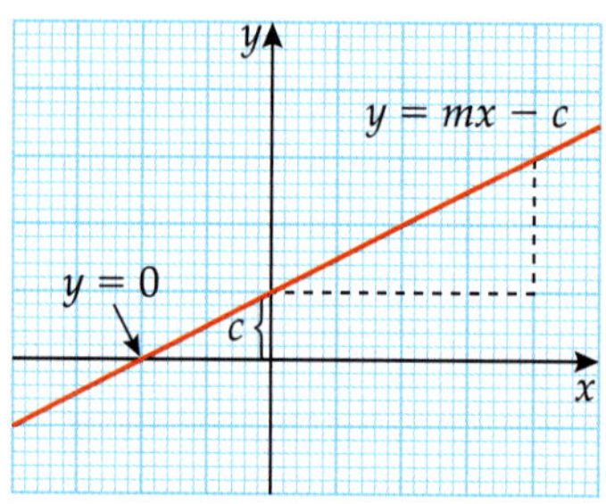
</td><td>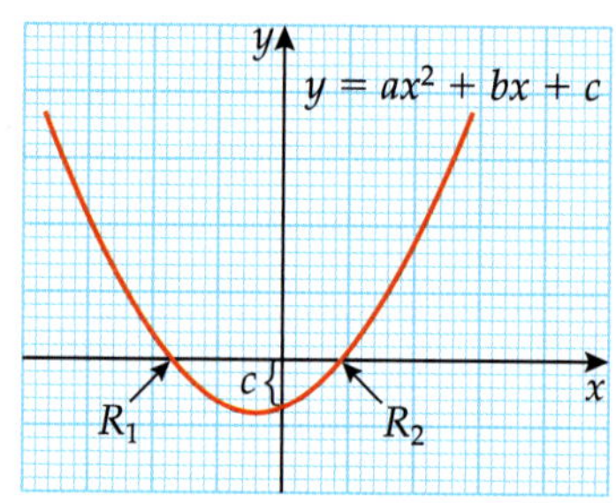
</td></tr>
<tr><td>Additional information</td><td>m = slope of straight line
$= \frac{(y_2 - y_1)}{(x_2 - x_1)}$</td><td>Factorised form of quadratic is
$y = (x - R_1)(x - R_2)$,
where R_1 and R_2 are the roots.
Roots are found when $y = 0$
$0 = (x - R_1)(x - R_2)$
$c = R_1R_2$</td></tr>
</table>

Worked examples

1. What is the function plotted in the graph? (Leave your answer in its expanded form.)

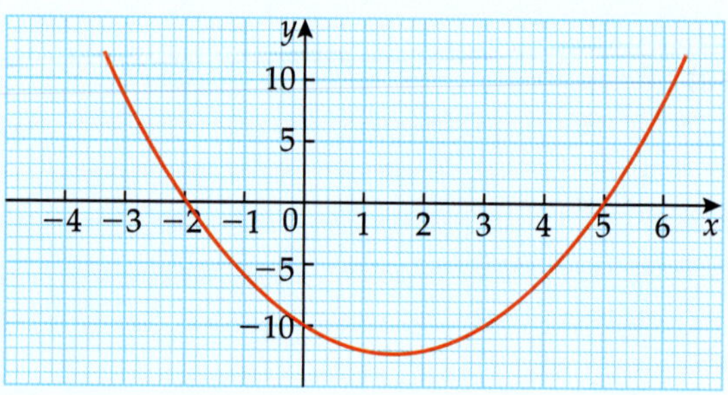

1. From the graph, the function intercepts the x-axis when $x = -2$ and $x = +5$, so the roots are

$R_1 = -2$ and $R_2 = +5$

So $y = (x + 2)(x - 5)$ (Substituting the roots into factorised form of quadratic equation.)

Check that this equation matches the graph.

Check 1: $c = R_1R_2$ (predicted value of y-Intercept from factorised quadratic equation)

$c = (-2)(+5)$ (Reading values from the graph.)

$c = -10$

This matches the graph (the y-intercept is $y = -10$).

Check 2: When $x = 1$, $y = (1 + 2)(1 - 5)$ (Substitute $x = 1$ into factorised equation)

$y = (3)(-4)$

$= -12$

The point $(1, -12)$ lies on the graph.

It is always a good idea to check your results by calculating or looking at 2 other points on the graph.

Exercise

1. Plot the following graphs for $-5 < x < 5$

a) $y = x^2$ **b)** $y = -3x + 5$ **c)** $y = (x - 4)(x - 2)$

d) $y = x(x + 3)$ **e)** $y = 4 - x$ **f)** $y = (x + 1)(x - 3)$

2. Write the functions that are plotted in the graphs below. You may leave any quadratic functions in their expanded form.

a)

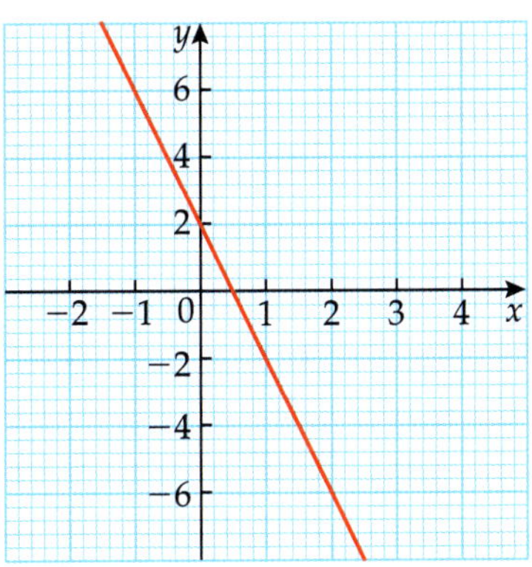

b)

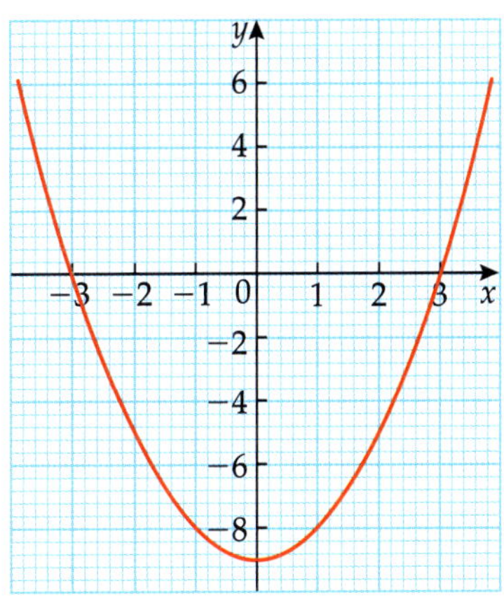

c)

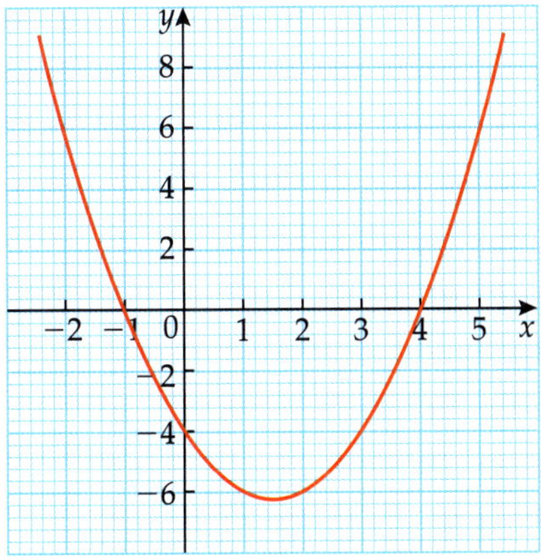

3. Sketch the following graphs. Clearly mark any roots and all intercepts.

a) $y = 2(x + 4)$

b) $y = 3x^2 - 2x^2 - 3x$

c) $y = x^2 - \frac{x^4}{x^2} + 15 - 3x$

Investigation

4. Sketch the graph of $y = x^3$ and determine whether it is a function. Draw the graph of $y = \sqrt{x}$. Is this a function?

Looking back

1. If $f: x \to x^2 + 3x - 4$, evaluate

a) $f(2)$ b) $f(4)$ c) $f(10)$ d) $f(-1)$

2. a) Construct a table for $f(x) = x^3$ for $-4 \leqslant x \leqslant 4$, $x \in \{I\}$.

b) Plot the graph of $f(x) = x^3$ for $-4 \leqslant x \leqslant 4$, $x \in \{R\}$.

3. Plot the graphs for $-5 \leqslant x \leqslant 5$

a) $y = 2x^2$ b) $y = 2 - 4x$ c) $y = (x + 1)(x - 2)$ d) $y = 6 - x^2$

4. Write the functions that are plotted in the graphs below. You may leave any quadratic functions in their expanded form.

a)

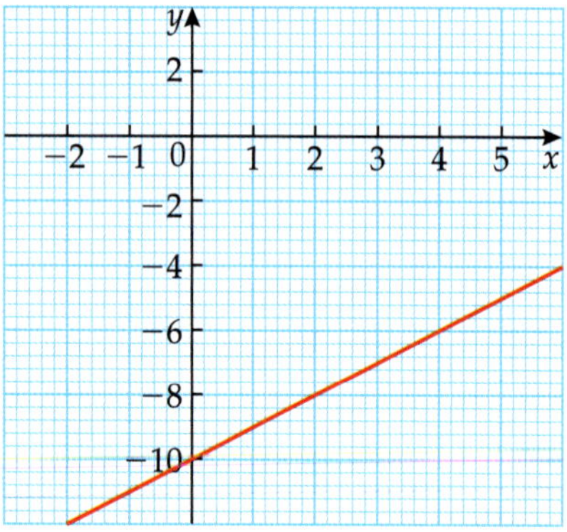

b)

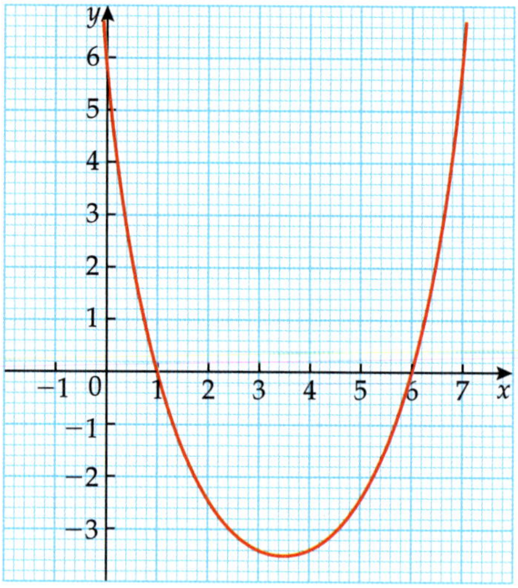

Key words
Inequality
Solution set
Set notation
Integer

Chapter 13

Inequalities

This chapter will show you how to use a range of different strategies to solve inequalities. You will practise using inequalities to solve real problems, and develop a set of rules for dealing with inequalities.

Unit 13.1 Inequalities on a number line

In this unit you will:

- use the signs $<$, $>$, $\leqslant$ and $\geqslant$
- write and solve simple inequalities
- show the solution of inequalities on a number line and in set notation.

An **inequality** is a **statement** that one expression is greater than or less than another in value.

$x > y$ means 'x is **greater than** y'
$x < y$ means 'x is **less than** y'
$x \geqslant y$ means 'x is **greater than or equal to** y'
$x \leqslant y$ means 'x is **less than or equal to** y'

The set of all answers which satisfy the original inequality is known as the **solution set**. You can show the solution set to an inequality in a number of ways.

Number line:
You can show an inequality on a number line.

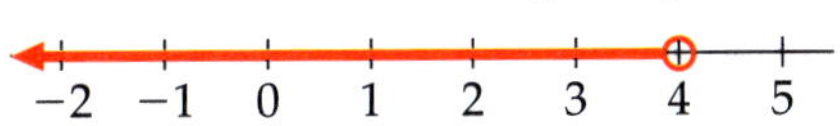

An open circle ○ means a number is not included. In this example, $x < 4$.

A closed circle ● means a number is included. In this example $x \geqslant -2$.

Sometimes there is more than one **condition**.

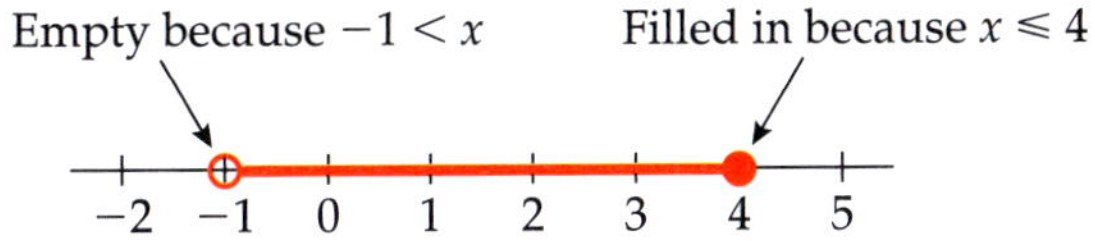

This tells us that $-1 < x$ and $x \leqslant 4$. This can be written as $-1 < x \leqslant 4$ by combining the two separate conditions.

Set notation:
The solution to an inequality is a set of numbers. For example, $x > 5$ describes the set of all numbers that are greater than 5. This set would include numbers like 6 and 5.01. One way of writing the solution to an inequality is by using **set notation**. To write $x > 5$ in set notation we would write $\{x \mid x > 5\}$. This is read as 'the set of x-values such that x is greater than 5'. The curly bracket is used to enclose a listed set. The vertical line is read as 'such that'. So:

$x < 1$	$\{x \mid x < 1\}$
$-4 < x \leqslant 10$	$\{x \mid -4 < x \leqslant 10\}$

Worked example

1. Find all the **integers** that satisfy each inequality:

a. $1 > b \geqslant -5$ **b.** $0 \geqslant p > -4$

a. $1 > b \geqslant -5$

$1 > b$ (A) and $b \geqslant -5$ (B)

For (A) to be true $b = 0, -1, -2, \ldots$

For (B) to be true $b = -4, -3, -2, \ldots$

So for (A) and (B) to be true: $b = 0, -1, -2, -3, -4$

SOLUTION SET $\{0, -1, -2, -3, -4\}$

b. $0 \geqslant p > -4$

$0 > p$ (A) and $p > -4$ (B)

For (A) to be true $p = 0, -1, -2, -3, \ldots$

For (B) to be true $p = -3, -2, -1, \ldots$

So for (A) and (B) to be true: $p = 0, -1, -2, -3$

SOLUTION SET $\{0, -1, -2, -3\}$

Exercise

1 Translate each phrase into algebraic notation.

a) x is no more than 11 **b)** y is at least -3

c) t is at most 27 **d)** m is not less than 6

2 Write down the integers between -10 and 10 which satisfy each inequality.

a) $d < 5$ **b)** $x > -5$ **c)** $y < 9.2$ **d)** $a \geqslant 0$ **e)** $x \leqslant 2.6$

3 Write down the integers that satisfy each diagram.

a)
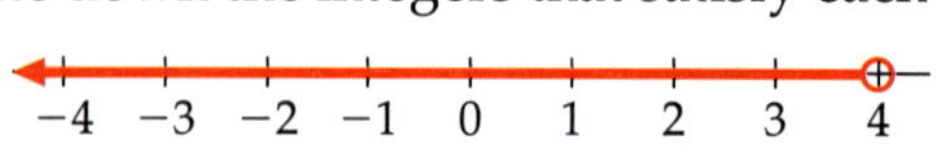

b)
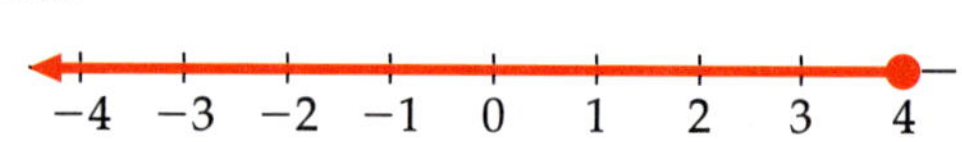

c)
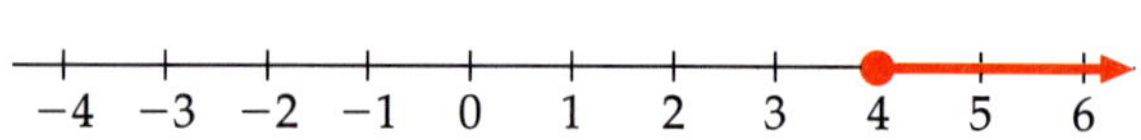

4 **a)** Write down the integers that satisfy each diagram.

b) Write down, using set notation, the inequality that is represented by the diagram.

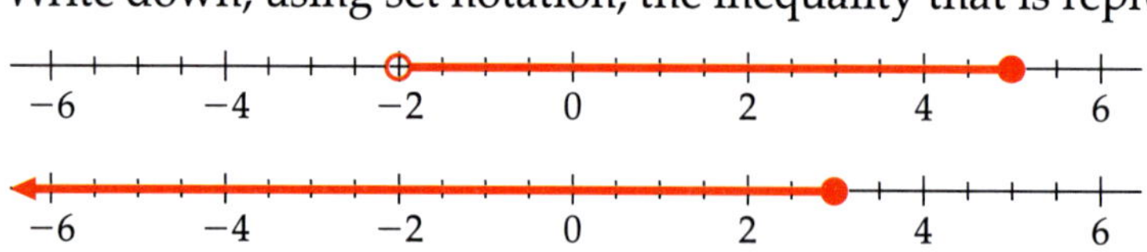

5 Write down the solution set to each diagram using inequalities.

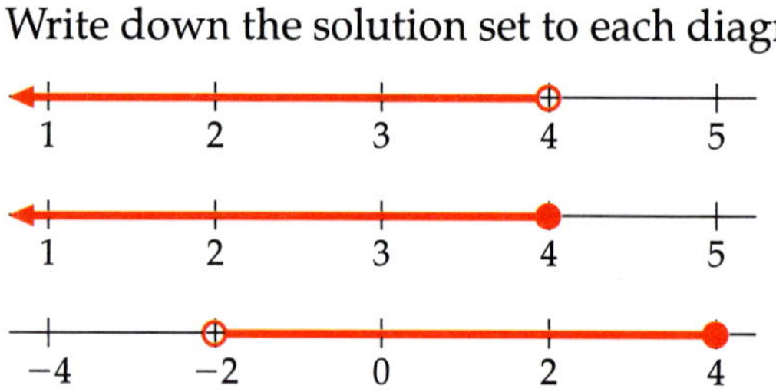

6 For the questions below:

i) draw the complete solution set on a number line.

ii) write down the integers, between -5 and 5, that satisfy each inequality.

a) $x \geqslant 3$ **b)** $x < 1$ **c)** $y \geqslant -2$ **d)** $x > -6$

e) $x \leqslant 4$ **f)** $y \leqslant -1$ **g)** $x < -3$ **h)** $x > -10$

7 The variable b satisfies each of these inequalities: $31.62 \leqslant b \leqslant 31.83$ and $31.95 > b > 31.74$. Mark the solution set for b on a number line.

8 Write down the integers that satisfy each set of inequalities.

a) $x > 2, x \leqslant 7$ **b)** $x \geqslant 1, x < 5$ **c)** $x \leqslant 0, x > -4$

d) $x \leqslant 6, x \geqslant 0$ **e)** $x < -9, x > -16$ **f)** $x \leqslant 4, x \geqslant -7$

g) $x \leqslant -6, x \geqslant -8$ **h)** $x \geqslant -12, x \leqslant 0$

9 Draw the complete solution set, on a number line, for these inequalities.

a) $x \geqslant 7, x < 2$ **b)** $x > 4, x \leqslant 1$ **c)** $x \geqslant 9, x < 7$

d) $x \geqslant 1, x \leqslant -1$ **e)** $x \geqslant 4, x \leqslant -4$ **f)** $x > -2, x < -8$

Why have you not been asked to find any integers that satisfy each set of inequalities?

10 For each inequality in question **9** draw the complete solution set on a number line.

Unit 13.2 Using graphs to solve inequalities

In this unit you will:

- determine the regions of the Cartesian plane that solve inequalities.

Key words
Greater than or equal to $\geqslant$
Less than or equal to $\leqslant$
Solution set

Remember:

- the set of points on a line satisfies the equation of the line
- a line divides the Cartesian plane into three regions or **solution sets**.

A

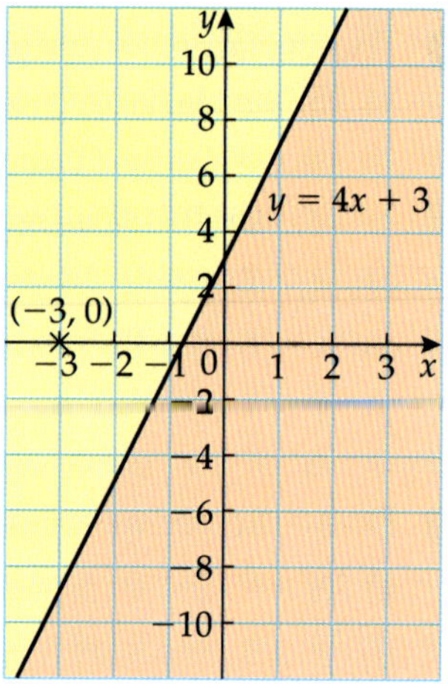

Graph A has points
i) on the line $y = 4x + 3$
ii) in the region $y > 4x + 3$
iii) in the region $y < 4x + 3$.

To locate the regions that are the solution set for the inequalities, test points in the regions on either side of the line.
For example, $(-3, 0)$ lies in one region of A (the yellow part).
When $x = -3$, $4x + 3 = -12 + 3 = -9$ and $y = 0 > -9$.
Therefore that region of A satisfies $y > 4x + 3$. The other side satisfies $y < 4x + 3$.

Sometimes a region is a solution for more than one inequality.

The shaded region A satisfies the inequalities $x \geqslant 0$ and $y \geqslant 0$.
The shaded region B satisfies the inequalities $-2 < x \leqslant 0$ and $0 \leqslant y \leqslant 4$.
The shaded region C satisfies the inequalities $x < -2$ and $y \leqslant 0$.

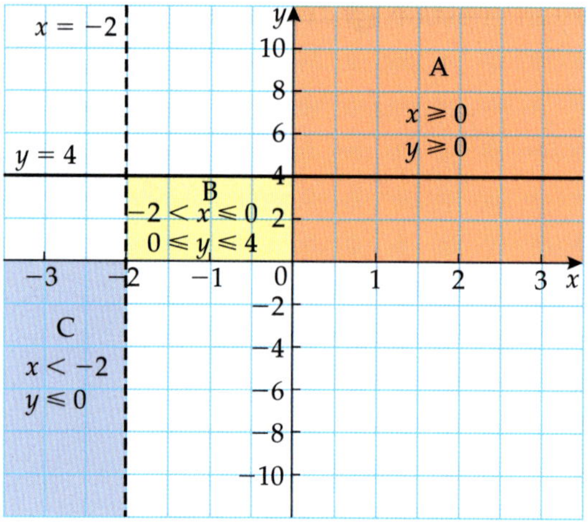

In region B x lies between -2 and 0, includes $x = 0$ but not $x = -2$ because that line is broken.

In region B y lies between 0 and 4 and includes both $y = 0$ and $y = 4$.

In this graph, the unshaded region represents $y < 4 - 2x$, and $x \geqslant -1$.

The equality is not included in this region, because $y = 4 - 2x$ is a broken line.

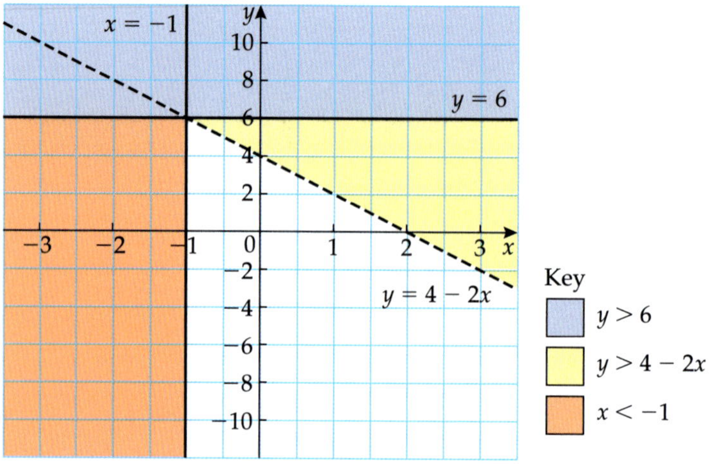

The point $(0, 0)$ lies in this region and satisfies the three inequalities. Other points in and out of this region may be used to test if the inequalities are true or false.

Exercise

1. Given the following graphs, write the inequalities that will satisfy the regions A to F.

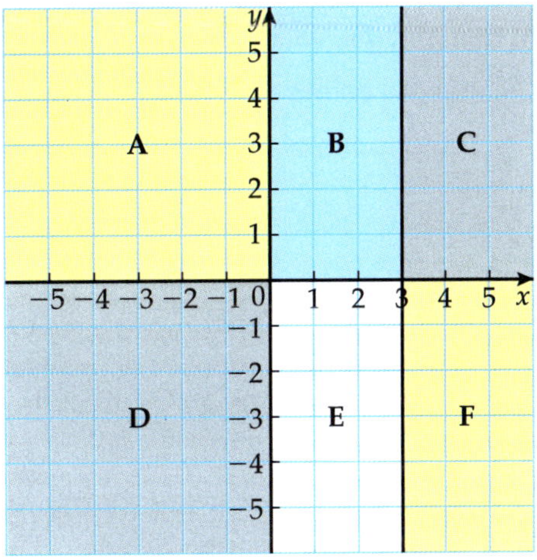

2. Use the graph in question 1 to copy and complete the table with one set of coordinates that lies in each region. Remember to check the x- and y-values in the inequalities you chose.

Regions	One set of coordinates
A	
B	
C	
D	
E	
F	

3. On a grid draw the lines $y = x$ and $y = 5 - x$ for values of x from -5 to 5.
Do a separate diagram for parts **a)**, **b)** and **c)**.
   a) Shade in the solution set for $x \geqslant 0$ and $y \geqslant 0$.
   b) Shade in the solution set for $x < 0$, $y < 0$ and $y > x$.
   c) Shade in the solution set for $x \geqslant 0$, $y \geqslant 0$, $y \geqslant x$ and $y \geqslant 5 - x$.

Remember the equality sign is not included when there is a broken line.

4. Draw the graph of $y = 6 + 2x$. Do a separate diagram for parts **a)**, **b)**, **c)** and **d)**.
   a) Shade in the solution set for $x \geqslant 0$, $y \leqslant 0$ and $y \leqslant 6 + 2x$.
   b) Shade in the solution set for $x \geqslant 0$ and $y < 6 + 2x$.
   c) Shade in the solution set for $x \leqslant 0$, $y \geqslant 0$ and $y \geqslant 6 + 2x$.
   d) Shade in the solution set for $x \geqslant 0$, $y \leqslant 0$ and $y \geqslant 6 + 2x$.

5. **a)** Sketch the region defined by the inequality $2x + 3y > 6$.
   b) On a separate graph, shade in the solution set for $2x + 3y \geqslant 6$ and $x \leqslant 4$.
   c) On a separate graph, shade in the solution set for $2x + 3y \geqslant 6$ and $y > x + 1$.

Unit 13.3 Solving real-world problems using graphs

In this unit you will:

- **identify points or regions on graphs that represent real-life situations**
- **draw graphs to represent real-life situations.**

Key words
Solution
Continuous
Discrete

Some of our problems encountered in our daily lives involve equations or inequalities.

Remember

- when equations are represented graphically the **solutions** lie on lines, in regions or at vertices
- the solution for more than one situation falls on the intersection of lines or regions
- a **discrete** variable does not allow fractions e.g. the number of people is discrete. If fractions are allowed then the variable is **continuous**.

Worked example

One day Roy had \$150 and decided to share lunch with friends. They wanted pizza at \$30 each and a bottle of drink at \$15 each. No fraction of a pizza or a bottle of drink is sold.

a. If each friend got a pizza and a bottle of drink, what is the most number of friends Roy may treat that day?

b. How much would he spend for himself and his friends?

c. Roy decides to spend all his money. What other item(s) could he buy?

Let the number of pizzas be x and the number of drinks be y.

Roy spent $30x + 15y$. It is an inequality because he can spend no more than \$150.

Therefore $30x + 15y \leqslant 150$.

He must buy at least one drink and one pizza, therefore $x \geqslant 1$ and $y \geqslant 1$.

In $30x + 15y \leqslant 150$, when $x = 0$, $y = 10$ and When $y = 0$, $x = 5$. Connecting the points (0, 10) and (5, 0) gives the line $30x + 15y = 150$.

Dividing by 15 gives $y = 10 - 2x$, y-intercept and gradient may be checked.

All the points represented by dots are possible solutions.

The values of x and y are discrete. No fraction of a bottle of drink or fraction of a pizza is sold.

Since each person should get a pizza and a drink then the solution narrows down to the dots on the red line (1, 1), Roy only, (2, 2), Roy and a friend, or (3, 3), Roy and two friends.

a. He can treat, at the most, two friends.

b. He would spend $\$(30 \times 3 + 15 \times 3) = \$135$.

c. He would have \$15 to purchase one other bottle of drink. This would be the point (3, 4) which lies in the region.

Exercise

1. Danny got \$600 to purchase T-shirts and shorts. A T-shirt costs \$100 and a pair of shorts costs \$90. He has to purchase at least one set.
   a) Write the information as inequalities.
   b) Represent the information graphically.
   c) Use your graph to find the most sets he can purchase.
   d) What will his change be, if any?

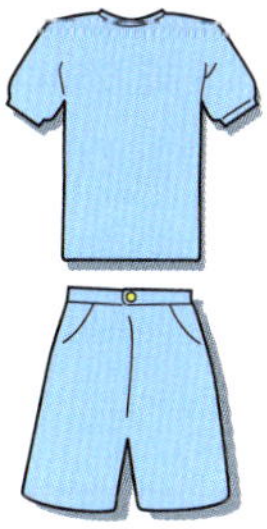

2. Mrs Prince and Mrs King met at a sale. Mrs Prince spent \$50 and bought 5 pens and 2 books. Mrs King spent \$45 and bought 6 of the same pens and 1 copy of the same book.
   a) Write the equations to show their purchases.
   b) Show this information on a graph.
   c) Use your graph to find the cost of a book and a pen.

3. The graph represents the number of hours per day that Jack and James are able to work on a site. James can work for no more than 10 hours per day. Jack can do 5 to 12 hours per day. x represents the number of hours per day for James and y represents that for Jack.
   a) Explain the meaning of the vertices of the shaded portion for their employer.
   b) Which point would give a maximum total time of work?
   c) What would be the total time?
   d) How many hours would each work?

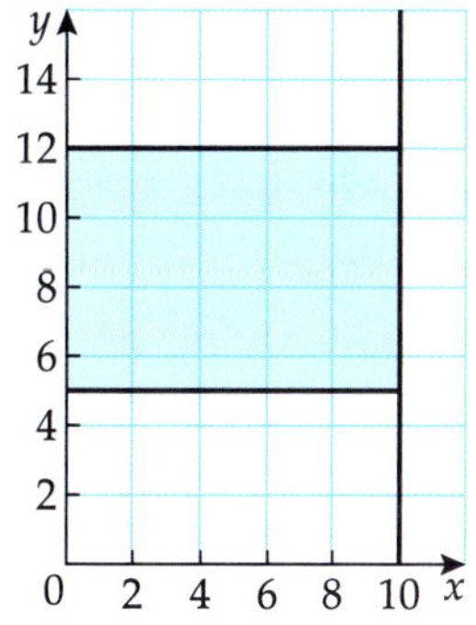

4. The manager of a dive shop wants to buy more snorkels and masks, but has only 120\$ to spare. Snorkels cost 20\$ each and masks cost 30\$ each. Snorkels take up 6 units of display shelving and masks take up 3 units. The shop has 24 units available. What combinations of masks and snorkels can be bought?

Show your answer as a solution set on a graph.

Unit 13.4 Further inequalities

In this unit you will:

- find the solutions to quadratic inequalities.

Key words
Linear inequality
Variable
Quadratic inequality
Inequality symbol

$x \geqslant 6$ $\qquad x < -6$ $\qquad x \leqslant 0$

The above inequalities are known as **linear inequalities.** For this unit we are going to look at another type of inequality, known as a **quadratic inequality**.

An inequality with a quadratic expression in one **variable** on one side and none on the other is called a quadratic inequality in one variable.

The following are all examples of quadratic inequalities.

$x^2 \geqslant 25$ $\qquad x^2 + 2 < 38$ $\qquad x^2 - 5x < -4$

Can you explain the difference between linear and quadratic inequalities?

To solve a quadratic inequality, replace the **inequality symbol** with an equals sign and solve the resulting equation. The solutions to the equation will allow you to establish intervals that will let you solve the inequality.

When solving quadratic equations, the first step involves factorisation.

The quadratic inequalities that we are going to investigate will already have been through the factorisation stage.

Worked examples

1. Solve the quadratic inequality $(x - 4)(x + 3) < 0$.
2. Solve the quadratic inequality $(x - 2)(x + 2) < 0$.

1. STEP 1: Change the inequality sign to an equal sign.
   $(x - 4)(x + 3) = 0$

   STEP 2: So, either $(x - 4) = 0$ or $(x + 3) = 0$.
   If $(x - 4) = 0$ $\qquad x = 4$
   If $(x + 3) = 0$ $\qquad x = -3$

   STEP 3: Plot the solutions $x = 4$ and $x = -3$ on a number line.

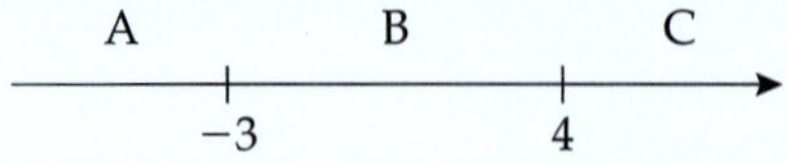

   We now have 3 intervals, A, B and C to investigate.

STEP 4: We need to decide which intervals produce a result that is <0 as shown by the original inequality. So, we pick a number from each interval and test it in the original inequality. If the result is true, that interval is a solution to the inequality.

Interval A: try $x = -4$ $(-4 - 4)(-4 + 3) = -8 \times -1 = 8 > 0$ does not satisfy

Interval B: try $x = 0$ $(0 - 4)(0 + 3) = -4 \times 3 = -12 < 0$ does satisfy

Interval C: try $x = 5$ $(5 - 4)(5 + 3) = 1 \times 8 = 8 > 0$ does not satisfy

So, only numbers in interval B satisfy the given inequality and so $x > -3$ and $x < 4$.

$x > -3$ and $x < 4$ OR $-3 < x < 4$

2. STEP 1: Change the inequality sign to an equal sign.

$(x - 2)(x + 2) = 0$

STEP 2: So, either $(x - 2) = 0$ or $(x + 2) = 0$.

If $(x - 2) = 0$ $x = 2$

If $(x + 2) = 0$ $x = -2$

STEP 3: Plot the solutions $x = 2$ and $x = -2$ on a number line.

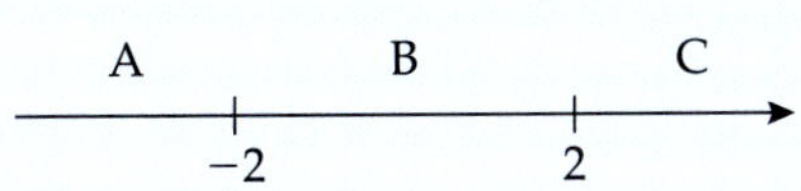

We now have 3 intervals, A, B and C to investigate.

STEP 4:

Interval A: try $x = -3$ $(-3 - 2)(-3 + 2) = -5 \times -1 = 5 > 0$ does not satisfy

Interval B: try $x = 0$ $(0 - 2)(0 + 2) = -2 \times 2 = -4 < 0$ does satisfy

Interval C: try $x = 3$ $(3 - 2)(3 + 2) = 1 \times 5 = 5 > 0$ does not satisfy

So, only numbers in interval B satisfy the given inequality and so $x > -2$ and $x < 2$.

$x > -2$ and $x < 2$ OR $-2 < x < 2$

Exercise

1 Solve the following quadratic inequalities:

a) $(x + 1)(-x + 4) < 0$ **b)** $(x + 3)(x - 2) < 0$ **c)** $(x - 5)(x - 9) \geqslant 0$

d) $(x + 3)(x - 4) < 0$ **e)** $(x + 7)(x - 5) < 0$ **f)** $(x - 5)(x + 1) \geqslant 0$

g) $(x + 3)(x - 2) > 0$ **h)** $(x - 1)(x - 4) < 0$ **i)** $(x - 3)(x + 2) > 0$

j) $(x - 2)(x - 3) > 0$ **k)** $(x - 2)(x - 3) < 0$ **l)** $(x + 2)(x - 1) > 0$

m) $(x + 2)(x - 2) < 0$

2 Show the solutions to the quadratic inequalities in question **1** on separate number lines.

3 Show the solutions to the quadratic inequalities in question **2** on separate number lines.

4 The following graph shows the solution to the quadratic inequality $(x-3)(x-1)<0$.

a) Use the graph to find the solution.

b) Check your solution using the written method.

c) Show the full solution on a number line.

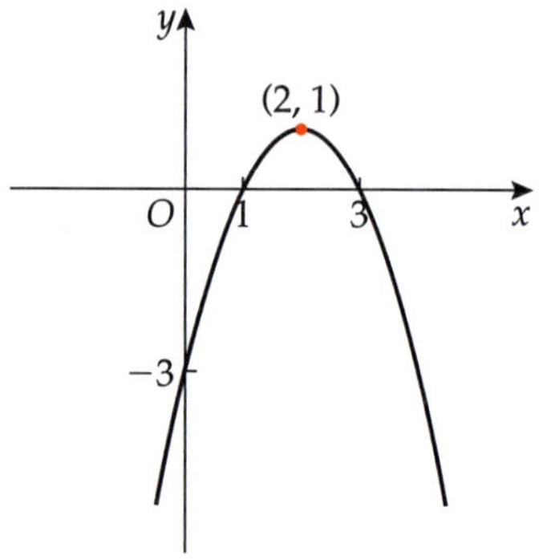

5 Here is a graph of the quadratic inequality $(x+1)(x-1)<0$.

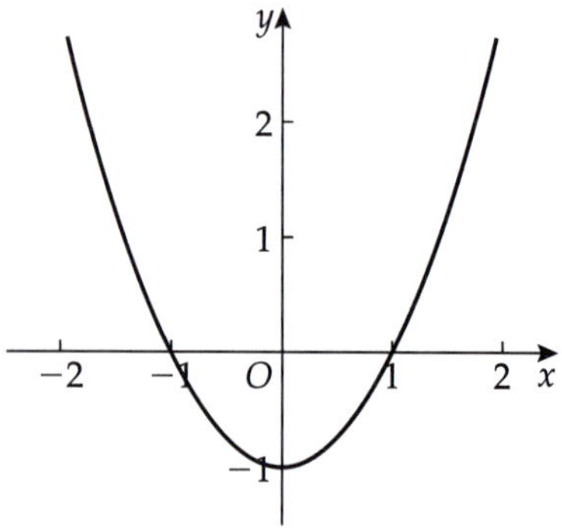

a) By looking at the graph write down the solution to the inequality.

b) Check your solution using the written method.

6 Solve the following quadratic inequalities:

a) $(x+3)(3x-2) \geqslant 0$ b) $(2x-5)(x+3)>0$

7 The following graph shows the solution to the quadratic inequality $(x+4)(x-2)>0$.

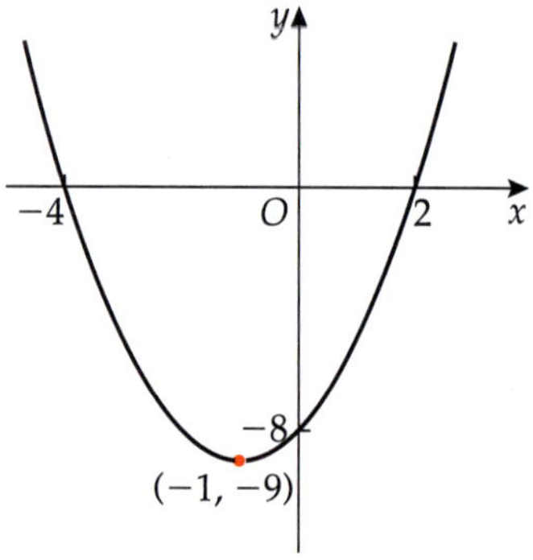

a) Use the graph to find the solution.

b) Check your solution using the written method.

c) Show the full solution on a number line.

Key words
Solve
Solution set

Unit 13.5 Rules of inequalities

In this unit you will:

- develop and test rules for operating with inequalities.

In solving inequalities we need to know whether we can treat them in the same way as equations. Let us investigate by considering the inequality $50 < 100$.

Positive numbers

1. Add 15 to both sides of the inequality. Is it still true?
   $50 + 15 < 100 + 15$
   $65 < 115$ STILL TRUE

2. Subtract 15 from both sides of the inequality. Is it still true?
   $50 + 15 < 100 - 15$
   $35 < 85$ STILL TRUE

3. Multiply both sides of the inequality by 10. Is it still true?
   $50 \times 10 < 100 \times 10$
   $500 < 1000$ STILL TRUE

2. Divide both sides of the inequality by 10. Is it still true?
   $50 \div 10 < 100 \div 10$
   $5 < 10$ STILL TRUE

Negative numbers

1. Add -15 to both sides of the inequality. Is it still true?
   $50 + -15 < 100 + -15$
   $35 < 85$ STILL TRUE

2. Subtract -15 from both sides of the inequality. Is it still true?
   $50 - -15 < 100 - -15$
   $65 < 115$ STILL TRUE

3. Multiply both sides of the inequality by -10. Is it still true?
   $50 \times -10 < 100 \times -10$
   $-500 < -1000$ NOT TRUE

4. Divide both sides of the inequality by -10. Is it still true?
   $50 \div -10 < 100 \div -10$
   $-5 < -10$ NOT TRUE

Let us collect our results together and put them into rules that we can follow.
To solve an inequality you can:

- add the same quantity to both sides
- subtract the same quantity from both sides
- multiply both sides by the same positive quantity
- divide both sides by the same positive quantity

To solve an inequality you cannot:

- multiply both sides by the same negative quantity
- divide both sides by the same negative quantity

Worked examples

1. Solve $2x - 5 < 3$ and show the answer on a number line.

$2x - 5 < 3$

Add 5 to both sides $\quad 2x - 5 + \mathbf{5} < 3 + \mathbf{5}$

$2x < 8$

Divide both sides by 2 $\quad x < 4$

In set notation: $\{x \mid x < 4\}$.

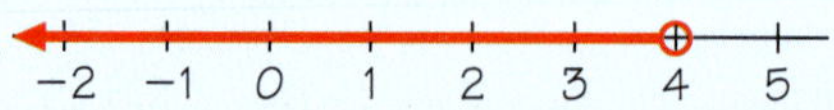

2. Find the solution set for $4x + 3 < 6x + 7$.

$4x + 3 \leqslant 6x + 7$

$(-6x) \quad 4x - \mathbf{6x} + 3 < 6x - \mathbf{6x} + 7$

$-2x + 3 \leqslant 7$

$(-3) \quad -2x + 3 - \mathbf{3} < 7 - \mathbf{3}$

$-2x \leqslant 4$

$(\div -2) \quad x \geqslant -2$

Remember, when dividing by a negative number the inequality sign is reversed.

The solution set is $\{2, 3, 4, 5, 6, 7, \ldots\}$.

In set notation: $(x \mid x > 22)$.

Exercise

1 Solve these inequalities:

a) $x + 3 < 6$ b) $5x > 15$ c) $2x + 1 < 5$

d) $x - 1 \geqslant 1$ e) $2(2x + 3) \geqslant 20$ f) $\dfrac{x}{5} \geqslant 3$

2 Show your solution sets for question **1** on number lines.

3 Solve the following inequalities and illustrate your solutions on a number line.

a) $3x - 5 < 7$ b) $2x + 1 > 2$ c) $4x - 3 > 9$

d) $5x + 4 > -1$ e) $8x - 3 > 7$

4 Find the solution sets for:

a) $4x + 1 > 2x + 5$ b) $3x - 1 < x + 5$ c) $5x + 1 \geqslant x - 7$

d) $2(5x - 4) < 32$ e) $6x + 2 > 3x + 5$ f) $2x + 2 > 5x + 5$

g) $3x - 1 \leqslant 7x + 7$ h) $7 - 2x \leqslant 4x + 10$

5 a) Rearrange the inequality $5n - 8 > 28 + n$ into the form $n >$ a number.

b) Write down the lowest whole number value of n which satisfies this inequality.

6 The length of a rectangle is $(x + 3)$ cm and its width is 4 cm. If the area is less than 60 cm^2, write an inequality and solve it. Show the solution on a number line.

7 When a number is subtracted from 11, then divided by 2, the result is always less than 5. Write this information as an inequality and solve it to show the possible values of the number.

8 The bridge across a river can carry a maximum load of 4 tonnes. Two trucks of weight 1 tonne, each carrying equal loads of x kg, pass each other on the bridge. Write an inequality statement to represent this information and solve it.
Explain your solution.

9 Solve the inequality $7x + 5 > 4x - 9$ and write down the smallest integer which satisfies it.

10 A social club want to hire a bus to take them out for the day. Company A charge $20 per hour. Company B charge $70 plus $10 per hour. Let x be the number of hours they want to hire the bus for.

a) Write down the inequalities satisfied by x where the cost of the number of hours charged by company A is less than the cost charged by company B.

b) Solve the inequality and explain what the solution tells you.

Looking back

1 Write down the integers that satisfy each set of inequalities.

a) $1 \leqslant x \leqslant 4$ **b)** $0 \leqslant x < 5$ **c)** $6 > x \geqslant 4$

d) $-3 \leqslant x \leqslant 2$ **e)** $-2 \leqslant x \leqslant 1$ **f)** $-5 < x < -1$

g) $3 < x \leqslant 5$ **h)** $2 \geqslant x \geqslant -2$ **i)** $-5 > x \geqslant -12$

2 Find all the integers which satisfy each inequality and show the solutions on a number line.

a) $4 \leqslant 2x \leqslant 8$ **b)** $-15 < 5x \leqslant 5$ **c)** $-16 < 4x \leqslant 0$

d) $-9 \leqslant 3x < 6$ **e)** $2 \leqslant 2x < 7$ **f)** $-5 < 2x < 5$

3 Draw a graph of $y = 5 + 3x$. Shade in the solution sets for

a) $x \geqslant 0$ and $y \leqslant 5 + 3x$

b) $x \geqslant 0$ and $y \geqslant 5 + 3x$

c) $x \leqslant 0, y \geqslant 0$ and $y \geqslant 5 + 3x$

Draw separate diagrams for each solution set.

4 A teacher spent $70 on 8 protractors and 10 set squares. Her colleague bought 6 protractors and 9 set squares for $57.

a) Write equations to show their purchases.

b) Show this on a graph.

c) Use your graph to find the cost of one of each.

Chapter 14

Frequency distribution

This chapter introduces you to discrete and continuous data, and the methods for analysing grouped and ungrouped data. You will review and extend your skills in representation of data, using frequency tables, histograms and frequency polygons.

Key words
Continuous
Data
Discrete

Unit 14.1 Discrete and continuous data

In this unit you will:

- differentiate between discrete and continuous data.

Discrete data has only exact and separate values.
Discrete data are usually numbers that can be counted, such as the numbers of students in classes (these can only be whole numbers).

Continuous data can be anywhere within a range of values. Continuous data are usually measurements such as lengths, masses, times.

The number of coins in Afiya's jar is discrete data.
Afiya's height is an example of continuous data.

Worked example

Determine whether the populations of different territories in the Caribbean are continuous or discrete data and give a reason.

Discrete, because populations are numbers of people and these are whole numbers.

Exercise

Determine whether the data described is discrete or continuous and give a reason.

1. Masses of coconuts harvested
2. Lengths of bananas in a hand
3. Numbers of coconuts harvested
4. Densities of households per hectare
5. Populations of jellyfish in lagoons
6. Shoe sizes
7. Ages of members of a cricket club
8. Lengths of feet

Key words
Pie chart
Bar chart

Unit 14.2 Frequency tables and charts for discrete data

In this unit you will:

- draw and interpret pie charts
- draw and interpret bar charts.

When handling large amounts of data, it is often useful to sort the data into groups, and also to display the data in diagram form, such as bar charts and pie charts.

Bar charts can be used to show patterns in data.

Pie charts are useful for showing proportions or parts of a whole.

Worked examples

1. The average contents of a box of tacks is said to be 67. As a check, the actual number is counted for a random sample of boxes. The results are shown in the table below.
   a. Decide whether to display the data as a bar chart or pie chart and give a reason.
   b. Use your chart to find
      i. the number of boxes in the sample. **ii.** the true average contents.
2. The types of land use for 2 islands are shown below. Draw suitable charts to display the data and use them to compare the islands.

1.

Number of tacks	64	65	66	67	68	69	70
Frequency	4	5	10	12	7	3	2

a. Bar chart – to show the distribution pattern.

b.

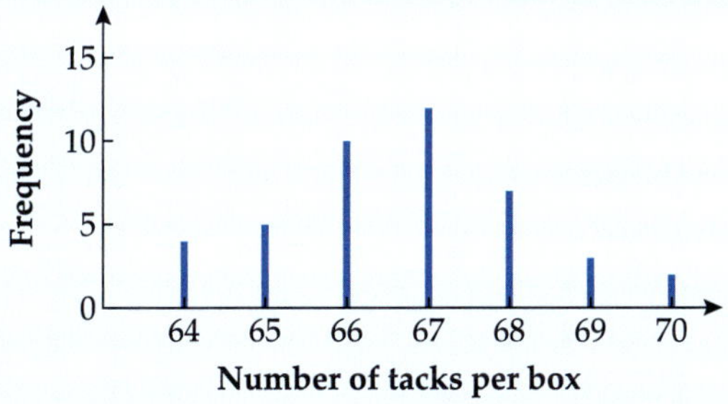

i. Number of boxes $= 4 + 5 + 10 + 12 + 7 + 3 + 2$
$= 43$

ii. Mean $= (4 \times 64 + 5 \times 65 + 10 \times 66 + 12 \times 67 + 7 \times 68 + 3 \times 69 + 2 \times 70) \div 43$
$= (256 + 325 + 660 + 804 + 476 + 207 + 140) \div 43$
$= 66.7$ (to 1 d.p.)

Mode $= 67$

Median $=$ 22nd value $= 67$

2.

Island	Crops	Grazing	Towns	Woodland	Coastal strip
St Ann's/km^2	360	45	60	25	50
St Bee's/km^2	390	600	100	2100	410

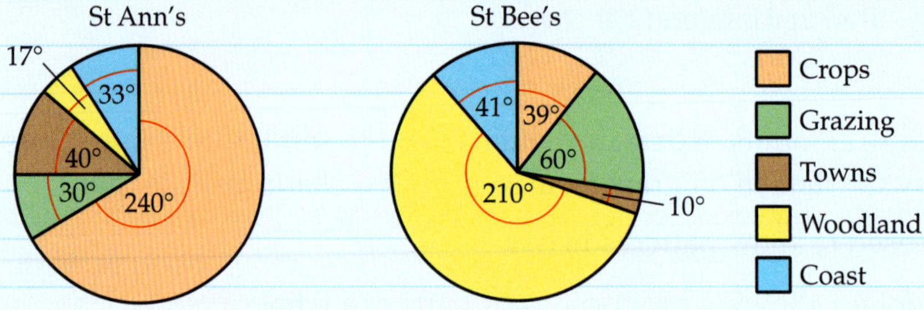

Pie charts are suitable for displaying these data, as we are comparing parts of a whole.

The pie charts show that St. Ann's has a lot of land devoted to growing crops, but St. Bee's is largely woodland. Although St. Bee's has a larger area taken up by towns, the proportion of town to country is higher on St. Ann's.

Exercise

1 The average contents of a packet of seeds is claimed as 43. The actual values in 40 packets are shown below. Draw a suitable chart to display the data, and find the mean, mode and median. Is the claim reasonable?

Number of seeds	41	42	43	44	45
Frequency	8	11	14	6	1

2 Estimates of the populations of several countries as at July 2006 are shown below. Draw a suitable chart and explain your reasons for choosing this type of chart.

Anguilla	13,477
Barbados	279,912
St Lucia	168,458
Trinidad and Tobago	1,304,981
Jamaica	2,758,124
St Kitts and Nevis	39,129
St Vincent and the Grenadines	117,848
Virgin Islands	108,605

3 The age structure for 2 countries is shown below. Draw comparative pie charts and comment on what they show.

Age range (years)	0–14	15–64	65 and over
Jamaica	912,000	1,644,000	201,000
Luxembourg	90,000	315,000	69,000

Find the total populations first.

Key words
Class interval
Class width
Frequency table
Frequency density
Histogram

Unit 14.3 Grouped data and histograms

In this unit you will:

- divide a continuous data into class intervals
- construct a frequency table for continuous data
- draw a histogram to represent continuous data.

Discrete data can be grouped and shown in a frequency table.
This table shows the numbers of students in different schools who walk to school.

Numbers	1–10	11–20	21–30	31–40
Frequency	6	15	18	9

The data can be shown on a bar chart:
There are gaps between the bars because there are no values between the end of one group and the start of the next one.

Pupils walking to school

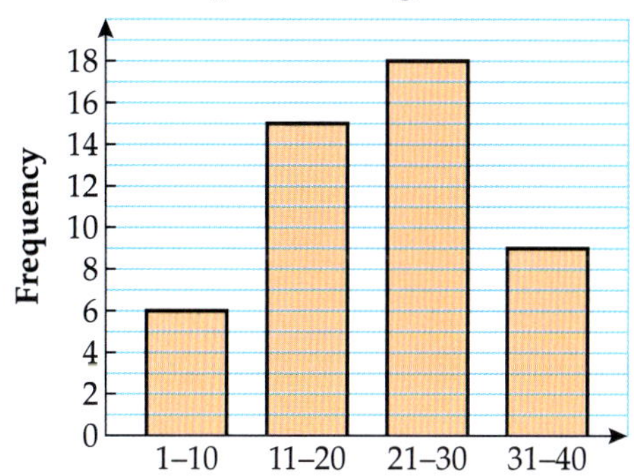

Continuous data lies within a range.
This range is then broken into smaller intervals called class intervals.
The difference between the top and bottom values in a **class interval** is called the **class width**.

For example, the heights, h m, of a group of students all lie in the range $1 < h \leqslant 2$.
This range can then be divided into equal width class intervals.
Dividing it into four gives $1 < h < 1.25$, $1.25 < h < 1.5$, $1.5 < h < 1.75$, $1.75 < h < 2$.
The heights of the students can then be shown in a **frequency table**:

Heights	$1 \leqslant h < 1.25$	$1.25 \leqslant h < 1.5$	$1.5 \leqslant h < 1.75$	$1.75 \leqslant h < 2$
Frequency	3	12	10	4

These are the class intervals; there is no gap between them.

A frequency table for continuous data can be illustrated by a **histogram**.
A histogram is similar to a bar chart except that the horizontal axis is a number line covering the range of values, there are no gaps between the bars and the vertical axis shows the **frequency density**.

$$\text{Frequency density} = \frac{\text{frequency}}{\text{class width}}$$

For the data in the table, the class width is 0.25, so the frequency densities are

$3 \div 0.25 = 12$, $12 \div 0.25 = 48$, $10 \div 0.25 = 40$ and $4 \div 0.25 = 16$.

This histogram illustrates the data in the frequency table.

Height of pupils

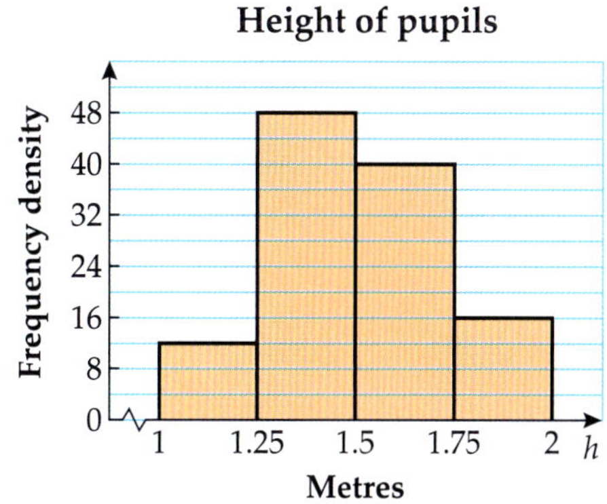

Worked example

These are the masses of some laptop computers.

1.22 kg, 1.32 kg, 1.55 kg, 1.57 kg, 1.60 kg 1.60 kg, 1.82 kg, 1.9 kg, 2.54 kg, 2.57 kg, 2.57 kg, 2.57 kg, 2.58 kg, 2.61 kg, 2.63 kg, 2.66 kg, 2.67 kg, 2.98 kg 2.98 kg, 2.99 kg, 3.00 kg, 3.10 kg, 3.15 kg, 3.25 kg, 3.38 kg, 3.45 kg, 3.80 kg, 3.90 kg, 4.00 kg, 4.10 kg.

a. Show these masses in a frequency table.

b. Draw a histogram to illustrate the data.

The masses range from 1.22 kg to 4.10 kg. Choose an interval that includes this and is easy to divide into smaller equal classes.

a.

Masses	$1 \leqslant m < 2$	$2 \leqslant m < 3$	$3 \leqslant m < 4$	$4 \leqslant m < 5$
Frequency	8	12	8	2
Frequency density	8	12	8	2

The mass 4.00 kg goes here.

Add another row to calculate frequency density. For this data, the class width is 1 so frequency density = frequency.

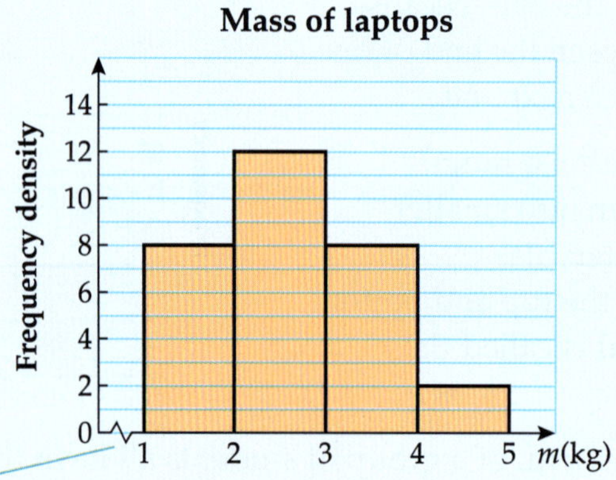

This shows that the horizontal axis does not start at zero.

Exercise

1 The frequency table shows the times taken by some students to complete a task.

Time, t minutes	$0 \leqslant t < 10$	$10 \leqslant t < 20$	$20 \leqslant t < 30$	$30 \leqslant t < 40$	$40 \leqslant t < 50$	$50 \leqslant t < 60$
Frequency	2	9	5	4	2	1

Construct a histogram to illustrate this data.

For questions **2** and **3**, show the data in a frequency table and illustrate it with a histogram.

2 The lengths of some rugs.

246 cm, 115 cm, 82 cm, 140 cm, 205 cm, 115 cm, 92 cm, 122 cm, 95 cm, 185 cm, 100 cm, 220 cm, 110 cm, 189 cm.

3 The masses of some books.

45 g	52 g	71 g	95 g	51 g	38 g	44 g	62 g	80 g	72 g
81 g	39 g	48 g	60 g	66 g	75 g	36 g	63 g	92 g	108 g
5 g	115 g	95 g	91 g	56 g	72 g	61 g	97 g	110 g	88 g

Key words
Frequency polygon
Mid-class value

Unit 14.4 Frequency polygons

In this unit you will:

- draw frequency polygons for grouped data
- interpret smooth frequency polygons.

We can construct a **frequency polygon** from a histogram by drawing straight lines from the bottom of the left-hand bar to the midpoint of the each bar, ending at the bottom of the right-hand bar.
This is the histogram from Unit 14.3 illustrating the distribution of the heights of some students.
The frequency polygon is drawn on top of the histogram.

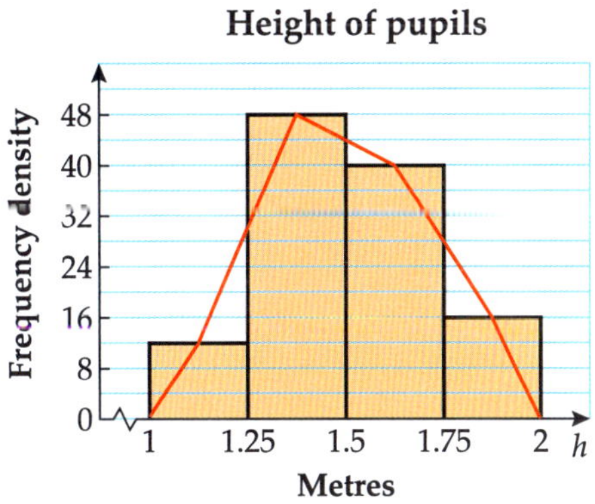

When we remove the bars we are left with the frequency polygon.

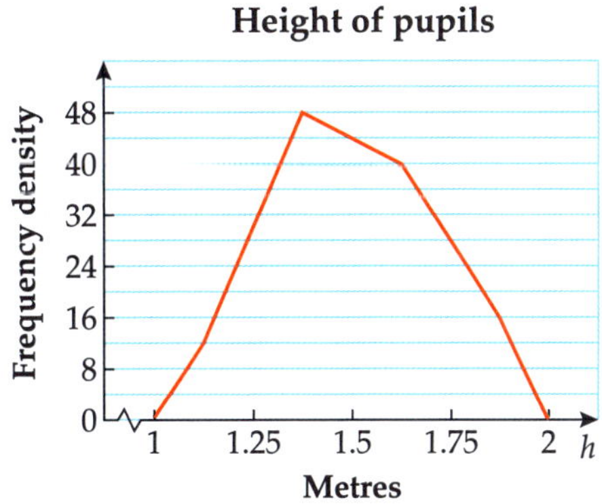

If the class intervals are smaller, the frequency polygon is smoother. When the class widths are very small, the curve becomes a smooth curve.

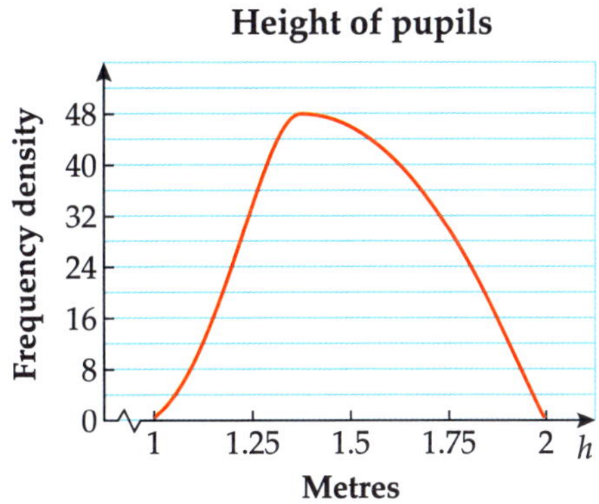

Worked example

This table shows the times taken by a group of students to travel to school.

Time (t minutes)	$10 \leqslant t < 20$	$20 \leqslant t < 30$	$30 \leqslant t < 40$	$40 \leqslant t < 50$	$50 \leqslant t < 60$
Frequency	10	18	5	3	2

Draw a frequency polygon for this data.

To draw a frequency polygon, we need to plot the frequency densities against the middle of the bars. We do not need to draw the bars to do this. The middle of a bar is the midpoint of the class interval, which we can find by averaging the upper and lower ends of the class. This is called the **mid-class value**.

Time (t minutes)	$10 \leqslant t < 20$	$20 \leqslant t < 30$	$30 \leqslant t < 40$	$40 \leqslant t < 50$	$50 \leqslant t < 60$
Frequency	10	18	5	3	2
Mid-class value	15	25	35	45	55
Frequency density	1	1.8	0.5	0.3	0.2

Add two rows to the table, one for the mid-class values and the other for the frequency density.

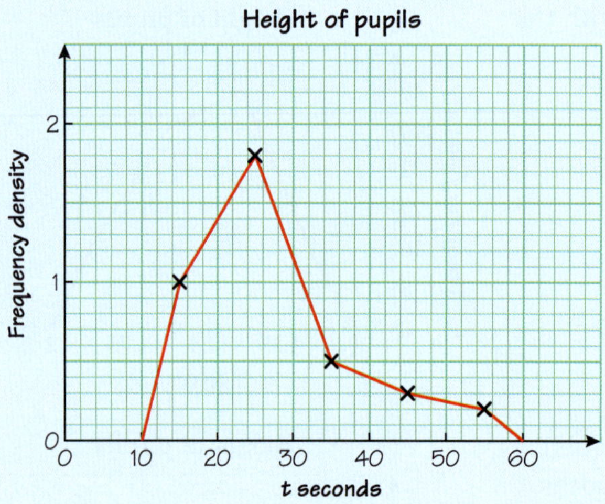

Exercise

1. This frequency table shows the times taken by a sample of students to travel home one evening.

Time (t minutes)	$0 \leqslant t < 10$	$10 \leqslant t < 20$	$20 \leqslant t < 30$	$30 \leqslant t < 40$	$40 \leqslant t < 50$
Frequency	2	8	6	3	1

a) Copy the table and add a column showing the mid-interval values.

b) Draw a frequency polygon to illustrate this data.

2 This table gives the lengths of 50 bean pods.

Length (l cm)	$3 \leqslant l < 5$	$5 \leqslant l < 7$	$7 \leqslant l < 9$	$9 \leqslant l < 11$
Frequency	10	27	24	12

Draw a frequency polygon to illustrate this data.

3 This frequency table shows the heights of a sample of seedlings.

Height, h cm	$0 \leqslant h < 1$	$1 \leqslant h < 2$	$2 \leqslant h < 3$	$3 \leqslant h < 4$	$4 \leqslant h < 5$
Frequency	8	15	32	20	4

The mid-interval values are not whole numbers.

Draw a frequency polygon to illustrate this data.

4 The frequency polygon shows the amounts of money spent in a shop by customers on Friday. The table shows the amounts spent by customers the following Monday.

Value of purchases on Friday

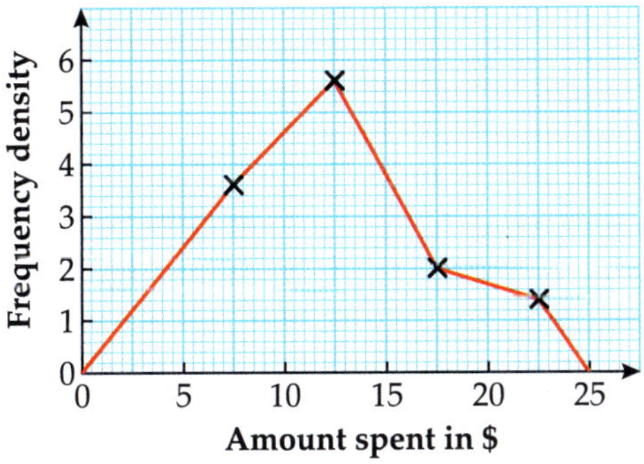

Value of purchases on Monday

Amount spent, $\$A$	Frequency
$0 \leqslant A < 5$	8
$5 \leqslant A < 10$	11
$10 \leqslant A < 15$	15
$15 \leqslant A < 20$	8
$20 \leqslant A < 25$	4

a) Calculate the number of customers on Friday who spent up to $5.

Use frequency = (frequency density) × (class width)

b) Calculate the total number of customers on Monday.

c) Explain why you cannot give the exact number of customers who spent between $10 and $12 on Monday.

d) Copy the frequency diagram above and draw a frequency polygon on it for the amounts spent by customers on Monday.

Unit 14.5 Mean of a grouped frequency distribution

Key words
Mean
Mid-class value

In this unit you will:

- review the method for finding the mean of an ungrouped frequency distribution
- find the mean of a grouped frequency distribution

We can find the **mean** from an ungrouped frequency table by multiplying each data value by its frequency to give the total. Then we divide by the total number of data values: this is the total of the frequencies.
This table shows the marks out of 5 scored by students in a test.

Mark, x	Frequency, f	$f \times x$
1	1	2
2	3	6
3	12	36
4	18	72
5	5	25
	Total: 39	Total: 141

The sum of the marks is 141.
There are 39 students so the mean mark $= 141 \div 39$
$= 3.6$ marks to 1 decimal place.

In a grouped frequency table, we do not know the individual data values.
We use the **mid-class value** as an estimate for the average data value for the class.
This means that the calculated mean value is an estimate.

This frequency table shows the heights of a sample of students:

Heights	$1 \leqslant h < 1.25$	$1.25 \leqslant h < 1.5$	$1.5 \leqslant h < 1.75$	$1.75 \leqslant h < 2$
Frequency	3	12	10	4

To estimate the mean value, we need to find each mid-class value and then multiply it by the frequency of that class. We add two rows to the table to help keep track of our work.

Heights	$1 \leqslant h < 1.25$	$1.25 \leqslant h < 1.5$	$1.5 \leqslant h < 1.75$	$1.75 \leqslant h < 2$	total
Frequency, f	3	12	10	4	29
Mid-class value, x	1.125	1.375	1.625	1.875	
fx	3.375	16.5	16.25	7.5	43.625

The estimated mean is then $43.625 \div 29 = 1.50$ to 2 decimal places.

The formula for the estimated mean value of a grouped frequency distribution is $\frac{\Sigma fx}{\Sigma f}$

where Σfx means the sum of all the (frequency $\times$ mid-class value) and Σf means the sum of all the frequencies.

Worked example

This table shows the times taken by a group of students to travel to school.

Time (t minutes)	$10 \leqslant t < 20$	$20 \leqslant t < 30$	$30 \leqslant t < 40$	$40 \leqslant t < 50$	$50 \leqslant t < 60$
Frequency, f	10	18	5	3	2

Find an estimate for the mean time taken.

Time (t minutes)	$10 \leqslant t < 20$	$20 \leqslant t < 30$	$30 \leqslant t < 40$	$40 \leqslant t < 50$	$50 \leqslant t < 60$	
Frequency, f	10	18	5	3	2	$\Sigma f = 38$
Mid-class value, x	15	25	35	45	55	
fx	150	450	175	135	110	$\Sigma fx = 1020$

$$\text{Mean} = \frac{\Sigma fx}{\Sigma f} = \frac{1020}{38} = 26.8 \text{ to 1 dp}$$

Exercise

1 This table gives the bowling figures for a club bowler one season.

Wickets per match	0	1	2	3	4	5
Frequency	2	3	5	4	3	1

Calculate the mean number of wickets taken per match.

2 Some pupils were asked how much they spent last week. The results are shown in the table. Find an estimate for the mean amount of money spent.

Amount ($)	0–24	25–49	50–74	75–99
Frequency, *f*	25	35	40	20

The data is discrete. Use the value in the middle of the group for the mid-class value.

3 Response times to a help line are given in this table.

Response time, (t minutes)	$5 \leqslant t < 10$	$10 \leqslant t < 15$	$15 \leqslant t < 20$	$20 \leqslant t < 25$	$25 \leqslant t < 30$
Frequency	3	6	27	18	6

Find an estimate for the mean response time.

4 Some seeds were planted and five weeks later, the heights of the resulting plants were measured. The results are shown in the table.

Height, h cm	$0 \leqslant h < 3$	$3 \leqslant h < 6$	$6 \leqslant h < 9$	$9 \leqslant h < 12$
Frequency	2	8	25	6

Find an estimate for the mean height of the seed1ings.

5 A sample of apprentices were timed to see how long it took them to complete a job. This table summarises the results. Estimate the mean time.

Time, *t* minutes	Number of apprentices
$0 \leqslant t < 5$	6
$5 \leqslant t < 10$	9
$10 \leqslant t < 15$	4
$15 \leqslant t < 20$	1

6 A questionnaire was trialed by asking some people to complete it. The time each person took was recorded and this frequency polygon summarises the results.

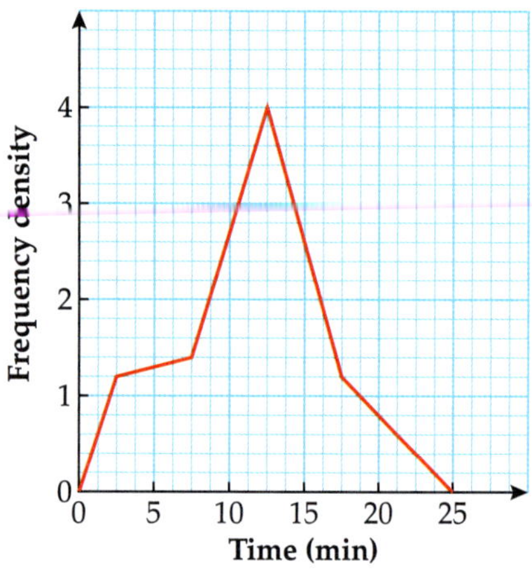

a) Copy and complete this table.

Mid-interval value, *x* (minutes)	Frequency, *f*	*fx*
2.5	5	12.5
7.5		
	Σf:	Σfx:

b) How many people were asked to complete the form?

c) Calculate an estimate for the mean time taken to complete this form.

7 The lengths of a sample of runner beans were measured and recorded in this table.

Height (*h* cm)	$0 \leqslant h < 5$	$5 \leqslant h < 10$	$10 \leqslant h < 15$	$15 \leqslant h < 20$
Frequency	3	6	11	4

Estimate the mean.

8 This frequency table shows the heights of the students in a class.

Height (*h* cm)	$148 \leqslant h < 150$	$150 \leqslant h < 152$	$152 \leqslant h < 154$	$154 \leqslant h < 156$	$156 \leqslant h < 158$
Frequency	4	7	8	7	5

Estimate the mean.

Key words
Central tendency
Mode
Median

Unit 14.6 Mode and median of grouped data

In this unit you will:

- review the mode and median of ungrouped data
- find the modal class and median class of grouped frequency distributions
- analyse frequency distributions using measures of central tendency.

The measures of **central tendency** are the mean, the mode and the median.
The **mode** is the data value that has the highest frequency.
The **median** is the middle data value when data is ordered. If there are n data values, the middle value is the $\frac{1}{2}(n+1)$th value.
We can find the mode and the median from an ungrouped frequency table or diagram.

These are the marks out of 5 scored by students in a test.

Mark, x	1	2	3	4	5
Frequency, f	1	3	12	18	5

The mark with the highest frequency is 3, so 3 is the modal mark.
The sum of the frequencies is 39, so the middle mark is the 20th mark in ascending order.
There are 16 marks below 4 and 34 marks below 5 so the median mark is 4.

In a grouped frequency distribution we can identify the class or group with the highest frequency. This is called the modal class.
We can also identify the class that contains the median. This is the median class.

This table shows the heights of a sample of students:

Heights	$1 \leqslant h < 1.25$	$1.25 \leqslant h < 1.5$	$1.5 \leqslant h < 1.75$	$1.75 \leqslant h < 2$
Frequency	3	12	10	4

The modal class is $1.25 \leqslant h < 1.5$.
There are 29 heights so the middle height is the 15th. This is also in the class $1.25 \leqslant h < 1.5$.

We can analyse data using the three measures of central tendency.

Worked examples

1. The distance jumped on the long jump by competitors at an athletics meeting are recorded in the table.

Distance, d m	$7.8 \leqslant d < 8.0$	$8.0 \leqslant d < 8.2$	$8.2 \leqslant d < 8.4$	$8.4 \leqslant d < 8.6$	$8.6 \leqslant d < 8.8$
Frequency	10	18	5	3	2

   a. Calculate the mean, modal and median class.
   b. Which is the most suitable measure of central tendency to use for this data?

1. a.

Distance, *d* m	$7.8 \leq d < 8.0$	$8.0 \leq d < 8.2$	$8.2 \leq d < 8.4$	$8.4 \leq d < 8.6$	$8.6 \leq d < 8.8$
Frequency	7	12	14	25	4
Mid-class value, *x*	7.9	8.1	8.3	8.5	8.7
fx	55.3	97.2	116.2	212.5	43.8

$\Sigma f = 62$, $\Sigma fx = 525$ so the mean distance $= 525 \div 62$ m $= 8.47$ m to 2 dp

The modal class is $8.4 \leq d < 8.6$.

$\Sigma f = 62$ so the median distance is the $\frac{1}{2}(62 + 1)$th distance $= 31\frac{1}{2}$th distance.

This in the class $8.2 \leq d < 8.4$

b. The mean is the most useful because it includes all the distances.

If there were a few very short or very long distances, the mean would not be a good choice. The median or the mode would be better choices because they are not affected by the outlying values.
The mode is better than the median when there is one class that has a much larger frequency than any of the others.

2. Mr Khan wants to buy a digital camera.
How useful is each of these pieces of information? Give reasons for your answers.
   a. The average price of a digital camera is \$895. This is the mean price.
   b. A middle range digital costs about \$580. This is the median price.
   c. The most common cost is about \$420. This is the modal price.

2. a. The mean price involves adding up all the prices so it takes account of very expensive cameras. If Mr Khan does not want an expensive machine, the mean price is not much use.
   b. The median price is quite useful because it shows that half of the models cost about \$580 or less.
   c. The modal price shows that there is more than one camera costing about \$420 so there is a choice of cameras at this price.

Exercise

1 This table records the marks of a group of students in an examination.

Mark, *x*	0–19	20–39	40–59	60–79	80–99
Frequency, *f*	1	3	37	45	24

a) Calculate the mean and find the median and modal groups.
b) Which measure of central tendency is the most representative of the distribution?

2 The table shows the times taken by a group of students to complete a multiple choice test.

Time, *t* minutes	$5 \leqslant t < 10$	$10 \leqslant t < 15$	$15 \leqslant t < 20$	$20 \leqslant t < 25$	$25 \leqslant t < 30$
Frequency	2	8	8	8	6

a) Calculate the mean and find the median and modal groups.
b) Which measure of central tendency is the most representative of the distribution?

3 The lengths of a sample of cocoa beans were measured and recorded in this table.

Length (*l* mm)	$1 \leqslant l < 3$	$3 \leqslant l < 5$	$5 \leqslant l < 7$	$7 \leqslant l < 9$
Frequency	7	6	0	2

a) Calculate the mean and find the median and modal groups.
b) Which measure of central tendency is the most representative of the distribution?

4 This frequency table shows the distances a sample of students in a travel to school.

Distance, *d* km	$0 \leqslant d < 0.5$	$0.5 \leqslant d < 1$	$1 \leqslant d < 1.5$	$1.5 \leqslant d < 2$	$2 \leqslant d < 2.5$
Frequency	20	45	10	5	2

a) Calculate the mean and find the median and modal groups.
b) Which measure of central tendency is the most representative of the distribution?

5 Mrs Joseph would like to enrol her daughter in a swimming club.
She wants to know the average age of the members.
How useful would each measure of central tendency be? Give reasons for your answers.

Looking back

1 These are the batting averages for a group of cricketers:

32.9	5.7	27.4	31.0	20.6	36.1	40.2	19.5	14.1	33.5
10.0	6.3	43.1	57.9	14.0	39.8	46.3	52.4	38.2	6.2
22.4	28.8	29.3	32.0	24.6	22.8	12.8	35.7	38.3	25.0

a) Is this data discrete or continuous?
b) Choose appropriate class intervals, and show this data in a frequency table.
c) Draw a histogram to illustrate this data.
d) On the same axes, draw a frequency polygon.

2 The top speeds of different motorbikes are recorded in a frequency table:

Top speed, *v* km/h	$100 \leqslant v < 120$	$120 \leqslant v < 140$	$140 \leqslant v < 160$	$160 \leqslant v < 180$	$180 \leqslant v < 200$
Frequency	4	5	21	8	2

a) Estimate the mean and find the median and modal groups for this data.
b) Which measure of central tendency best represents the distribution?

Project

Statistics can be used to test a theory.
In this project we are going to test if reaction times can be improved with practice.
This experiment can be done in pairs by the whole class.
It needs at least fifteen people to take part.

To do this, ask a person to catch a ruler when it is dropped.
Hold the ruler touching the person's open hand, then release the ruler.
Record the measurement above the index finger on a data collection sheet like this one. This is trial 1.

Repeat this experiment 10 times without recording the measurement. This is the practice.

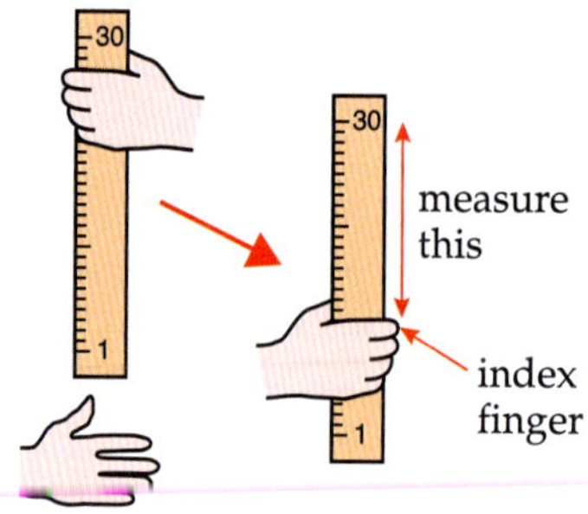

Now repeat the experiment again and record the measurement.
This is trial 2.

Go through the whole process again for trial 3.
Record the measurements on a data collection sheet like this one.

Name	Trial 1 (mm)	Trial 2 (mm)	Trial 3 (mm)

When the data has been collected, it needs sorting and analysing.
The data can be rewritten in order and the measures of central tendency calculated directly from the data.

The data can be entered into a spreadsheet like this.

	A	B	C	D	E	F	G	H	I	J	K	L	M	N	O	P	Q
1																	
2																	
3	Trial 1	5.8	9	3.2	7.8	9.4	8.7	6.7	9.7	8.3	9.8	7.3	8.2	5.3	7.7	8.3	
4	Trial 2	6.1	9.3	4.9	7.8	9.1	8.8	8.2	8.3	8.5	7.5	9.8	7.5	5.9	7.2	7.6	
5	Trial 3	5.2	6.2	4.7	7.2	8.3	8.1	6	6.2	7.3	8.3	6.1	7.1	5.2	6.8	6.9	
6																	
7	Sorted data.								median								mean
8	Trial 1	3.2	5.3	5.8	6.7	7.3	7.7	7.8	8.2	8.3	8.3	8.7	9	9.4	9.7	9.8	7.68
9	Trial 2	4.9	5.9	6.1	7.2	7.5	7.5	7.6	7.8	8.2	8.3	8.5	8.8	9.1	9.3	9.8	7.77
10	Trial 3	4.7	5.2	5.2	6	6.1	6.2	6.2	6.8	6.9	7.1	7.2	7.3	8.1	8.3	8.3	6.64

The tools in the spreadsheet can be used to order the data and to calculate the measures of central tendency.

Alternatively, the data can be grouped, for example:

Distance caught in cm (D)	$3 \leqslant D < 5$	$5 \leqslant D < 7$	$7 \leqslant D < 9$	$9 \leqslant D < 11$
Trial 1	1	3	7	4
Trial 2	1	2	9	3
Trial 3	1	8	6	0

The measures of central tendency can be calculated from the grouped frequency table.

The data also need illustrating.
The choices are pie charts, bar charts, frequency polygons or histograms.

You could combine illustrations for the three trials on one diagram.
This diagram shows the data above as a compound histogram.

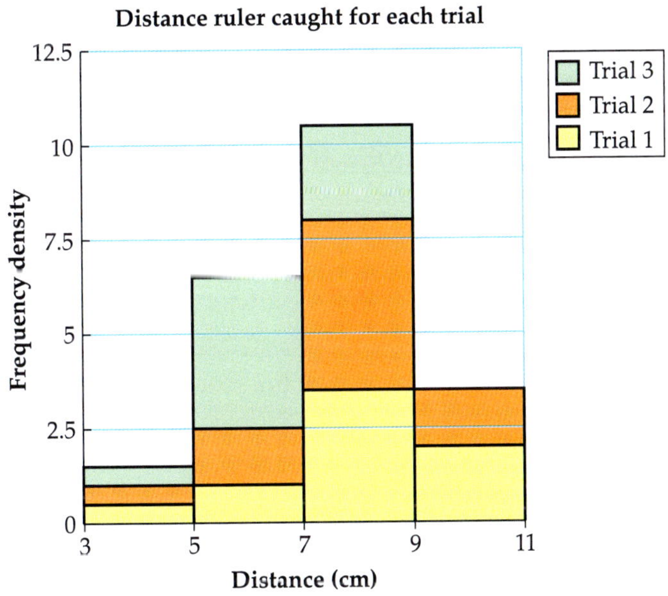

Now decide whether the evidence shows that practice improves reaction time and give reasons for the conclusion.

When the project is complete, make a poster explaining the project and justifying the choices made for the calculations of measures of central tendency and for the diagrams used to illustrate the data.

R3

Revise and consolidate 3

1 Write down all the integers between $^-5$ and $^+5$ inclusive, which satisfy the following inequalities.

a) $m > {}^-2$ **b)** $1.2 > t$ **c)** ${}^-3 \geqslant p$ **d)** $t \leqslant 1$ and $t > {}^-4$

2 14 boys in class 3T take tests in geography and science, both scored out of 50.

Geography	25	40	46	34	30	38	29	41	37	43	38	28	36	35	45
Science	21	32	40	26	21	31	25	33	32	37	29	24	32	29	40

a) Draw a scatter diagram for their marks.

b) Draw a stem-and-leaf graph for each subject.

c) Use the shape of the two diagrams to help decide which test was the more difficult. Explain your answer.

d) If a pupil got a mark of 32 in geography, can you estimate his score in science? Explain your answer.

3 A group of 15- to 20-year-olds were asked to press a bell as soon as they saw a flash on a screen. The reaction times in seconds were:

0.63 0.58 0.54 0.47 0.67 0.53 0.49 0.62 0.49 0.64

a) Using an assumed mean, find the mean.

b) Find the mode, median and range of the data.

The results for 25- to 30-year-olds were:

Mean: 0.63 median 0.62 mode 0.67 range 0.3

c) Which group had the faster reaction time? Explain your answer.

4 The longest 100 throws for the mens' javelin at a sports club were:

Distance (m)	$60 \leqslant D < 63$	$63 \leqslant D < 66$	$66 \leqslant D < 69$	$69 \leqslant D < 72$	$72 \leqslant D < 75$	$75 \leqslant D < 78$	$78 \leqslant D < 81$
Frequency	4	6	26	37	15	11	1

a) Explain in words what $60 \leqslant D < 63$ means. Give an example of the smallest value and the largest value that fit into this group.

b) Draw a frequency diagram of this data.

c) What is the modal class?

d) Why can we not say the range is $81 - 60 = 21$ m?

5 This table shows the results of a test.

Mark	1–10	11–20	21–30	31–40	41–50	51–60	61–70	71–80
Frequency	1	3	9	14	19	13	8	7

a) Draw a cumulative frequency diagram.

b) Use your diagram to estimate the median and interquartile range.

c) 10 people achieved an A grade in the test. How many marks did you need for an A grade?

6 In a shooting competition, two competitors achieved the following results after 50 shots.

Score	1	2	3	4	5
Competitor A	3	10	18	10	9
Competitor B	2	6	22	16	4

a) Draw superimposed frequency polygons for the competitors and comment on the results.

b) Which competitor had the highest average score?

7 The results of a Year 2 long jump competition were:

Length (cm)	$180 \leqslant l < 210$	$210 \leqslant l < 240$	$240 \leqslant l < 270$	$270 \leqslant l < 300$	$300 \leqslant l < 330$
Frequency	13	27	30	19	12

a) Calculate an estimate of the mean.

b) What is the modal class?

c) If a pupil were picked at random, what is the probability that he or she will jump less than 240 cm?

8 **a)** Draw a pair of axes where $^-7 \leqslant x \leqslant 7$ and $^-10 \leqslant y \leqslant 10$.
On the axes plot the following graphs.
i) $y = 2x^2 - 1$ **ii)** $y = x^3$ **iii)** $y = x$

b) Use your graph to solve the following pairs of simultaneous equations:
i) $y = 2x^2 - 1$ and $y = x$ **ii)** $y = x^3$ and $y = x$

c) Make a comment about the number of solutions to your equations.

9 Represent the following on number lines. Draw a new number line from $^-5$ to 5 for each part.

a) $a \geqslant {}^-4$ **b)** $^-2 < b < 5$ **c)** $^-2.5 < c \leqslant 3$ **d)** $0 \leqslant d < 2$

10 Mrs Brown says your memory is better when you are younger.

a) Plan an investigation to test Mrs Brown's theory. Include a suggested hypothesis. State how you will collect the data, how much data you will collect and what steps you will take to minimise bias.

b) Design a data collection sheet.

11 **a)** Draw a pair of axes where $^-5 \leqslant x \leqslant 5$ and $^-5 \leqslant y \leqslant 5$.

b) On the axes, draw the graph of $x = {}^-2$.
Shade in the area which satisfies the inequality $x > {}^-2$.

c) On the same axes, draw the graph of $y = 2x - 1$.
Shade in the area which satisfies the inequality $y \geqslant 2x - 1$.

d) State whether the following points are in an unshaded area, a single shaded area or a double shaded area.

(1, 4) (3, $^-2$) ($^-1$, $^-3$) ($^-2$, $^-3$)

12 Solve the following inequalities.

a) $^-3 < a + 5$ **b)** $\frac{3b}{4} \geqslant 9$ **c)** $2(2x - 6) > 4$ **d)** $7 - 2x \leqslant 1$

Test your knowledge

Test 1 (Chapters 1 and 2)

1 $\frac{4}{9} \times \frac{3}{5} =$

a) $\frac{43}{95}$ **b)** $\frac{1}{2}$ **c)** $\frac{4}{15}$ **d)** $\frac{8}{45}$

2 The fraction $\frac{5}{6}$ written as a recurring decimal is

a) 0.8333 **b)** $0.\dot{8}\dot{3}$ **c)** $0.8\dot{3}$ **d)** $0.\dot{8}3$

3 24 million expressed in scientific notation is

a) 24×10^7 **b)** 24×10^6 **c)** 2.4×10^7 **d)** 2.4×10^6

4 How many of the following statements are true?

$\sqrt{50} \approx 22.4$ $\sqrt{500} \approx 22.4$ $\sqrt{5000} \approx 224$

a) 1 **b)** 2 **c)** 3 **d)** 0

5 The value of m in $x^m = \sqrt[4]{x}$ is

a) 4 **b)** -4 **c)** $\frac{1}{4}$ **d)** $\frac{-1}{4}$

6 Mr McIntyre has $7\frac{1}{2}$ watermelons in his grocery shop. He cuts the melon into slices that are $\frac{1}{16}$ of a watermelon each. How many slices does he get?

7 Work out:

a) $1\frac{1}{8} + \frac{6}{8}$ **b)** $\frac{1}{2} + \frac{1}{4}$ **c)** $\frac{3}{5} + \frac{2}{3}$

d) $3\frac{1}{9} - \frac{4}{5} + 1\frac{1}{3}$ **e)** $\frac{3}{4} - \frac{1}{3} + 4\frac{5}{6}$ **f)** $1\frac{11}{15} + \frac{3}{5} - 1\frac{1}{3}$

8 For a staff bonus, a company divides $192 260 equally between 36 employees, rounding down to the nearest dollar.

a) How much does each employee receive?

b) How much money is left over from the original amount?

9 A property developer has bought a section of coastline 27.3 km long. He divides it into plots 0.7 km long to sell to hotel chains. How many plots can he sell?

10 $17^2 = 289$, $23^2 = 529$ and $31^2 = 961$. Use these facts to find the value of:

a) 17×18 **b)** 23×22 **c)** 30×31 **d)** 16×17 **e)** 29×31

11 Simplify, giving answers using index notation:

a) $2^3 \times 2^2$ **b)** $3^5 \times 3^{-2}$ **c)** $2^4 \times 3^2 \times 2^{-5}$ **d)** $5^4 \times 5^3$ **e)** $7^4 \div 7^{-2}$

Test 2 (Chapters 3 and 4)

1 The perimeter of this semicircle is

a) 132 cm **b)** 108 cm

c) 66 cm **d)** 63 cm

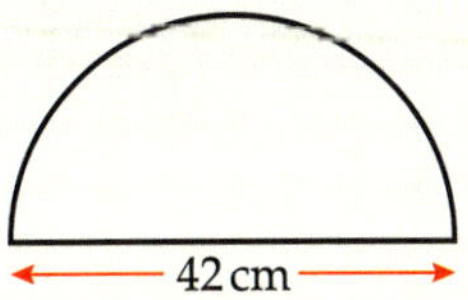

2 The length of an arc with central angle 21° on a circle of radius 15 m is

a) 5.5 m **b)** 2.7 m **c)** 41.2 m **d)** 8.2 m

3 The area of a sector with central angle 307° on a circle of radius 6 cm is

a) 32.1 cm^2 **b)** 96.4 cm^2 **c)** 64.1 cm^2 **d)** 192.9 cm^2

4 The area of this shape is

a) 10.1 cm^2

b) 8.6 cm^2

c) 11.2 cm^2

d) 13.4 cm^2

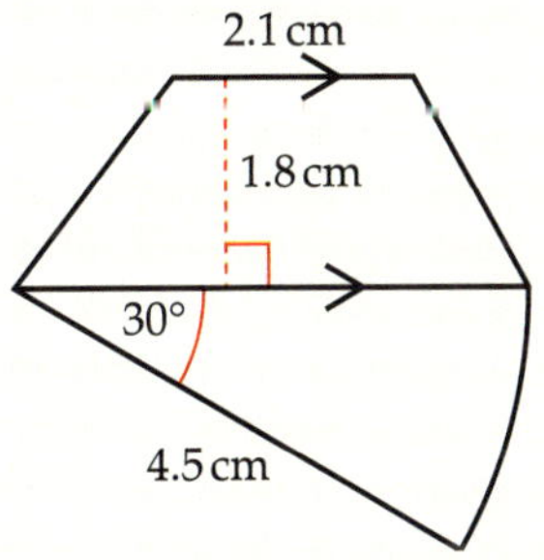

5 How many vertices does a cube have?

a) 4 **b)** 6 **c)** 8 **d)** 16

6 A parallelogram has vertices at (0, 0), (1, 3), (8, 3).

a) What are the coordinates of the fourth vertex?

b) Calculate the area of the parallelogram.

7 Five bricks of volume 3 cm × 5 cm × 6 cm are placed in a wooden box 40 cm long, 20 cm wide and 9 cm high.
What is the volume of space left in the box?

8 Find the surface area and volumes of these cuboids:

a) length = 10 cm, width = 6 cm, height = 2 cm

b) length = 212 mm, width = 96 mm, height = 11 mm

c) length = 0.8 m, width = 9.2 m, height = 1.5 m

d) length = 18.6 cm, width = 0.3 cm, height = 6.5 cm

9 Bottle caps are made by cutting as many circles as possible of diameter 4 cm out of a strip of metal 300 cm long and 4 cm wide.

a) How many bottle caps can be made from the strip of metal?

b) What percentage of the metal is wasted?

10 A one litre orange juice carton is in the shape of a cuboid. The base of the carton is 8 cm long and 5 cm wide. What is the height of the carton?

Test 3 (Chapters 5 and 6)

1 The expression $\frac{1}{20} - \frac{k}{10}$ is equivalent to

a) $\frac{1-k}{20}$ **b)** $\frac{-k}{2}$ **c)** $\frac{1-2k}{20}$ **d)** $\frac{1-k}{10}$

2 The least common multiple of $3xy$ and $4x^2$ is

a) $12x^2y$ **b)** $12xy$ **c)** $12x^2y^2$ **d)** $12xy^2$

3 The expression $(x + y)^2 - (x - y)^2$ written as a product of factors is

a) 0 **b)** $2x \times 2y$ **c)** $(x + y)(x - y)$ **d)** $2(x + y)(x - y)$

4 The roots of the equation $y = x^2 - x - 6$ are

a) $x = 3$ and $x = -4$ **b)** $x = 3$ and $x = -2$

c) $x = 2$ and $x = -3$ **d)** $x = 1$ and $x = 6$

5 Which of these points lies on the graph of $y + x^2 = 18$.

a) $(5, {}^-7)$ **b)** $(4, 3)$ **c)** $(6, {}^-9)$ **d)** $(3, 9)$

6 Simplify the following expressions:

a) $\frac{p}{10} + \frac{3}{q}$ **b)** $\frac{7}{s} + \frac{12}{t}$ **c)** $\frac{20}{j} + \frac{k}{10}$ **d)** $\frac{2x}{5} - \frac{1}{y}$ **e)** $\frac{x}{y} + \frac{y}{x}$ **f)** $\frac{2p}{q} - \frac{1}{2q}$

7 Expand the brackets and simplify each expression.

a) $4(f + 2g) + 2(3f + g)$ **b)** $-5(m - 8n) + 2(6m + n)$

c) $2x(x + y) - 4y(3x - y)$ **d)** $j(k^2 + j) + 3j^2(k - 12)$

8 Plot the graph of $y = x^2$ for values of x between 0 and 10. Use your graph to find:

a) the square of 3.6 **b)** the square of 7.8

c) the square root of 70 **d)** the square root of 94

9 Solve the simultaneous equations $3a - b = 12$ and $a - b = 2$.

10 This is the graph of a quadratic equation.

a) What are the roots of the equation?

b) Write down the equation in the form $y = x^2 + bx + c$.

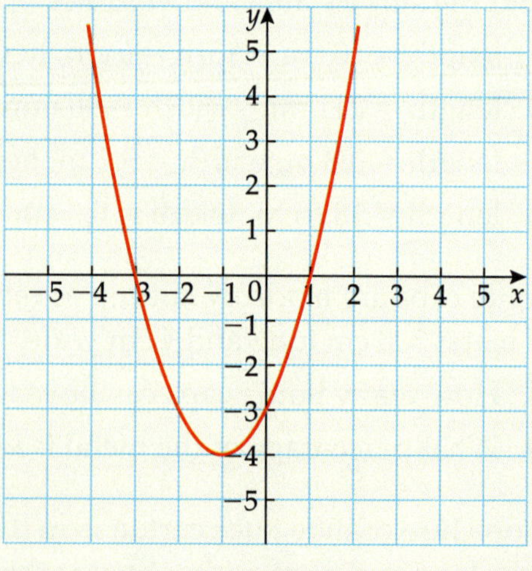

Test 4 (Major test A: Chapters 1–7) ....

1. The value of $(\sqrt{49} - \sqrt{25})^2$ is

   a) 4 b) 24 c) 48 d) 576

2. The expression $\frac{1}{x^3} \div x^{-6}$ is equivalent to

   a) x^{18} b) x^3 c) x^{-3} d) x^{-9}

3. How many edges does a hemisphere have?

   a) 0 b) 1 c) 2 d) 3

4. The roots of the equation $y = x^2 + 13x + 36$ are

   a) $x = 13$ and $x = 36$
   b) $x = -4$ and $x = -9$
   c) $x = 3$ and $x = 12$
   d) $x = 18$ and $x = -2$

5. Mr Lee buys ten MP3 players for $1200. He sells each one for $174. His percentage profit is

   a) 54% b) 4.5% c) 14.5% d) 45%

6. Express in scientific notation

   a) 436 b) 3 million c) 24 000 d) 0.0053 e) 0.708

7. Find the highest common factor of each pair of terms:

   a) $12x$ and 4
   b) 10 and $20y$
   c) z^3 and $2z$
   d) $9m^2$ and $3m$
   e) k^3 and k^2
   f) $8s^3$ and $4s^2$

8. Calculate the area of a sector with central angle 175° in a circle of radius 33 cm.

9. Solve the simultaneous equations $3j + k = 50$ and $2j - k = -30$.

10. Draw a graph to find the solutions of the simultaneous equations $y = 8 - x^2$ and $y = \frac{1}{2}x$.

11. Calculate:

    a) the simple interest on $800 for 3 years at $5\frac{1}{2}$% per annum
    b) the compound interest on $3000 for 2 years at 9% per annum.

12. Mr Sullivan buys 900 shares in a retail company, costing $22.50 each.

    a) What is the value of Mr Sullivan's investment?
    b) After one year the share price has increased to $24. Calculate Mr Sullivan's profit, as an amount of money and as a percentage of his principal.

Test 5 (Chapters 8 and 9)

1 Which one of these triangles is right-angled?

a)
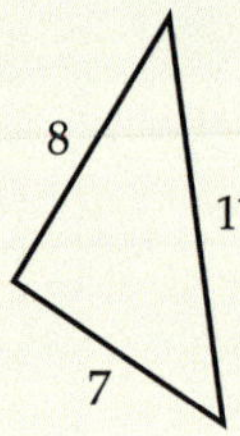

b)
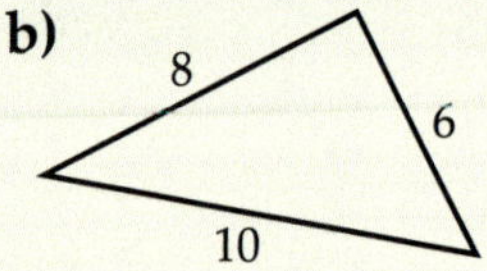

c)
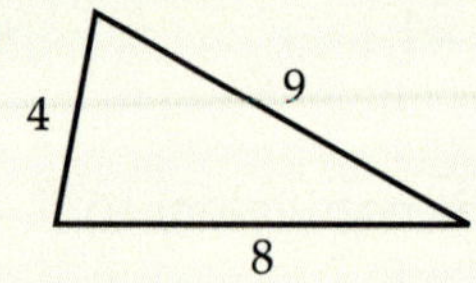

d)
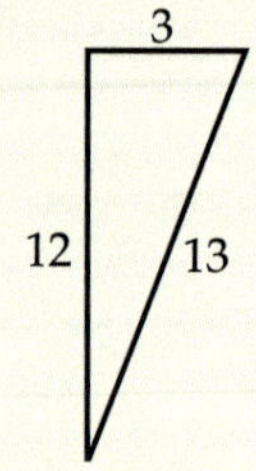

2 A regular polygon has an external angle of 40°. How many sides does it have?

a) 3 **b)** 18 **c)** 9 **d)** 6

3 The cosine of 83° is

a) 0.12 **b)** 0.99 **c)** 0.25 **d)** 8.14

4 Which of the following angles is a solution to the equation $\sin x = \cos x$?

a) 90° **b)** 30° **c)** 60° **d)** 45°

5 Find the length of the missing side in each of the following triangles, correct to one decimal place.

a)
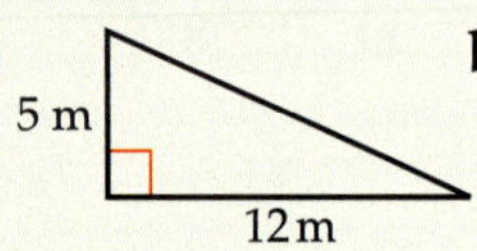

b)
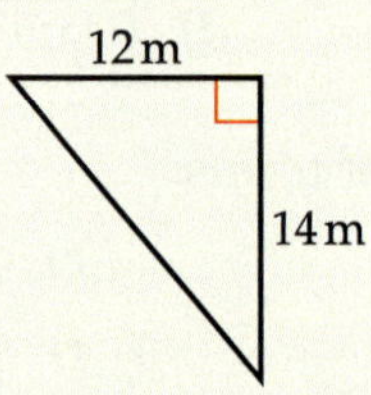

c)
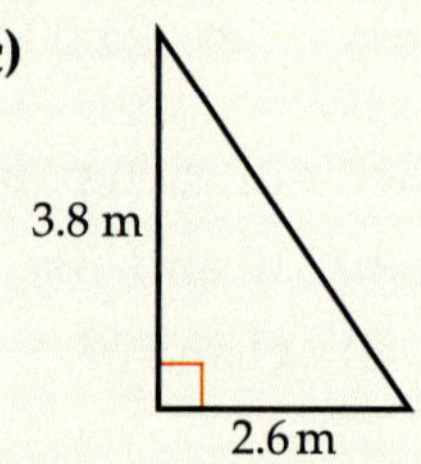

d)
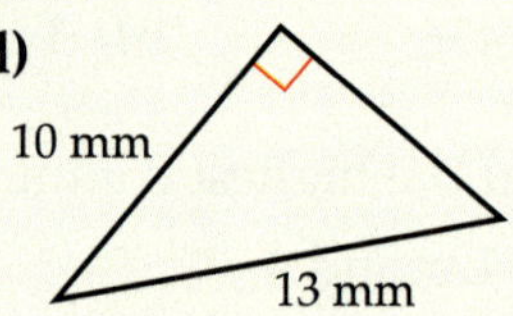

6 Find the size of the shaded angle in each of the following triangles, correct to one decimal place:

a)
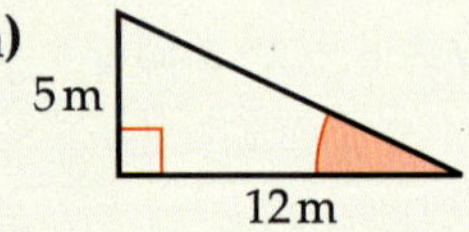

b)
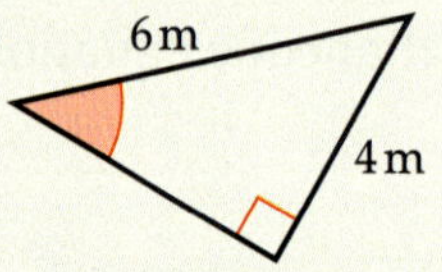

c)
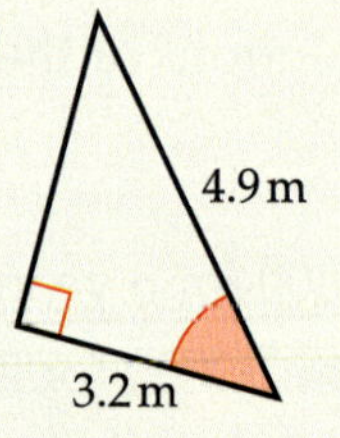

d)
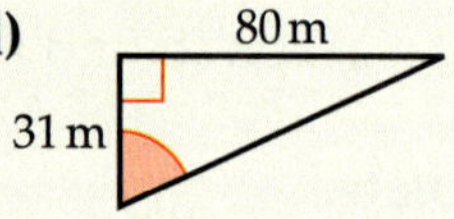

7 Winston is standing 15 m from the base of a tree. He measures the angle of elevation of the top of a tree as 40°. Calculate the height of the tree.

8 A ladder is leaning against a wall. The base of the ladder is 2.4 m from the wall, and the ladder reaches 6.5 m up the wall. Calculate the length of the ladder.

9 A sailing boat travels 130 km due North. It then turns and travels 60 km due East. How far is the sailing boat from its starting position?

10 Construct the parallelogram ABCD in which AB = 3.5 cm , BC = 5.5 cm and ∠B = 120°. Measure the diagonal AC.

Test 6 (Chapters 10 and 11)

1 The letter H has rotational symmetry of order

a) 1 **b)** 2 **c)** 4 **d)** 8

2 The scale factor of this enlargement is

a) 2

b) 2.5

c) 5

d) −5

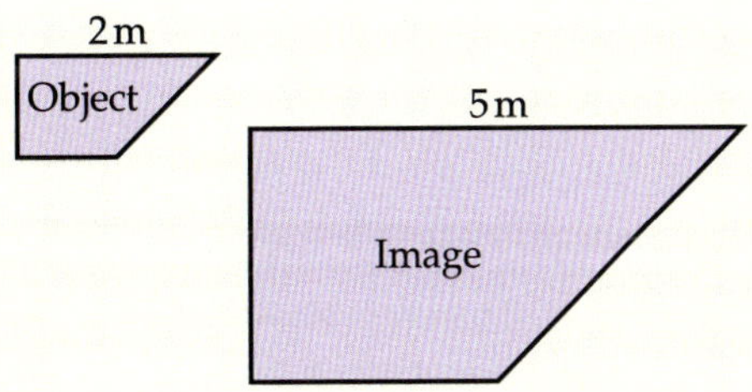

3 When a shape is enlarged with a scale factor of −2, the image is

a) the same size as the original shape but upside down

b) rotated through 180°

c) a reflection of the original shape

d) smaller than the original shape

4 How many of these triangles are similar?

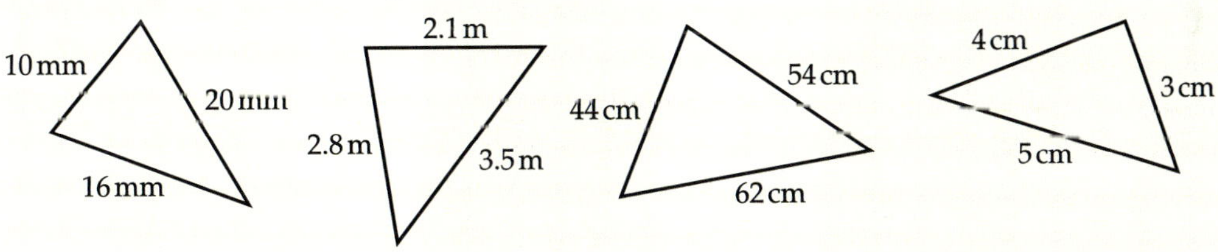

a) 0 **b)** 2 **c)** 3 **d)** 4

5 Draw a coordinate grid on graph or squared paper, with axes from −12 to 12 in both directions. Draw triangle PQR with P(1, 3), Q(5, 9) and R(11, 3). Find the coordinates of the images of P, Q and R after:

a) a rotation of 90° clockwise, centre (0, 0)

b) a reflection in the y-axis

6 Write down the order of rotational symmetry of each of the following patterns.

a) 

b) 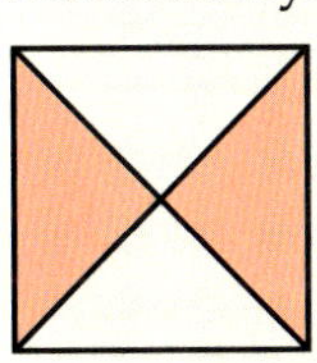

c) 

d) 

7. Fully describe this rotation.

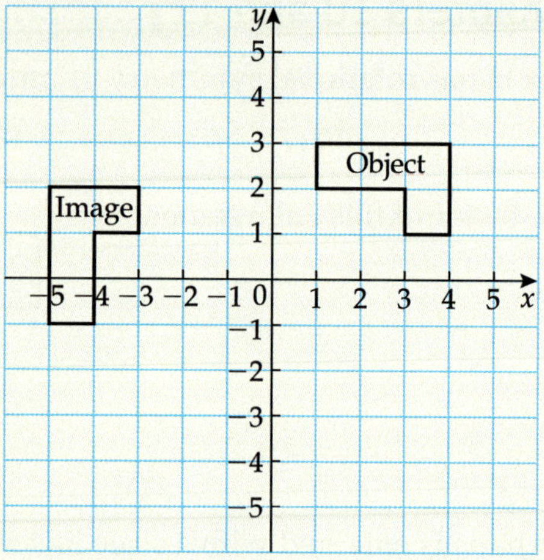

8. The Spitfire aeroplane has a wingspan of 11 m. Keela builds a model of the Spitfire. Her model is in proportion to the actual aeroplane. It has a wingspan of 44 cm and is 36 cm long.

   a) How long is the actual Spitfire?

   b) What is the scale factor of the enlargement from the model to the actual Spitfire?

Test 7 (Chapters 12 and 13)

1 The shaded region in the Venn diagram is the set

a) (A ∪ B) **b)** (A ∩ B)

c) (A ∪ B)′ **d)** (A ∩ B)′

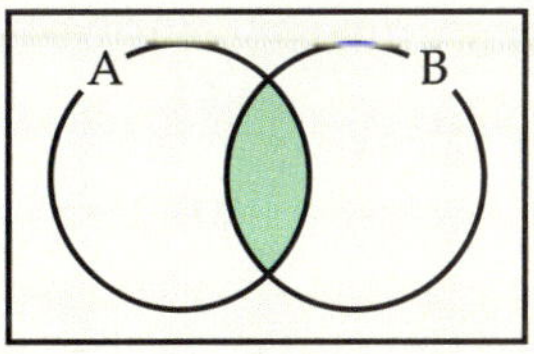

2 If **N** = {natural numbers}, **R** = {real numbers} and **Q** = {rational numbers} then

a) **R** ⊂ **N** ⊂ **Q** **b)** **N** ⊂ **R** ⊂ **Q** **c)** **Q** ⊂ **N** ⊂ **R** **d)** **N** ⊂ **Q** ⊂ **R**

3 If $x > y$ then which of the following statements is **not** true?

a) $2x > 2y$ **b)** $x - 3 > y - 3$ **c)** $5x > 5y$ **d)** $\frac{1}{2}x + 1 > \frac{1}{2}y + 1$

4 Which of the relations shown in the arrow diagrams is **not** a function?

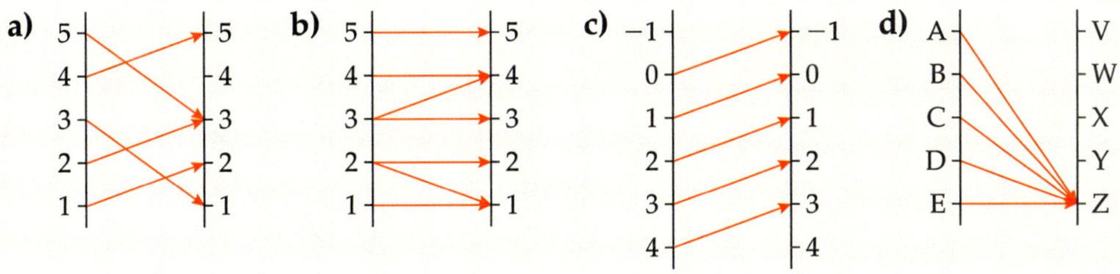

5 On a copy of this diagram, shade the region (P ∪ Q)′

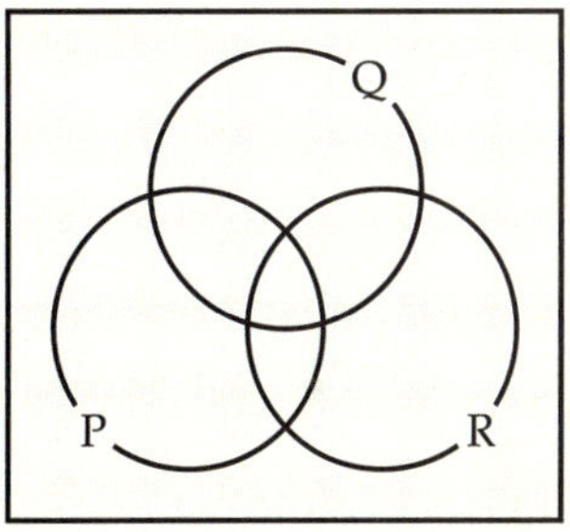

6 If $g: \to 2x + 3$ and $h: \to x^2 - 1$ write down

a) $g(2)$ **b)** $h(-2)$ **c)** $h(0)$ **d)** $g(-3)$

7 Draw a mapping diagram for the function
$f(x) = 2x - 1$ for $-3 \leqslant x \leqslant 3, x \in$ {natural numbers}

8 Find all the integers that satisfy the inequality $5 \leqslant 2x < 18$.

9 Solve these quadratic inequalities

a) $(x - 3)(x + 4) \leqslant 0$ **b)** $(x - 2)(x - 9) > 0$

c) $x(x + 3) \geqslant 0$ **d)** $(x + 5)(x - 1) < 0$

10 Draw a graph of $y = \frac{1}{2}x + 3$. Shade in the solution set for $x \geqslant 0$ and $y < \frac{1}{2}x + 3$.
Write down the coordinates of two points inside your solution set.

Test 8 (Major test B: Chapters 1–14)

1 How many of the following are recurring decimals?

$0.\dot{1}$ $\quad$ 4.222 $\quad$ $3.0\dot{2}\dot{1}$

a) 0 $\quad$ **b)** 1 $\quad$ **c)** 2 $\quad$ **d)** 3

2 The roots of the equation $y = x^2 + 6x - 16$ are

a) $x = 8$ and $x = -2$ $\quad$ **b)** $x = -8$ and $x = -2$

c) $x = 2$ and $x = -8$ $\quad$ **d)** $x = 2$ and $x = 8$

3 The centre of this enlargement is

a) $(-2, -1)$

b) $(0, 0)$

c) $(-1, -1)$

d) $(-1, 1)$

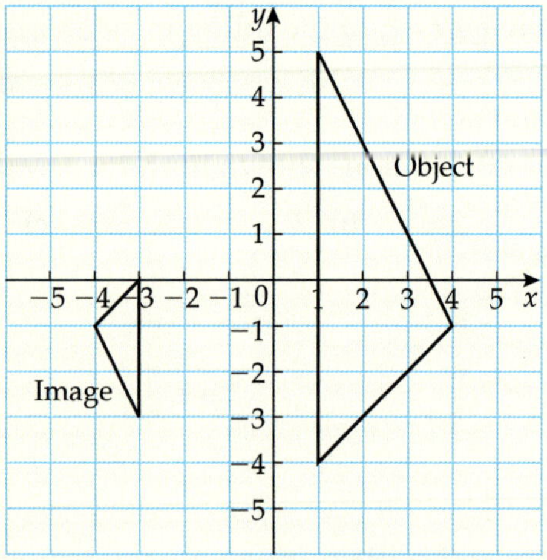

4 Calculate the value of:

a) $\left(\frac{2}{3}\right)^3$ $\quad$ **b)** $\left(\frac{1}{4}\right)^2$ $\quad$ **c)** $\left(1\frac{2}{5}\right)^2$ $\quad$ **d)** $\sqrt{\frac{25}{49}}$

5 Christina is painting her house. Each pot of paint will cover 13.24 m^2 of wall. Christina has 100 m^2 of walls to paint in total. How many pots of paint will she need?

6 Calculate the volume of this prism:

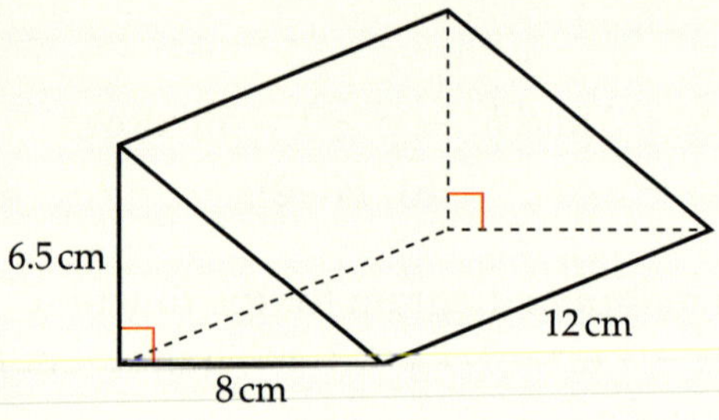

7 Solve the simultaneous equations $3m - n = 35$ and $m + n = 25$.

8 A company's share price increases by 8% each year for two years. Find the value of a \$15 000 investment at the end of two years.

9 The Leaning Tower of Pisa in Italy leans to one side at an angle of 6° to the vertical. A stone dropped from the top of the tower lands 5.8 m from the base of the tower. Calculate the vertical distance from the top of the tower to the ground.

10 Solve the inequality $3x - 4 < 2x + 1$ and show your solution on a number line.

Test 9 (Major test C: Chapters 1–14) ....

1 Written in standard form 0.073 =

a) 7.3×10^{-2} **b)** 7.3×10^{-1} **c)** 7.3×10^{2} **d)** 73×10^{3}

2 The expression $\frac{6}{x} - \frac{2}{y}$ is equivalent to

a) $\frac{4}{x-y}$ **b)** $\frac{2(3y-x)}{xy}$ **c)** $\frac{4xy}{x-y}$ **d)** $\frac{2(y-3x)}{xy}$

3 The tangent of 50° is

a) 0.5 **b)** 0.64 **c)** 1.19 **d)** 0.77

4 Which of these is a realistic estimate for the volume of a shoebox?

a) 1000 cm^3 **b)** 10 000 cm^3 **c)** 1 m^3 **d)** 0.2 m^3

5 How many 0.12 m lengths of fabric can you cut from a 4.08 m length of fabric?

a) 32 **b)** 34 **c)** 36 **d)** 38

6 Nick eats $\frac{3}{11}$ of his chocolate bar on Monday and $\frac{1}{3}$ of it on Tuesday.
What fraction of the chocolate bar is left for him to eat on Wednesday?

7 Find the length of the side marked x in each of the following triangles, correct to one decimal place:

a)

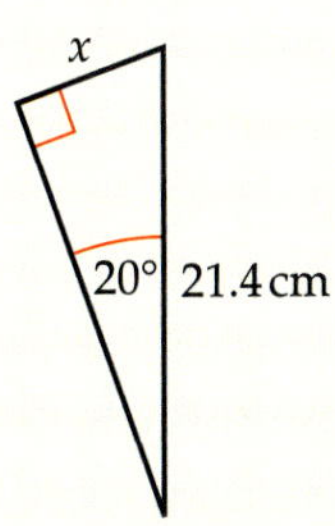

b)

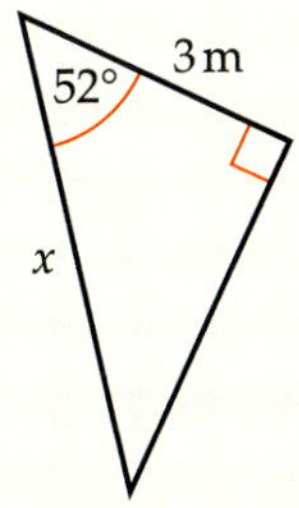

c)

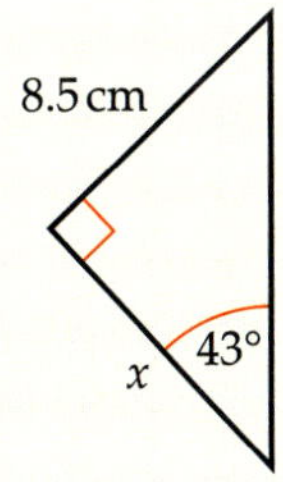

d)

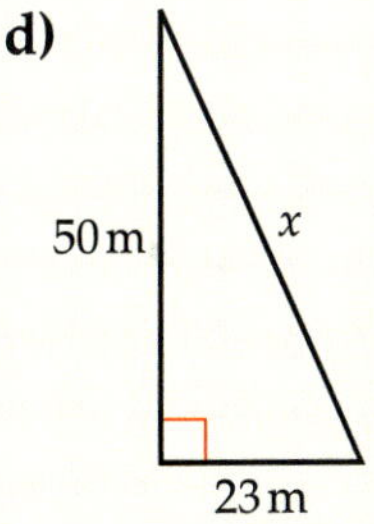

8 A photograph is enlarged with scale factor 5.5. A house in the original photograph is 1.8 cm wide. How wide is the house in the enlarged image?

9 Find all the integers that satisfy the inequality $-5 < 3x + 1 \leqslant 20$

10 This table shows the 100 m sprint times of a group of third-year students:

Time in seconds (t)	$12 \leqslant t < 14$	$14 \leqslant t < 16$	$16 \leqslant t < 18$	$18 \leqslant t < 20$
Number of people	2	6	7	3

Estimate the mean and find the median and modal groups for this data.

Test 10 (Major test D: Chapters 1–14) ....

1. The recurring decimal $0.\dot{5}\dot{6}$ written as a fraction is

   a) $\frac{4}{11}$ **b)** $\frac{9}{25}$ **c)** $\frac{3}{9}$ **d)** $\frac{5}{33}$

2. The solutions of the simultaneous equations $x + y = 0$ and $6x - y = 49$ are

   a) $x = y = 9.8$ **b)** $x = 7, y = -7$ **c)** $x = y = -7$ **d)** $x = -7, y = 7$

3. The sine of 60° is

   a) $\frac{\sqrt{2}}{3}$ **b)** $\frac{1}{2}$ **c)** $\frac{1}{\sqrt{2}}$ **d)** $\frac{\sqrt{3}}{2}$

4. The set of all integers that satisfy the inequality $(x + 3)(x + 7) < 0$ is

   a) $\{3, 4, 5, 6, 7\}$ **b)** $\{4, 5, 6\}$

   c) $\{-3, -4, -5, -6, -7\}$ **d)** $\{-4, -5, -6\}$

5. Simplify the following expressions:

   a) $\left(\frac{a}{2}\right)^2$ **b)** $\left(\frac{4}{2c}\right)^3$ **c)** $\left(\frac{2b}{4ab}\right)^2$

6. A speedboat costs $190 000 new and its value depreciates by 14% every year. Calculate its value after

   a) 1 year **b)** 3 years.

7. Solve these quadratic equations:

   a) $x^2 + 6x - 27 = 0$ **b)** $m^2 + 7m + 10 = 0$

   c) $t^2 - 13t = -30$ **d)** $n^2 = 25$

8. Using a ruler and a pair of compasses only, construct a right-angled triangle ABC with AB = 6.2 cm, BC = 9 cm and ∠B = 90°. Measure and write down the length of AC and the sizes of ∠A and ∠C.

9. Draw a diagram to show that the following statement is not true: 'A reflection in the y-axis followed by a reflection in the x-axis is equivalent to a reflection in the line $y = -x$.'

10. Alison has taken a photograph of the Big Ben clock tower in London. The lengths on her photograph are proportional to the actual lengths. She measures the width of the actual clock tower as 12 m. In her photograph the tower is 0.6 cm wide and 4.6 cm tall. Use these measurements to estimate the height of the actual clock tower.

11. Raul measured the heights in cm of 10 of his friends:

    155 136 139 141 150 148 151 132 162 155

    a) Is this data discrete or continuous?

    b) Choose appropriate class intervals, and show this data in a frequency table.

    c) Draw a histogram to illustrate this data.

    d) On the same axes, draw a frequency polygon.

Answers

Chapter 1

Page 2

1 **a)** 32 **b)** 17, 41 **c)** 6 **d)** 49 **e)** 294

2 36

3 402 300, four-hundred-and-two-thousand, three hundred

4 2, 3, 5, 7, 11, 13, 17, 19, 23, 29, 31, 37, 41, 43, 47, 53, 59

5 122

6 1, 19 – 19 is a prime number

7 −\$88.40

8 **a)** 27 **b)** −54 **c)** 69 **d)** −6

Page 4

1 **a) i)** $\frac{3}{4}$ **ii)** $\frac{5}{6}$ **iii)** $\frac{2}{3}$ **iv)** $\frac{4}{9}$ **v)** $\frac{3}{8}$ **vi)** $\frac{2}{5}$
b) i) $\frac{1}{4}$ **ii)** $\frac{1}{6}$ **iii)** $\frac{1}{3}$ **iv)** $\frac{5}{9}$ **v)** $\frac{5}{8}$ **vi)** $\frac{3}{5}$

2 **a)** $\frac{11}{4}$ **b)** $\frac{10}{3}$ **c)** $\frac{27}{5}$ **d)** $\frac{25}{6}$
e) $\frac{13}{8}$ **f)** $\frac{11}{7}$ **g)** $\frac{9}{2}$ **h)** $\frac{23}{10}$
i) $\frac{56}{15}$ **j)** $\frac{20}{3}$

3 **a)** $2\frac{1}{5}$ **b)** $7\frac{1}{3}$ **c)** $1\frac{3}{4}$ **d)** $2\frac{4}{7}$
e) $4\frac{1}{2}$ **f)** $1\frac{5}{6}$ **g)** $3\frac{5}{9}$ **h)** $3\frac{10}{11}$
i) $3\frac{3}{8}$ **j)** $3\frac{4}{12} = 3\frac{1}{3}$

4 **a)** $1\frac{1}{2}$ (or equivalent) **b)** $1\frac{3}{7}$
c) $2\frac{10}{60} = 2\frac{1}{6}$ **d)** $1\frac{30}{100}$ or $1\frac{3}{10}$

5 **a)** $\frac{11}{2} > \frac{25}{7}$ **b)** $2\frac{1}{5} < 3\frac{1}{5}$ **c)** $6\frac{3}{8} < \frac{53}{8}$
d) $\frac{17}{4} < 4\frac{3}{4}$ **e)** $5\frac{4}{5} > \frac{27}{5}$ **f)** $\frac{31}{12} = 2\frac{7}{12}$
g) $6\frac{2}{3} < \frac{22}{3}$ **h)** $\frac{46}{10} = 4\frac{3}{5}$ **i)** $8\frac{1}{2} > \frac{33}{4}$
j) $\frac{23}{11} < 2\frac{3}{11}$

7 $\frac{15}{4} = 3\frac{3}{4}$ oranges. 6 whole oranges left, and $\frac{1}{4}$ orange

Page 6

1 **a)** $\frac{2}{3} = \frac{4}{6}$ **b)** $\frac{3}{5} = \frac{6}{10}$ **c)** $\frac{4}{12} = \frac{2}{6}$ **d)** $\frac{5}{6} = \frac{20}{24}$
e) $\frac{3}{4} = \frac{9}{12}$ **f)** $\frac{5}{8} = \frac{10}{16}$ **g)** $\frac{7}{10} = \frac{21}{30}$ **h)** $\frac{6}{7} = \frac{18}{21}$
i) $\frac{5}{9} = \frac{15}{27}$ **j)** $\frac{7}{20} = \frac{35}{100}$

2 **a)** $\frac{3}{5}$ **b)** $\frac{3}{4}$ **c)** $\frac{2}{3}$ **d)** $\frac{4}{5}$ **e)** $\frac{5}{6}$
f) $\frac{3}{10}$ **g)** $\frac{4}{7}$ **h)** $\frac{1}{3}$ **i)** $\frac{7}{9}$ **j)** $\frac{5}{8}$

3 **a)** $\frac{15}{24}$ **b)** $\frac{14}{27}$ **c)** $\frac{24}{35}$ **d)** $\frac{15}{20}$
e) $\frac{56}{70}$ **f)** $\frac{15}{28}$ **g)** $\frac{16}{54}$ **h)** $\frac{24}{72}$

5 **a)** $\frac{9}{12}$ **b)** $\frac{20}{25}$ **c)** $\frac{6}{9}$ **d)** $\frac{28}{48}$

6 RH Fraction with a denominator of:
2: none
3: $\frac{1}{3} = \frac{8}{24}$; $\frac{1}{3} = \frac{9}{27}$
4: $\frac{1}{4} = \frac{7}{28}$; $\frac{1}{4} = \frac{8}{32}$; $\frac{1}{4} = \frac{9}{36}$; $\frac{3}{4} = \frac{9}{12}$
5: none
6: $\frac{1}{6} = \frac{7}{42}$; $\frac{1}{6} = \frac{9}{54}$; $\frac{3}{6} = \frac{9}{18}$; $\frac{4}{6} = \frac{8}{12}$
7: $\frac{1}{7} = \frac{4}{28}$; $\frac{1}{7} = \frac{6}{42}$; $\frac{1}{7} = \frac{8}{56}$; $\frac{1}{7} = \frac{9}{63}$; $\frac{3}{7} = \frac{6}{14}$; $\frac{3}{7} = \frac{9}{21}$; $\frac{4}{7} = \frac{8}{14}$
8: $\frac{1}{8} = \frac{3}{24}$; $\frac{1}{8} = \frac{4}{32}$; $\frac{1}{8} = \frac{7}{56}$; $\frac{1}{8} = \frac{9}{72}$; $\frac{2}{8} = \frac{4}{16}$; $\frac{3}{8} = \frac{9}{24}$
9: $\frac{1}{9} = \frac{3}{27}$; $\frac{1}{9} = \frac{4}{36}$; $\frac{1}{9} = \frac{6}{54}$; $\frac{1}{9} = \frac{7}{63}$; $\frac{1}{9} = \frac{8}{72}$; $\frac{2}{9} = \frac{4}{18}$; $\frac{2}{9} = \frac{8}{36}$; $\frac{3}{9} = \frac{6}{18}$
Altogether: 31 possibilities

Page 7

1 **a)** $\frac{4}{5}$ **b)** $\frac{6}{8} = \frac{3}{4}$ **c)** $\frac{9}{10}$
d) $\frac{10}{6} = \frac{5}{3} = 1\frac{2}{3}$ **e)** $1\frac{4}{8} = 1\frac{1}{2}$ **f)** $2\frac{4}{10} = 2\frac{2}{5}$
g) $4\frac{4}{6} = 4\frac{2}{3}$ **h)** $3\frac{1}{2}$ **i)** $\frac{4}{5}$
j) $4\frac{2}{4} = 4\frac{1}{2}$

2 **a)** $\frac{1}{4}$ **b)** $\frac{1}{10}$ **c)** $\frac{5}{8}$
d) $\frac{5}{12}$ **e)** $1\frac{2}{5}$ **f)** $1\frac{5}{12}$
g) $3\frac{1}{4}$ **h)** $4\frac{1}{3}$

3 **a)** $1\frac{1}{4}$ **b)** $\frac{5}{6}$ **c)** $\frac{3}{8}$
d) $\frac{5}{10} = \frac{1}{2}$ **e)** $\frac{9}{10}$ **f)** 2
g) $2\frac{1}{4}$ **h)** $1\frac{7}{8}$ **i)** $1\frac{1}{6}$
j) $1\frac{1}{4}$

4 $\frac{1}{2}$ 5 $\frac{1}{6}$ 6 $\frac{3}{10}$

7 **a)** $\frac{5}{6}$ **b)** $\frac{3}{8}$ **c)** $\frac{3}{5}$

8 Numerous additions are possible. They include:
Adding two fractions: quarters
$1\frac{1}{4} + \frac{2}{4}\ (1\frac{1}{4} + \frac{1}{2})$ $1 + \frac{3}{4}$
$1\frac{2}{4} + \frac{1}{4}\ (1\frac{1}{2} + \frac{1}{4})$
Adding two fractions: eighths
$\frac{1}{8} + 1\frac{5}{8}$ $\frac{3}{8} + 1\frac{3}{8}$
$\frac{5}{8} + 1\frac{1}{8}$ $\frac{7}{8} + \frac{7}{8}$
Adding three fractions: quarters
$\frac{1}{4} + \frac{1}{4} + 1\frac{1}{4}$ $\frac{1}{4} + \frac{2}{4} + 1\ (\frac{1}{4} + \frac{1}{2} + 1)$
$\frac{1}{4} + \frac{3}{4} + \frac{3}{4}$ $\frac{2}{4} + \frac{2}{4} + \frac{3}{4}\ (\frac{1}{2} + \frac{1}{2} + \frac{3}{4})$
Adding three fractions: eighths
$\frac{1}{8} + \frac{1}{8} + 1\frac{4}{8}\ (\frac{1}{8} + \frac{1}{8} + 1\frac{1}{2})$ $\frac{1}{8} + \frac{2}{8} + 1\frac{3}{8}\ (\frac{1}{8} + \frac{1}{4} + 1\frac{3}{8})$
$\frac{1}{8} + \frac{3}{8} + 1\frac{2}{8}\ (\frac{1}{8} + \frac{3}{8} + 1\frac{1}{4})$ $\frac{1}{8} + \frac{4}{8} + 1\frac{1}{8}\ (\frac{1}{8} + \frac{1}{2} + 1\frac{1}{8})$
$\frac{1}{8} + \frac{5}{8} + 1$ $\frac{1}{8} + \frac{6}{8} + \frac{7}{8}\ (\frac{1}{8} + \frac{3}{4} + \frac{7}{8})$
$\frac{2}{8} + \frac{2}{8} + 1\frac{2}{8}\ (\frac{1}{4} + \frac{1}{4} + 1\frac{1}{4})$ $\frac{2}{8} + \frac{3}{8} + 1\frac{1}{8}\ (\frac{1}{4} + \frac{1}{4} + 1\frac{1}{8})$
$\frac{2}{8} + \frac{4}{8} + 1\ (\frac{1}{4} + \frac{1}{2} + 1)$ $\frac{2}{8} + \frac{5}{8} + \frac{7}{8}\ (\frac{1}{4} + \frac{5}{8} + \frac{7}{8})$
$\frac{2}{8} + \frac{6}{8} + \frac{6}{8}\ (\frac{1}{4} + \frac{3}{4} + \frac{3}{4})$ $\frac{3}{8} + \frac{3}{8} + 1$
$\frac{3}{8} + \frac{4}{8} + \frac{7}{8}\ (\frac{3}{8} + \frac{1}{2} + \frac{7}{8})$ $\frac{3}{8} + \frac{5}{8} + \frac{6}{8}\ (\frac{3}{8} + \frac{5}{8} + \frac{3}{4})$
$\frac{4}{8} + \frac{4}{8} + \frac{6}{8}\ (\frac{1}{2} + \frac{1}{2} + \frac{3}{4})$ $\frac{4}{8} + \frac{5}{8} + \frac{5}{8}\ (\frac{1}{2} + \frac{5}{8} + \frac{5}{8})$
Other possibilities include: Adding two fractions sixteenths, Adding three fractions: sixteenths, Adding four fractions, etc

9 Many pairs of fractions are possible. They include:
thirds: $(\frac{1}{3}, \frac{2}{3})$, $(\frac{2}{3}, 1)$, $(1, 1\frac{1}{3})$, $(1\frac{2}{3}, 1\frac{1}{3})$, …
sixths: $(\frac{1}{6}, \frac{1}{2})$, $(\frac{5}{6}, \frac{1}{2})$, $(1\frac{1}{6}, \frac{5}{6})$, $(1\frac{1}{2}, 1\frac{1}{6})$, …
ninths: $(\frac{1}{9}, \frac{4}{9})$, $(\frac{2}{9}, \frac{5}{9})$, $(\frac{4}{9}, \frac{7}{9})$, $(\frac{5}{9}, \frac{8}{9})$, …
twelfths: $(\frac{1}{12}, \frac{5}{12})$, …

Page 10

1 **a)** $\frac{4}{5}$ **b)** $\frac{16}{25}$ **c)** $\frac{19}{25}$ **d)** $\frac{97}{200}$
e) $\frac{129}{250}$ **f)** $\frac{3}{50}$ **g)** $\frac{3}{8}$ **h)** $\frac{41}{50}$
i) $\frac{76}{125}$ **j)** $\frac{21}{250}$

2 **a)** 0.8 **b)** 0.75 **c)** 0.6 **d)** 0.85
e) 0.62 **f)** 0.36 **g)** 0.444 **h)** 0.395
i) 0.17 **j)** 6.25

3 **a)** 1.6 **b)** 0.375 **c)** 2.75 **d)** 1.625

4 a) 0.666 666 6 b) 0.222 222 2
c) 0.888 888 8 d) 0.166 666 6
e) 0.777 777 7 f) 0.833 333 3
g) 0.444 444 4 h) 0.333 333 3

5

	1	2	3	4	5	6	8	9	10
1	1	2	3	4	5	6	8	9	10
2	0.5	1	1.5	2	2.5	3	4	4.5	5
3	$0.3\dot{3}$	$0.6\dot{6}$	1	$1.3\dot{3}$	$1.6\dot{6}$	2	$2.6\dot{6}$	3	$10.3\dot{3}$
4	0.25	0.5	0.75	1	1.25	1.5	2	2.25	2.5
5	0.2	0.4	0.6	0.8	1	1.2	1.6	1.8	2
6	$0.1\dot{6}$	$0.3\dot{3}$	0.5	$0.6\dot{6}$	$0.8\dot{3}$	1	$1.3\dot{3}$	1.5	$1.6\dot{6}$
8	0.13	0.25	0.38	0.5	0.63	0.75	1	1.13	1.25
9	$0.1\dot{1}$	$0.2\dot{2}$	$0.3\dot{3}$	$0.4\dot{4}$	$0.5\dot{5}$	$0.6\dot{6}$	$0.8\dot{8}$	1	$1.1\dot{1}$
10	0.1	0.2	0.3	0.4	0.5	0.6	0.8	0.9	1

6 $\frac{3}{110}$

7

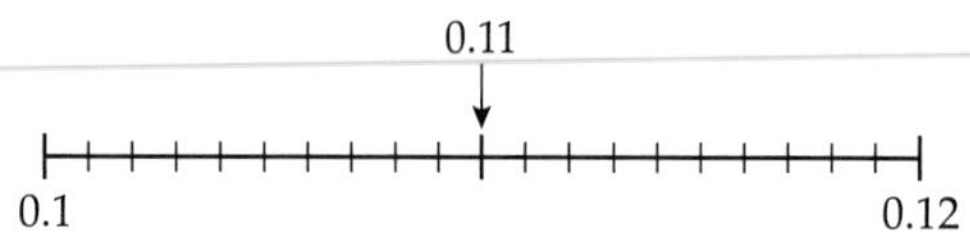

8 $\frac{1}{9} = 0.111\,111\,1$, $\frac{2}{9} = 0.222\,222\,2$, $\frac{3}{9} = 0.333\,333\,3$,
$\frac{4}{9} = 0.444\,444\,4$, $\frac{5}{9} = 0.555\,555\,5$, $\frac{6}{9} = 0.666\,666\,6$,
$\frac{7}{9} = 0.777\,777\,7$, $\frac{8}{9} = 0.888\,888\,8$, $\frac{9}{9} = 1.0$,
$\frac{10}{9} = 1.111\,111\,1$, $\frac{11}{9} = 1.222\,222\,2$
$\frac{1}{11} = 0.090\,909$, $\frac{2}{11} = 0.181\,818$, $\frac{3}{11} = 0.272\,727$,
$\frac{4}{11} = 0.363\,636$, $\frac{5}{11} = 0.454\,545$, $\frac{6}{11} = 545\,454$,
$\frac{7}{11} = 0.636\,363$, $\frac{8}{11} = 0.727\,272$, $\frac{9}{11} = 0.818\,181$,
$\frac{10}{11} = 0.909\,090\,9$
Note that the digits in the ninths link to the multiples of 11, and the digits in the elevenths link to the multiples of 9

9 Answers will vary.
If the denominator is 3, and the numerator is not a multiple of 3, then the fraction is equivalent to a recurring decimal with 1 digit recurring.
If the denominator is 11, and the numerator is not a multiple of 11, then the fraction is equivalent to a recurring decimal with 2 digits recurring.

Page 12

1 a) $0.\dot{1}$ b) $3.6\dot{2}$ c) $19.\dot{3}\dot{2}$
d) $7.6\dot{1}\dot{9}$ e) $0.\dot{1}\dot{8}$ f) $0.\dot{1}4285\dot{7}$

2 a) $\frac{2}{5}$ b) $\frac{8}{25}$ c) $\frac{27}{100}$
d) $\frac{341}{500}$ e) $\frac{31}{250}$ f) $\frac{3}{4}$
g) $\frac{7}{25}$ h) $\frac{3}{50}$ i) $\frac{1}{40}$

3 a) $0.\dot{2}$ b) $0.8\dot{3}$ c) $0.\dot{3}$
d) $0.\dot{8}$ e) 0.214…

4 a) $\frac{2}{3}$ b) $\frac{1}{9}$ c) $\frac{5}{9}$
d) $\frac{7}{9}$ e) $\frac{2}{11}$ f) $\frac{91}{99}$
g) $\frac{513}{999}$ h) $\frac{125}{999}$

Page 14

1 a) $1\frac{1}{2}$ b) $\frac{1}{3}$ c) 5 d) $1\frac{1}{2}$ e) $2\frac{1}{2}$

2 a) 40, $\frac{9}{40}$ b) 42, $1\frac{17}{42}$ c) 45, $1\frac{5}{9}$
d) 48, $\frac{1}{8}$ e) 60, $\frac{37}{60}$

3 a) $4\frac{1}{35}$ b) $1\frac{43}{45}$ c) $\frac{3}{10}$ d) $5\frac{49}{88}$ e) $5\frac{43}{72}$

4 a) $3\frac{19}{24}$ b) $1\frac{19}{60}$ c) $\frac{13}{40}$ d) $4\frac{9}{20}$ e) $\frac{43}{60}$

5 $30\frac{5}{12}$ inches

6 $\frac{1}{8}$

7 a) John b) $\frac{3}{40}$

8 Possible answers include: $\frac{1}{2} + \frac{1}{12}$

9 94.50$

Page 16

1 a) $\frac{2}{3}$ b) $\frac{8}{15}$ c) $\frac{4}{9}$ d) $\frac{7}{18}$ e) $\frac{5}{8}$ f) $\frac{5}{16}$

2 a) 3 b) 14 c) 16 d) 5 e) 3 f) $8\frac{4}{5}$

3 a) $\frac{1}{8}$ b) $\frac{16}{25}$ c) $\frac{1}{3}$ d) $4\frac{10}{11}$ e) $1\frac{61}{64}$

4 9 m^2

5 $\frac{1}{4}$

6 a) $\frac{1}{9}$ b) $\frac{1}{10}$ c) $4\frac{17}{27}$

7 $\frac{1}{4}$ m^2, $\frac{1}{16}$ m^2

8 $38\frac{1}{2}$ cm^2

9 a) $10\frac{1}{2}$ $inch^3$ b) 35 g

10 a) $1\frac{3}{4}$ m b) 3

Page 18

1 a) $\frac{5}{3}$ b) $\frac{4}{5}$ c) $\frac{3}{4}$

2 a) $\frac{2}{3}$ b) $1\frac{1}{3}$ c) $\frac{2}{3}$ d) $\frac{6}{7}$ e) $\frac{4}{5}$

3 a) 6 b) 6 c) $1\frac{1}{2}$ d) $4\frac{1}{2}$ e) $2\frac{1}{2}$

4 6

5 $1\frac{4}{5}$ cm

6 a) 3 b) $\frac{5}{7}$ c) $\frac{1}{2}$

7 a) $\frac{1}{4}$ b) $\frac{3}{5}$ c) $1\frac{1}{3}$ d) $\frac{2}{9}$

8 a) 16 b) $\frac{1}{4}$

9 $1\frac{1}{2}$ cm b) $\frac{3}{4}$ cm

Page 19

1 a) Multiplying by 0.1 is the same as dividing by 10.
b) Dividing by 0.1 is the same as multiplying by 10.
c) Multiplying by 0.01 is the same as dividing by 100.
d) Dividing by 0.01 is the same as multiplying by 100.
e) Multiplying by 0.08 is the same as multiplying by 0.01, then multiplying by 8.
f) Dividing by 0.6 is the same as dividing by 0.1 then dividing by 6.

2

	×1000	÷10	×0.1	×0.01	÷0.1	÷0.01
4.29	4290	0.429	0.429	0.0429	42.9	429
13.6	13600	1.36	1.36	0.136	136	1360
4.3	4300	0.43	0.43	0.043	43	430
7	7000	0.7	0.7	0.07	70	700

3 a) smaller b) greater c) greater
d) smaller e) smaller f) greater

4 a) 12 b) 20.79 c) 60
d) 2.24 e) 8.42 f) 13.5

5 a) 0.396 b) 150 c) 4.424
d) 82 e) 1.95 f) 1.68

6 5.64 mm

7 400

8 167

9 Answers will vary.

10 a) Answers will vary. b) No

Page 22

1 a) 0.12 b) 0.072 c) 0.0252
d) 0.069 e) 0.27 f) 0.028
g) 0.081 h) 0.384 i) 0.015
j) 0.8 k) 0.9 l) 17
m) 0.8 n) 16 o) 12
p) 0.7 q) 6 r) 0.25

2

	0.027	
0.18	0.15	
0.3	0.6	0.25

3

	0.75	
0.6	0.8	
0.24	0.4	0.5

4 a) 0.075 m^2 b) 12 c) 16

5 Answers will vary.

Page 24

1 a) 0.4, 0.6, 0.7, 0.9 b) 0.7, 1.1, 1.3, 1.7
c) 0.64, 0.66, 0.68, 0.70 d) 0.58, 0.67, 0.76, 0.85
e) $0.8\dot{3}$, 1.0, $1.1\dot{6}$, $1.\dot{3}$

2 a) $\frac{4}{10}, \frac{6}{10}, \frac{7}{10}, \frac{9}{10}$ b) $\frac{7}{10}, \frac{11}{10}, \frac{13}{10}, \frac{17}{10}$
c) $\frac{64}{100}, \frac{66}{100}, \frac{68}{100}, \frac{70}{100}$ d) $\frac{58}{100}, \frac{67}{100}, \frac{76}{100}, \frac{85}{100}$
e) $\frac{5}{6}, 1, \frac{7}{6}, \frac{4}{3}$

3 0.4, 0.5, 0.6

4 $\frac{5}{6}$

5 $\frac{4}{5}$ and $\frac{7}{25}$

6 a) $\frac{6}{11}$, 0.57, $\frac{4}{7}$, $0.\dot{6}$, 0.71 b) $\frac{9}{16}$, 0.58, $\frac{9}{13}$, 0.7
c) 0.8, 0.89, 0.9, $\frac{15}{16}, \frac{20}{21}$ d) $\frac{3}{20}$, 0.17, 0.19, $\frac{1}{5}, \frac{3}{8}$
e) $\frac{3}{10}$, $0.\dot{3}$, $\frac{3}{8}$, 0.38, 0.4

Looking back (page 25)

1 a) $0.41\dot{6}$ b) $0.041\dot{6}$ c) $0.4\dot{6}$
d) $0.3\dot{6}$ e) $0.9\dot{3}$

2 a) $\frac{11}{15}$ b) $\frac{4}{21}$ c) $4\frac{11}{36}$

3 22.185

4 6

5 a) $0.375 < \frac{3}{6}$ b) $\frac{3}{7} > 0.41$ c) $\frac{2}{3} < 0.74$
d) $0.625 = \frac{5}{8}$ e) $0.1 < \frac{1}{6}$ f) $\frac{11}{12} > 0.9$
g) $0.79 > \frac{7}{9}$

6 a) $0.\dot{6}$ $\frac{13}{22}$ 0.59 $\frac{8}{15}$ $\frac{9}{17}$ b) $\frac{1}{8}$ 0.1 0.08 $\frac{1}{13}$ 0.05
c) $\frac{7}{9}$ $\frac{14}{19}$ 0.7 0.68 $\frac{13}{20}$ d) $0.\dot{8}$ $\frac{7}{8}$ 0.86 $\frac{9}{11}$ 0.8
e) 0.45 $\frac{4}{9}$ 0.4 0.39 $\frac{5}{13}$

8 a) $0.\dot{2}\dot{7}$ b) 0.23 c) 0.125
d) 0.8 e) $0.\dot{4}2857\dot{1}$ f) 0.24
g) 0.1875 h) 0.85 i) 0.272

9 $0.\dot{0}7692\dot{3}$, $0.\dot{1}5384\dot{6}$, $0.\dot{2}3076\dot{9}$, $0.769230\dot{7}6923\dot{0}$,
$0.\dot{8}4615\dot{3}$, $0.\dot{9}2307\dot{6}$

10 a) $1 + \frac{1}{2} + \frac{1}{4} + \frac{1}{8} + \frac{1}{16} = 1\frac{15}{16}$
$1 + \frac{1}{2} + \frac{1}{4} + \frac{1}{8} + \frac{1}{16} + \frac{1}{32} = 1\frac{31}{32}$
$1 + \frac{1}{2} + \frac{1}{4} + \frac{1}{8} + \frac{1}{16} + \frac{1}{32} + \frac{1}{64} = 1\frac{63}{64}$
$1 + \frac{1}{2} + \frac{1}{4} + \frac{1}{8} + \frac{1}{16} + \frac{1}{32} + \frac{1}{64} + \frac{1}{128} = 1\frac{127}{128}$

b) Answer to next line = $1\frac{(2^n - 1)}{2^n}$ where n is the term number.

c) 2

Chapter 2

Page 27

1 a) base 11; index 5 b) base 2; index 4
c) base 3; index 1 d) base 7; index 0
e) base 5; index −3 f) base 4; index 8
g) base 1; index 9 h) base $\frac{3}{4}$; index 6
i) base −4; index 2 j) base $\frac{1}{3}$; index −5

2 a) $5 \times 5 \times 5$
b) $4 \times 4 \times 4 \times 4$
c) $2 \times 2 \times 2 \times 2 \times 2 \times 2$
d) $3 \times 3 \times 3 \times 3 \times 3 \times 3 \times 3$
e) 8×8
f) $1 \times 1 \times 1 \times 1 \times 1$
g) $2 \times 2 \times 2 \times 2 \times 2 \times 2 \times 2 \times 2$
h) $10 \times 10 \times 10$
i) $11 \times 11 \times 11 \times 11$
j) 6×6

3 a) 6^2 b) 3^4 c) 7^5 d) 2^3 e) 5^4
f) 4^5 g) 8^1 h) 2^7 i) 9^6 j) 4^3

4 a) 16 b) 81 c) 64 d) 144 e) 32
f) 125 g) 1 h) 128 i) 75 j) −72

Page 29

1 a) $3x^4$ b) $2m^7$ c) $20z^6$
d) $10n^5$ e) $8q^3$ f) $8a^6$

2 a) x b) $4a^4$ c) $\frac{6x^4}{w^3}$
d) $2n$ e) $\frac{2}{y^3}$ f) 7

3 a) t^{12} b) k^3 c) p^2g^2
d) y^3z^2 e) $8a^{x+y}$ f) $\frac{q^3r^4}{4}$

4 a) 36 b) 4 096 c) 4 608
d) 729 e) 144 f) 118 098

5 a) 5^0 b) 42^0 c) a^0 d) y^0

6 a) 1 b) 1 c) 1 d) 1

7 1

8 a) 9 b) 8 c) h

Page 31

1 a) 2 b) 11 c) 6 d) 7
e) 9 f) 12 g) 3 h) 4
i) 8 j) 100

2 a) 20 b) 18 c) 60 d) 21
e) 60 f) 80 g) 22 h) 180
i) 210 j) 600

3 **a)** 30 **b)** 20 **c)** 50 **d)** 14
e) 70 **f)** 600

4 **a)** $\frac{9}{8}$ **b)** $\frac{2}{5}$ **c)** $\frac{1}{3}$ **d)** $\frac{5}{4}$ **e)** $\frac{10}{11}$

5 7 cm

6 100 m

7 **a)** **b)** and **d)** are square.

8 **a)** 272 **b)** 1640 **c)** 156 **d)** 240
e) 1560 **f)** 132

Page 32

1 **a)** 2 **b)** 125 **c)** 5 **d)** 5 **e)** $\sqrt{64}$

2 **a)** 6 **b)** 7 **c)** 2 **d)** 5 **e)** 3

Page 34

1 **a)** 60 million **b)** 9.9 million
c) 12.89 million **d)** 1.342 million
e) $\frac{1}{2}$ million **f)** 0.234 million

2 **a)** 10^6 **b)** 10^2 **c)** 10^4
d) 10^5 **e)** 10^1

3 **a)** 4.9×10 **b)** 5×10^2
c) 1.8×10^3 **d)** 1.42×10
e) 1.65×10^5 **f)** 7×10^{-1}
g) 9.5×10^{-3} **h)** 4×10^{-2}
i) 8.32×10^{-4} **j)** 7.6×10^{-5}
k) 6×10^9 **l)** 3×10^6

4 **a)** 1200 **b)** 300
c) 800 000 **d)** 2 590 000
e) 91 000 **f)** 0.5
g) 0.000 473 **h)** 0.036
i) 0.0018 **j)** 0.000 0207

Page 38

1 **a)** 1 **b)** 2 **c)** 5
d) 5 **e)** 0

2 **a)** 3 **b)** 4 **c)** 5
d) 3 **e)** 2 **f)** 5

3 **a)** 4.57 **b)** 4.57345 **c)** 4.5735
d) 4.573453 **e)** 4.6

4 **a)** 0.04 **b)** 0.043827 **c)** 0.0438
d) 0.044 **e)** 0.0

5 **a)** 126.35 **b)** 31.43 **c)** 0.32
d) 0.05

6 **a)** 6.514 **b)** 0.394 **c)** 231.146
d) 0.003

7 **a)** 10 000 **b)** 13 000 **c)** 12 600
d) 12 623.4 **e)** 12 623.37

8 **a)** 0.043827 **b)** 0.043827 **c)** 0.04383
d) 0.0438 **e)** 0.044 **f)** 0.04

9 **a)** 1370 **b)** 128 **c)** 0.054

10 **a)** 0.1 **b)** 4.9 **c)** 0.3

Page 40

1 **a)** ± 4 **b)** ± 40 **c)** ± 2
d) ± 20 **e)** ± 13

2 **a)** 2 **b)** 5 **c)** 4
d) 6 **e)** 8

3 **a)** ± 6.325 **b)** ± 12.649 **c)** ± 2.530
d) ± 1.3 **e)** ± 4.111

4 **a)** $243^{\frac{1}{5}}$ **b)** $b^{\frac{1}{10}}$ **c)** $g^{\frac{1}{2}}$
d) $h^{\frac{1}{3}}$ **e)** $100^{\frac{1}{6}}$

5 **a)** $\frac{1}{a^5}$ **b)** p^2q^3 **c)** $\frac{1}{c^3}$
d) $\frac{1}{49^{\frac{1}{2}}}$ **e)** $\frac{c^4}{b^3}$

6 **a)** d^5 **b)** $\frac{1}{f^{22}}$ **c)** $\frac{1}{y^8}$
d) q^3 **e)** r^5s **f)** $128u^7$

Looking back (page 41)

1 **a)** x^{11} **b)** y^5 **c)** $4x^5$
d) $2a^6$ **e)** $30d^6$ **f)** $45s^7$
g) $3x^5y^4$ **h)** $36a^4b^5$ **i)** $x^2y^2z^2$
j) $24x^4yz^5$ **k)** $21m^2n^2$ **l)** x^8
m) $27a^6$ **n)** $64x^{12}$ **o)** $2m^9$

2 **a)** x^2 **b)** $2a$ **c)** $5xy^2$
d) a^7 **e)** $\frac{4}{x^2}$

3 **a)** $m = \frac{1}{2}$ **b)** $m = 0$ **c)** $m = -3$
d) $m = \frac{2}{3}$ **e)** $m = 3$

4 **a)** 16 **b)** x^2 **c)** $m^{\frac{5}{2}}n^{\frac{3}{2}}$
d) $3x$ **e)** $2b^{\frac{1}{3}}$

Chapter 3

Page 43

1 **a)** 175.8 cm **b)** 31.4 cm **c)** 62.8 cm
d) 241.8 cm

2 **a)** 43 cm **b)** 21 cm **c)** 85.98 m
d) 64.5 cm **e)** 41.1 m **f)** 30.9 cm

3 **a)** Johnathan 188.4 m; Jeremy 157 m
b) Johnathan, by 628 m

4 64.3 cm

Page 46

1 **a)** 10.5 cm **b)** 18.3 cm
c) 11.0 m **d)** 11.3 cm

2 **a)** The length of an arc with central angle 240° is $\frac{2}{3}$ of the circumference.
b) The length of an arc with central angle 18° is $\frac{1}{20}$ of the circumference.

3 **a)** 383.5 cm **b)** 29.1 m **c)** 222.6 cm

4 31.4 m

5 144°

Page 48

1 **a)** 5 m² **b)** 25 m² **c)** 3.57 m²
d) 0.825 m² **e)** 0.0575 m²

2 **a)** 5 cm² **b)** 25 cm² **c)** 725 cm²
d) 82.54 cm² **e)** 3.75 cm²

3 **a)** 500 mm² **b)** 6 500 mm² **c)** 275 mm²
d) 670 mm² **e)** 22 mm² **f)** 30 mm²

4 **a)** 70 000 cm² **b)** 750 000 cm² **c)** 35 000 cm²
d) 5 000 cm² **e)** 3 475 cm²

5 **a)** 25 mm, 18 mm
b) **i)** 4.5 cm² **ii)** 450 mm²

6 **a)** 52 cm² **b)** 48 cm² **c)** 11.25 cm²

7 **a)** 33 cm² **b)** 75 cm² **c)** 59.99 cm²
d) 323.96 cm²

8 128

Page 51

1 6.5 cm^2

2 a) $\frac{1}{5}$ b) $\frac{1}{3}$ c) $\frac{5}{18}$ d) $\frac{1}{6}$

3 a) 39.3 cm^2 b) 9.4 cm^2 c) 0.5 m^2
d) 4817.1 mm^2 e) 116.1 cm^2 f) 16.7 m^2

4 772.7 km^2

Page 53

1 a) 30 b) 72 c) 48

2 a) 252 cm^3 b) 253 cm^3 c) 393.6 cm^3

3 a) 768 cm^3 b) 168 000 cm^3

4 5 m^3

5 Students' own answers

6 a) 3 cm^3 b) 5 cm^3 c) 4 cm^3
d) 8 cm^3 e) 20 cm^3 f) 16 cm^3

7 22 m^3 8 216 cm^3

9 6144 cm^3 10 245 cm^3

11 135

Page 56

1 a) i) 280 ml ii) 150 ml iii) 335 ml
iv) 29 ml v) 500 ml vi) 2.5 l
vii) 24 000 l viii) 200 l
b) i, ii, iii, iv c) vi, vii, viii d) 80 e) 120

2 32 3 10 cm

4 125 5 62.5 cm

6 12

Page 58

1

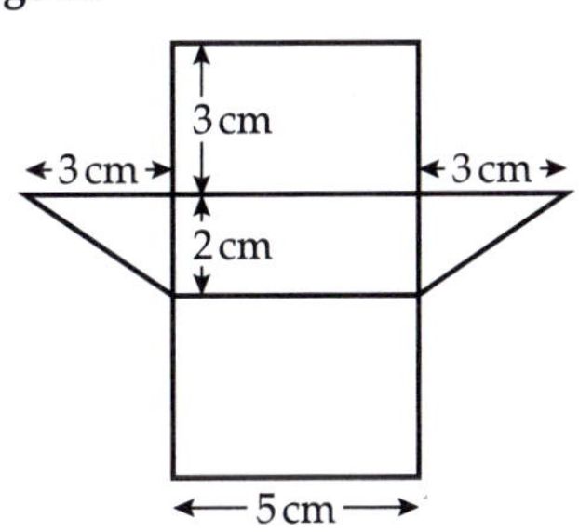

2

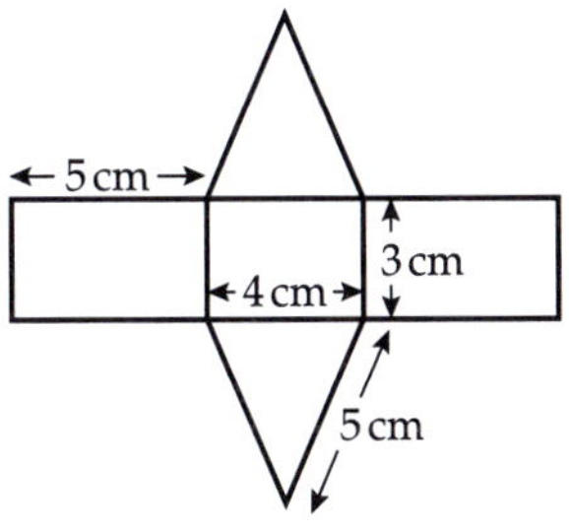

3 Right-angled triangular prism

4

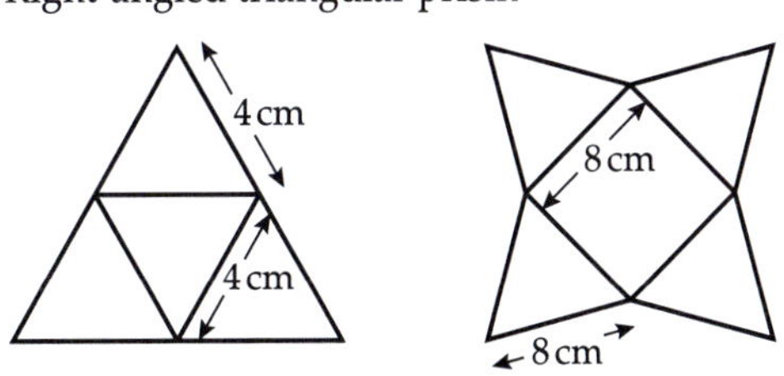

5 Answers will vary.

6 Answers will vary.

Page 60

1 Volume = 540 cm^3, surface area = 600 cm^2

2 a) i) 1600 cm^3,
ii) 70 cm^3.
iii) 3780 cm^3
iv) 182 cm^3
b) i) 880 cm^2
ii) 125 cm^2
iii) 1536 cm^2
iv) Surface area = 213 cm^2

3 a) 8.75 m^3 b) 20.75 m^2

4 942.5 cm^3 Surface area = 2 × circumference × height + 2 × area of ends

5 Base 20 cm square, height 10 cm

6–8 Answers will vary.

Looking back (page 61)

1 39.2 cm

2 a) 103.6 m b) 15.9 cm

3 120.0 m^2

4 a) 209.2 m^2 b) 44.51 mm^2
c) 493.3 cm^2 d) 2241.7 mm^2

5 0.168 m^3 6 0.0747 m^3

Chapter 4

Page 64

1 a) $12x + 20$ b) $42x - 18$
c) $nx + 2n$ d) $a^2 + 4a$
e) $b^2 - 2b$ f) $3r^2 - 5r$
g) $3y^2 + zy - 3y$ h) $2p^2 + 2pr + 4p$

2 a) $x + 12$ b) $5x + 2$ c) $x^2 + 9x$
d) $11x - 18$ e) $x^2 - x + 2$ f) $2x^2 - 7$

3 b) and d)

4 Each row, column and diagonal sums to $9a^2$.

5 a) The width is five more than the height
b) The height is three less than the width
c) $a(a + 5)$ d) $a(a - 3)$
e) $a(a + 5) - a(a - 3)$ f) $8a$
g) 32 cm^2

Page 66

1 a) $\frac{11}{21}$ b) $\frac{42}{55}$ c) $\frac{7}{30}$
d) $\frac{61}{63}$ e) $\frac{38}{120} = \frac{19}{60}$ f) $\frac{7}{30}$

2 a) $\frac{2 + x}{2x}$ b) $\frac{8 + 3m}{4m}$ c) $\frac{45 - p}{5p}$
d) $\frac{8y - 18}{9y}$ e) $\frac{j + 6}{2j}$ f) $\frac{90 - 9a}{10a}$

3 a) $\frac{5a + 12}{15}$ b) $\frac{9b - 21}{63}$ c) $\frac{5c - 36}{45}$
d) $\frac{63 - 12d}{84}$ e) $\frac{11 + 5e}{55}$ f) $\frac{f - 1}{3}$

4 a) $\frac{3x + 2y}{6}$ b) $\frac{3a - 5p}{15}$ c) $\frac{4q + 7w}{28}$
d) $\frac{3b + a}{ab}$ e) $\frac{4d - 5c}{cd}$ f) $\frac{11f - 12e}{ef}$
g) $\frac{6 + xw}{2w}$ h) $\frac{qm - 45}{5q}$ i) $\frac{8q + 7p}{7q}$

5 a) $\frac{1}{x}$ b) $\frac{1}{y}$ c) $\frac{y+x}{xy}$ d) $\frac{19}{60}$

6 a) $\frac{x}{2}$ b) x c) $\frac{x}{4}$ d) $\frac{5x}{4}$

7 The correct answer is Carla's.
Alexandra adds the tops and the bottoms together.
James multiplies the tops together rather than adding.
Glyn multiplies the tops and adds the bottoms.

Page 68

1 a) $6a + 9 = 3(2a + 3)$
b) $5(3b - 4) = 15b - 20$
c) $32c - 16 = 16(2c - 1)$
d) $18d + 24 = 6(3d + 4)$
e) $11(e + 5) = 11e + 55$
f) $24f - 60 = 12(2f - 5)$
g) $20(2g - 3) = 40g - 60$
h) $2h^2 + 5h = h(2h + 5)$

2 a) $10a + 25 = 5(2a + 5)$
b) $22c - 110 = 11(2c - 10)$
c) $4(5e + 9) = 20e + 36$
d) $f(2f - 3) = 2f^2 - 3f$
e) $e(e - 4) = e^2 - 4e$
f) $m(3m^2 + m) = 3m^3 + m^2$

3 a) $3(x + 2)$ b) $4(q + 2)$
c) $5(r - 4)$ d) $10(p + 12)$
e) $3(2n + 3)$ f) $5(2m - 5)$
g) $7(3w - 10)$ h) $4(2 + 3x)$
i) $8(3 + 10t)$ j) $11(7 - y)$

4 a) $a(3a^2 + 5)$ b) $b(4b - 7)$
c) $c(1 + 9c)$ d) $d^2(3 + 4d^2)$
e) $e(7e^4 + 1)$ f) $f(2f^2 - 1)$
g) $g(15 - 17g^2)$ h) $h(h^2 + 1)$

5 a) $2(x^2 + 2x)$ or $x(2x + 4)$ or $2x(x + 2)$
b) 2 and $x^2 + 2x$ or x and $2x + 4$ or $2x$ and $x + 2$

6 a) i) $n + 1$ ii) $n + 2$ b) $3n + 3$
c) $3(n + 1)$ d) 7, 8 and 9

Page 70

1 a) $(x + 4)(x + 7)$ b) $(x + 3)(x + 6)$
c) $(x - 4)(x - 5)$ d) $(x - 7)(x + 11)$
e) $(x + 6)(x - 5)$

2 a) $(a + b)(a + 7) = a^2 + ab + 7a + 7b$
b) $m^2 - 6$
c) $9x - 9 = 9(x - 1)$

3 4 cm × 9 cm

Looking back (page 71)

1 a) $10x + 6$ b) $35x - 5$
c) $y^2 + 3y$ d) $a^3 + 3ab - 4a$

2 a) $\frac{3+n}{3n}$ b) $\frac{5x-21}{7x}$
c) $\frac{4b+7a}{ab}$ d) $\frac{2q+5p}{5q}$

3 a) $5(x + 3)$ b) $7(4n - 9)$
c) $a(3a - 5)$ d) $m(\frac{m}{2} + 3)$

4 a) $(e + 3)(e + 4)$ b) $(x + 9)(x + 2)$
c) $(a + 7)(a - 2)$ d) $(g - 3)(g - 3)$

Revise and Consolidate 1

Page 72

1 a) $1\frac{3}{20}$ b) $\frac{1}{18}$ c) $3\frac{14}{15}$ d) $1\frac{11}{12}$

2 a) $\frac{4}{7}$ b) $1\frac{1}{2}$ c) $\frac{2}{3}$ d) $\frac{1}{3}$

3 a) True b) True c) False
d) False e) True

4 a) i) 66 cm ii) 181.5 cm^2 iii) $166\frac{3}{8}$ cm^2
b) 138 cm^3

5 i) 10^6 ii) 10^3 iii) 10^{-2}

6 i) 100 ii) 0.1 iii) 7200

7 a) 4 000 000 b) 0.027
c) 0.0005 d) 93

8 a) $8a^5$ b) $5b$ c) $2t^{-1}$ d) $24m^6$

9 a) 125 b) 71 c) 2 d) 400

10 a) 154 cm^2 b) 28.3 cm^2

11 90 cm

12 a) i) 0.5 ii) $0.\dot{5}1\dot{8}$
iii) 0.04 iv) 3.14 to 2 d.p.

13 a) 1306.24 mm^2 b) 1698.112 cm^3
c) i) 1700 cm^3 ii) 1698.1 cm^3
d) 2020 g

14 a) i) $6mp \times 10^8$ ii) $2p \times 10^3$
b) i) 900 000 000 ii) 1000

15 a) 3 b) $3(x + 2)$ c) $5(3y - 4)$
d) $12(4 + x)$ e) $a(a + 2)$ f) $t(3 + 4t)$

16 a) $\frac{3+2x}{3x}$ b) $\frac{3d-5}{6}$
c) $\frac{4x+3y}{12}$ d) $\frac{b-3a}{ab}$
e) $\frac{24+pr}{6r}$ f) $\frac{ac+2b}{bc}$

17 a) $3\frac{7}{12}$ b) $\frac{3a^4+4ab^2}{a^2b}$
c) 12 d) ab
c) $\frac{1}{3}$ f) $\frac{y^2+2x}{2y}$

18 a) ii) and iii)
b) i) $0.\dot{1}$ ii) $0.\dot{2}\dot{3}$ iii) $0.3\dot{4}$
c) i) $\frac{1}{9}$ ii) $\frac{23}{99}$ iii) $\frac{31}{90}$

19 a) $24a^9$ b) 36^{-3} c) 2^{12}
d) t^{-1} e) $3a^2b^6$

20 a) $a^2 + 9a + 18$ b) $b^2 - 2b - 15$
c) $2c^2 - 11c + 12$ d) $9d^2 - 24d + 16$

21 a) $3(x + 3y)$ b) $2x(2x + 3)$
c) $x^2(5x + 7)$ d) $2x^2(x + 5x^2 + 2)$

Chapter 5

Page 75

1 a) $x = 2, y = 7$ b) $x = -2, y = -4$
c) $x = \frac{5}{7}, y = 3\frac{1}{7}$ d) $x = \frac{2}{3}, y = 2\frac{1}{3}$
e) $x = 3, y = -3$ f) $x = \frac{3}{7}, y = -\frac{9}{7}$

2 A: $y = 2, y = 1 - x$
B: $x = 3, y = 2$
C: $y = 2, y = x - 2$
D: $y = 1 - x, y = x - 2$
E: $x = 0, y = x - 2$
F: $x = 3, y = 1 - x$

3 $x = -2, y = 1$ and $x = 1, y = 4$

Page 77

1 a) $x = 9, y = 6$ b) $x = -7, y = 10$
c) $x = 1, y = 2$ d) $x = -7, y = 6$
e) $x = 3, y = 2$ f) $x = 1, y = 3$
g) $x = -\frac{2}{7}, y = 2\frac{1}{7}$ h) $x = -\frac{5}{4}, y = -2$

2 Substitution, because the terms have a common factor.

Page 79

1 d, f, g

2 a)

x	−3	−2	−1	0	1	2	3
$x^2 + 1$	10	5	2	1	2	5	10
x^2	9	4	1	0	1	4	9
$x^2 - 1$	8	3	0	−1	0	3	8

b)

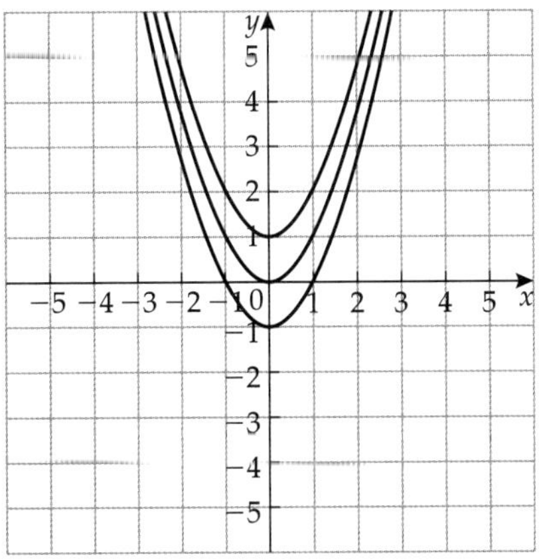

c) 0, 1, 2

3

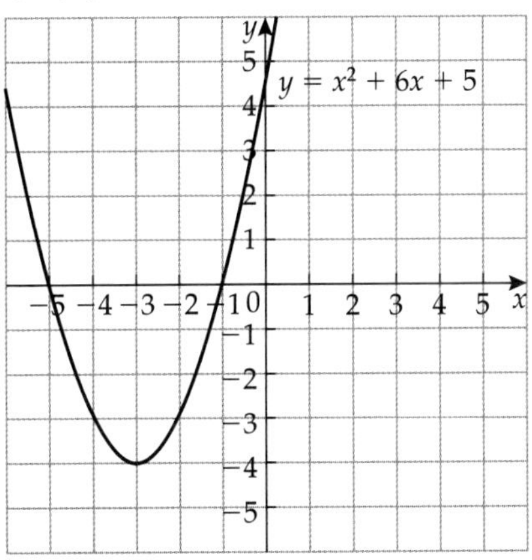

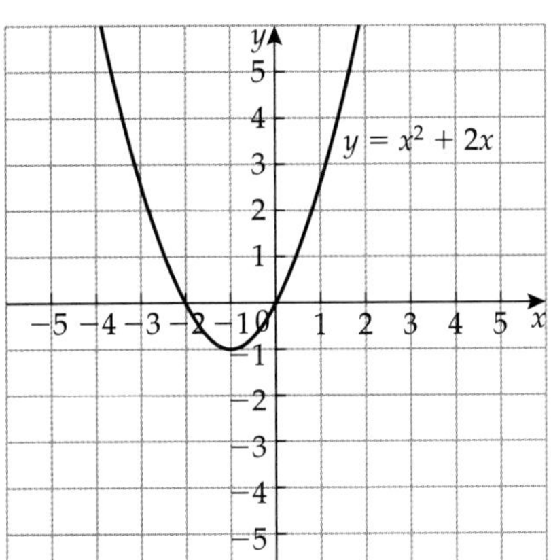

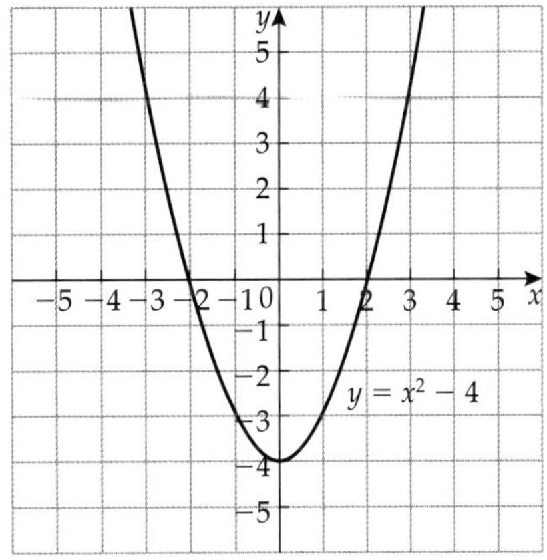

Page 82

1 a) $(x - 3)^2$ b) $x(x - 12)$
c) $x(x - 9)$ d) $x(x + 10)$

2 a) 0, 2 b) 0, −5 c) 0, 1 d) 1, −1

3 a)

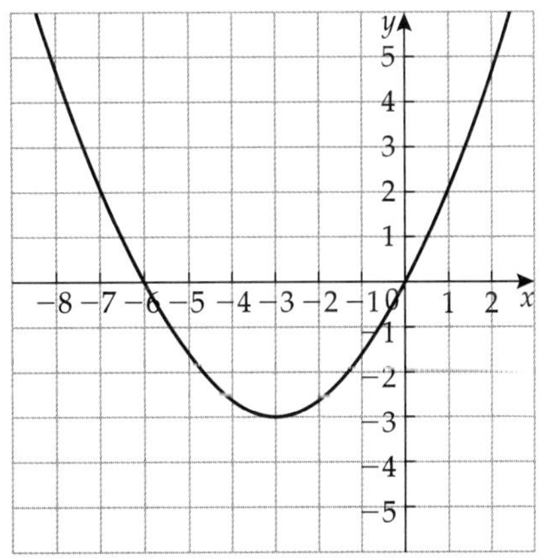

b)

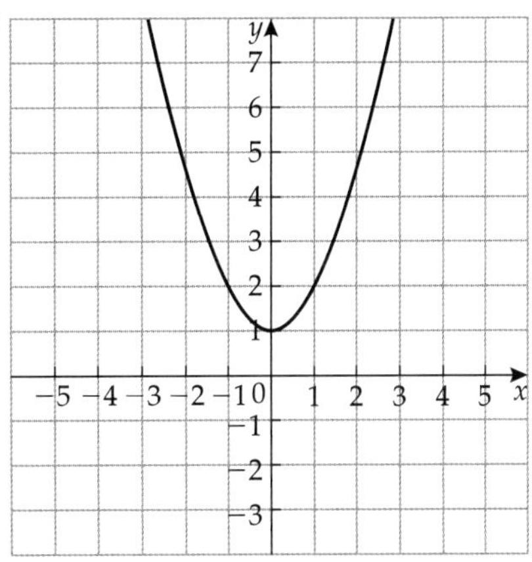

c)

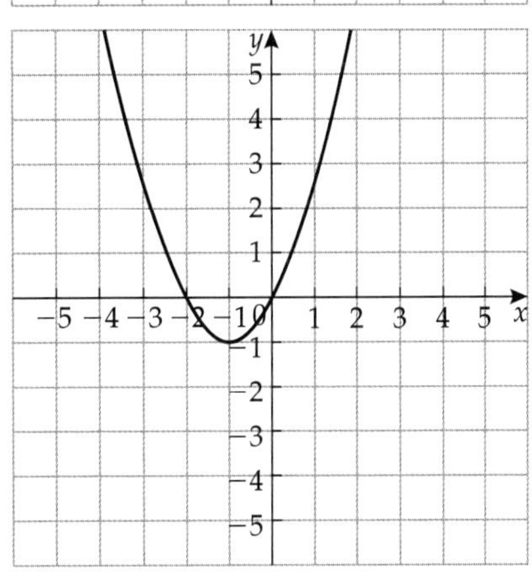

4 **a)** $y = x^2 - 2x$ **b)** $y = x^2 + 3x$ **c)** $y = x^2 - 4$

Looking back (page 83)

1 **a)** $x = 2, y = 13$ **b)** $x = -1, y = 4$
c) $x = -24, y = -6$
2 $9.40
3 **a)** 1, 10 **b)** 5, −2 **c)** 2, 6 **d)** 0, −3
4 x-intercepts at
a) 4 **b)** 3 **c)** −4 **d)** −2 **e)** 9

Chapter 6

Page 86

1 **a)** actual profit = $5, percentage profit = 12.5%
b) actual profit = $5, percentage profit = 10%
c) actual profit = $0.05, percentage profit = 2.86%
d) actual profit = 25 cents, percentage profit = 50%
2 **a)** 20% **b)** 20% **c)** 30% **d)** 83.28%
3 100% profit
4 $540
5 $532
6 $425
7 $283.48
8 **a)** $300 **b)** $400 **c)** $1 020 **d)** $1 800

Page 88

1 **a)** $12 **b)** $247.5 **c)** $216
d) $704 **e)** $67 500
2 5 years
3 7%
4 **a) i)** $2 550 **ii)** $7 140 **iii)** $14 152.50
b) i) $348.61 **ii)** $976.11 **iii)** $1 934.79
5 **a)** 1 year = $8 569.92, 2 years = $9 119.04,
3 years = $9690.12, 4 years = $10 282.56,
5 years = $11 020.20, 6 years = $11 684.88,
7 years = $12 371.52
b) 1 year = $569.92, 2 years = $1 119.04,
3 years = $1690.12, 4 years = $2 282.56,
5 years = $3 020.20, 6 years = $3 684.88,
7 years = $4 371.52
c) Students' own answers
d) Students' own answers

Page 91

1 $33 868.80
2 **a)** $53.42 **b)** $307.94 **c)** $742.34
3 **a)** $601.14 **b)** $1 458.66 **c)** $1 458.66
d) Students' own answers

Page 93

1 **a)** $80 960
b) $62 695 to the nearest dollar
2 **a)** $823 to the nearest dollar
b) $697 to the nearest dollar
c) 9 years
3 83.1%
4 20%
5 **a)** The bike depreciates by 20% of its value at the start of each year. Its value will never depreciate to zero.
b) $295 to the nearest dollar

Page 95

1 **a)** $28 800 **b)** $3600; 12.5%
2 **a)** 2500 **b)** $95 000 **c)** 5% loss
3 James's investment is worth the same as Lewis's; because they have both invested the same amount of money.
4 End of year 1: $24 700; End of year 2: $23 465
5 $32 620

Looking back (page 96)

1 **a)** $2040
b) $1253 to the nearest dollar
c) 3 years and 6 months
2 5% per year
3 **a)** $84 000 **b)** $15 960 **c)** 19%

Chapter 7

Page 99

1 **a)** 5 m **b)** 25 cm **c)** 5 cm
d) 9 cm **e)** 3.5 m
2 14.14 m
3 25 cm
4 12.5 cm
5 **a)** $1^2 + 2^2 < 3^2$
c) $4^2 + 4^2 < 8^2$
d) hypotenuse not the longest side
6 **a)** 5.16 m **b)** 23.6 cm
7 **b)** 5.29 m
8 **b)** 5 m
c) i) 5.83 m **ii)** 7.68 m

Page 104

1 **a)** 0.344 **b)** 3.487 **c)** 0.466
d) 1.732 **e)** 1.664 **f)** 1
2 **a)** 35.7° **b)** 89.1° **c)** 54.5°
d) 40.2° **e)** 0.4°
3 **a)** 0.203 **b)** 1.103 **c)** 1.477
d) 0.479 **e)** 0.139
4 $a = 51.3°, b = 79°, c = 82.5°$
5 $n = 8.34$ m, $p = 18.81$ m, $q = 68.69$ m

Page 106

1 **a)** 0.766 **b)** 0.707 **c)** 0.951
d) 0.866 **e)** 0.999 **f)** 1
2 **a)** 0.579 **b)** 0.462 **c)** 0.955
d) 0.876 **e)** 0.148
3 **a)** 0.956 **b)** 0.996 **c)** 0.515
d) 0 **e)** 0.866 **f)** 0.707
4 **a)** 0.340 **b)** 0.995 **c)** 0.540
d) 0.805 **e)** 0.698
5 $a = 48.2°, b = 23.6°, c = 23.6°, d = 36.9°, e = 48.6°$
6 $p = 74.31$ m, $q = 7.51$ cm, $r = 6.18$ m, $s = 6.18$ cm, $t = 46.82$ cm
7 **a)** 10.6 cm **b)** 21.2 cm
8 **a)** 3.38 cm **b)** 7.25 cm

Page 110

1 1.93 m
2 25.73 m
3 4.663 m

4 a) 10.72 m b) 41.41 m
5 a) 2.5 km b) 41.1°
6 a) 8 m b) 13.86 m
7 Sissy 73.86 m, Joan 77.67 m
8 AB = AD = 7.62 cm, DC = BC = 4.24 cm, ∠ABC = ∠ADC = 111.7°, ∠DCB = 90°, ∠DAB = 46.4°

Looking back (page 112)

1 $a = 7.7$ m, $b = 103.3$ mm, $c = 111.2$ m, $d = 17.8$ cm, $e = 14.7$ cm
2 48.1 cm
3 a) $\frac{b}{a}$ b) $\frac{c}{a}$ c) $\frac{b}{c}$
d) $\frac{c}{a}$ e) $\frac{b}{a}$ f) $\frac{c}{b}$
4 a) 8.7 cm b) 5 cm
5 a) 30° b) 60°
6 84.46 cm (2 d.p.) 7 b) 49.81 m

Chapter 8

Page 114

1 a) pentagon b) triangle
c) 7-sided polygon d) quadrilateral
e) square
2 a) yes, all sides and angles equal
b) no, angles not marked equal
c) yes, equilateral triangle
d) no, angles not equal
3 a) 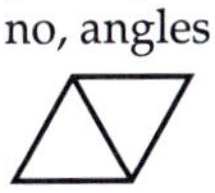 b) 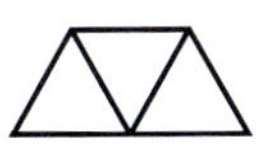
c) a straight line d) regular hexagon
4 a) isosceles trapezium
b) Joining E to A and E to D gives three equilateral triangles.

Page 116

1 a) $b = 110°$, angles on a straight line sum to 180°, $a = 35°$, angles in a quadrilateral sum to 360°
b) $c = 115°$, angles in a pentagon sum to 540°, $d = 65°$, angles on a straight line sum to 180°
c) $e = 136°$, angles in a hexagon sum to 720°, after calculating interior angles as 95° and 137°, using angles on a straight line sum to 180°
2 a) 40°, angles on a straight line sum to 180°
b) 85°, angles of a triangle sum to 180°
c) 95°, angles on a straight line sum to 180°
d) 85°, corresponding angle to ∠ADE
e) 145°, angles of a quadrilateral sum to 360°
3 a) $x = 115°$, base angle of the isosceles triangle is $\frac{1}{2}(180°-50°) = 65°$, angles on a straight line sum to 180° so the angle opposite x is 115°, opposite angles of a parallelogram are equal so x is found.
b) $x = 70°$, top angle of the right-angled triangle = 40° because alternate angles are equal, the angle of the isosceles triangle adjacent to x is 70°, angles on a straight line sum to 180° so x is found.
4 a) $x = 145°$, find the base angle of the isosceles triangle, subtract from 90° angle, then supplementary parallelogram angles
b) $x = 255°$, top angle of the right-angled triangle = 30° because alternate angles are equal, the angle of the isosceles triangle adjacent to x is 75°, angles at a point sum to 360° so x is found.
5 Exterior angles always sum to 360°.
Proof: Interior angle sum = $180°(n - 2)$, so
exterior angle sum = $n\left(180° - \frac{\text{interior angle sum}}{n}\right)$,
which cancels to 360°.
6 Always two triangles less than the number of sides.

Page 119

1 135°
2 Exterior = 20°, interior = 160°
3 a) Exterior = 45°, interior = 135°
b) Exterior = 24°, interior = 156°
c) Exterior = 15°, interior = 165°
d) Exterior = 6°, interior = 174°
4 20 sides
5 18
6 12
7

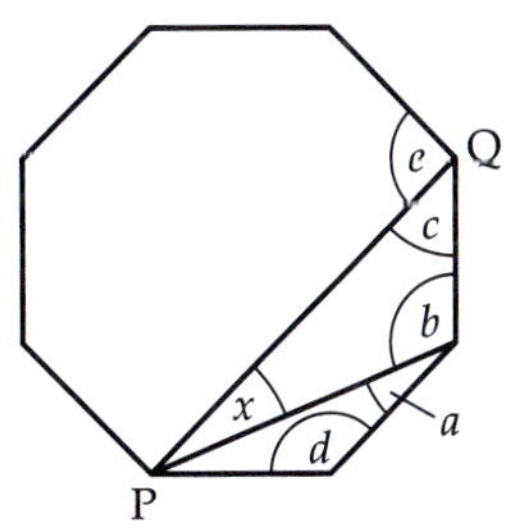

Interior angle of a regular octagon = $180 - \frac{360}{8} = 135°$
$\angle d = 135°$, interior angle
$\angle a = \frac{180° - 135°}{2} = 22.5°$, base angle of an isosceles triangle
$\angle e = 90°$, as this is a regular octagon
$\angle c = 135° - 90° = 45°$
$\angle b = 135° - 22.5° = 112.5°$
$\angle x = 180° - 112.5° - 45° = 22.5°$
It is possible to calculate all the angles at P.

Page 122

1 a) 3.9 cm, 56°, 84° b) 5 cm, 60°, 60°
c) 7.6 cm, 23°, 67°
2 student constructions
3 a) 3.4 cm b) 5.8 cm, 59° c) 7 cm
4 5.7 cm
5 13.6 cm

Page 124

1 7 cm 2 4.8 cm
3 6.1 cm
4 a) student construction b) 6 cm
c) mark a point on the circle then with radius set to 6 cm, mark equal divisions round the circle.
5 a) equilateral b) student construction
6 4.6 cm

Looking back (page 125)

1 **a)** 13.2 cm **b)** 55.7 m

2 Accurate construction, 7.1 cm; 9.7 cm; (allow +/−0.4 cm)

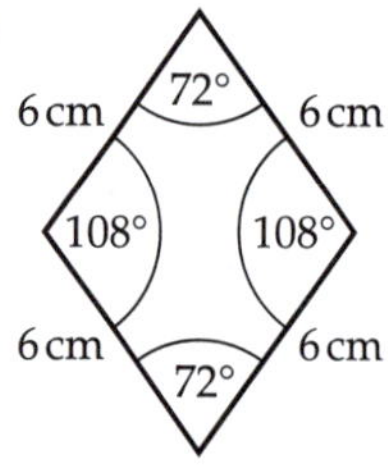

3

Number of sides (n)	Radius of circle (cm)	Perimeter of regular polygon (cm)
3	5	26.0
4	5	28.3
5	5	29.4
6	5	30.0
8	5	30.6

Chapter 9

Page 128

1

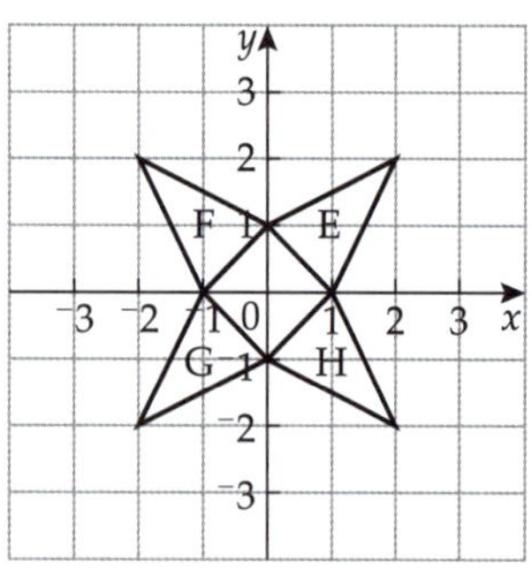

d) i) 90° **ii)** 180° **iii)** 90°

2 **a), b)**

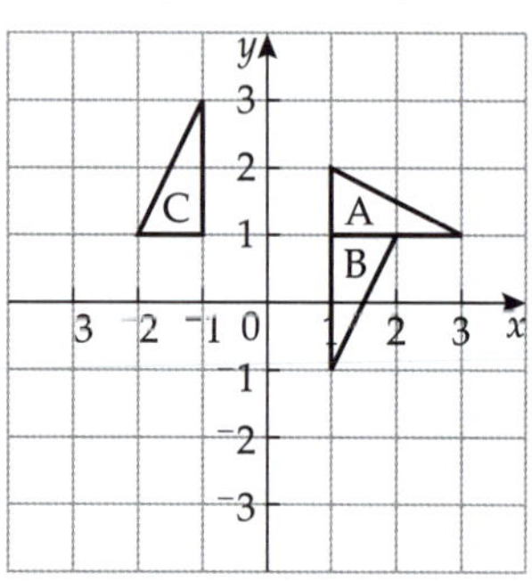

c) A(1, 1), (3, 1), (1, 2)
B(1, 1), (2, 1), (1, −1)
C(−1, 1), (−2, 1) (−1, 2)

3 **d)** star

Page 131

1 **a)** 3 **b)** 2 **c)** 1 **d)** 4 **e)** 3 **f)** 1
2 **a)** 4 **b)** 2 **c)** 1 **d)** 3 **e)** 1 **f)** 6
2 **a)** 1 **b)** 2
3 **a)** 3

b)

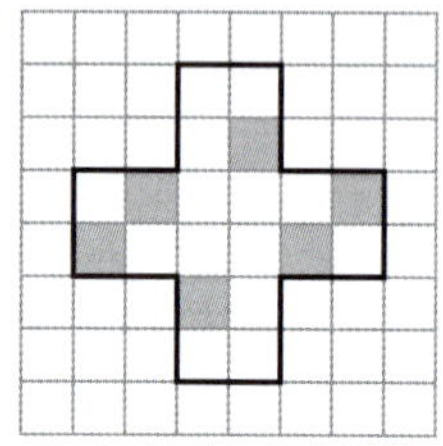

c)

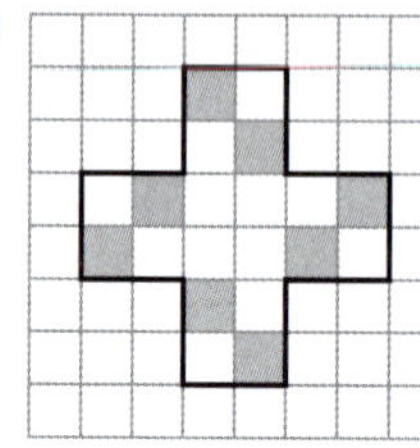

4 **a)**

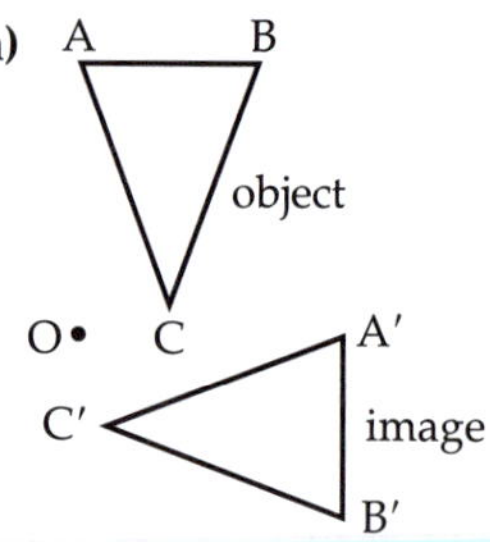

b)

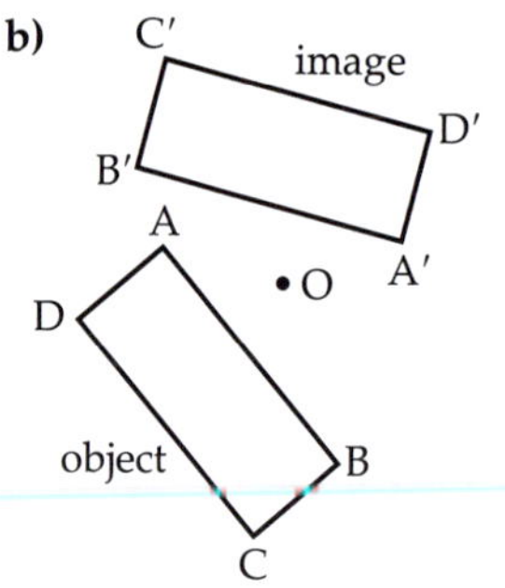

c)

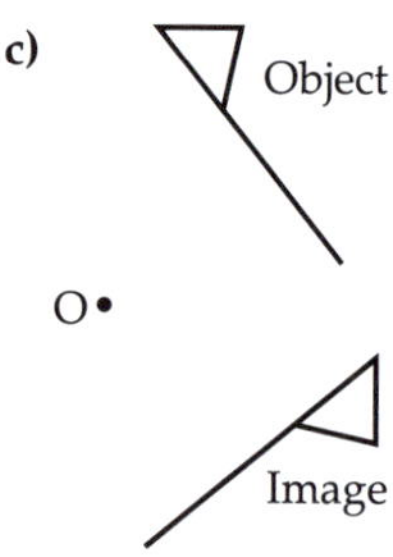

5 **a)** Rotation 270° (or 90° clockwise) with centre of rotation (4, 4)
b) Rotation 90° with centre of rotation (3, −1)
c) Rotation 180° with centre of rotation (1, 0)

Looking back (page 133)

1 **a)** 5 **b)** 6 **c)** 3
2 Rotation 270° (or 90° clockwise) with centre of rotation (2, 2)
3 **a)** parallelogram **b)** rectangle

Chapter 10

Page 135

2 Enlargement centre (0, 0), scale factor 2
3 Enlargement centre (0, 0), scale factor 4
4 **b)** They are in a straight line.
5 **a)** Centre A, scale factor 2
b) Centre J, scale factor 3

Page 138

1 **a)** A′B′ = 8 cm, B′C′ = 10 cm, C′D′ = 20 cm, D′A′ = 10 cm

2 **i) a)** 1 : 2
b) A = $\frac{1}{2}$B
c) scale factor 2
d) 12 cm

ii) a) 1 : 5
b) A = $\frac{1}{5}$B
c) scale factor 5
d) 25 cm

iii) a) 1 : 3
b) A = $\frac{1}{3}$B
c) scale factor 3
d) 60 cm

3 **i)** 32 cm **ii)** 24 cm **iii)** 10 cm

4 No, because one side is enlarged by a scale factor of 4, the other side is enlarged by a scale factor of 3.

5 **a)** Yes: Corresponding sides in same ratio
b) No: Corresponding sides not in same ratio
c) Yes: Angles of one triangle equal to angles of other
d) No: Corresponding sides not in same ratio
e) Yes: Two corresponding sides in same ratio and included angles equal

6 **a)** No: Only one angle always equal
b) Yes: Angles always equal and sides always in same ratio
c) Yes: Angles always equal and sides always in same ratio
d) No: Angles not always equal

Page 141

1 **a)** $\frac{1}{2}$ **b)** 1.5 **c)** -2 **d)** $-\frac{1}{3}$

2

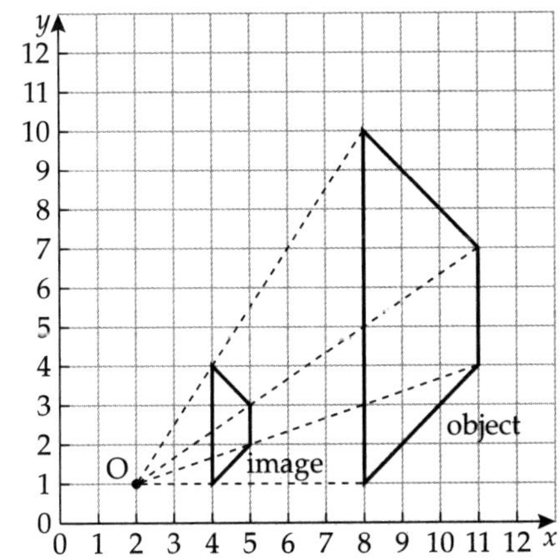

3

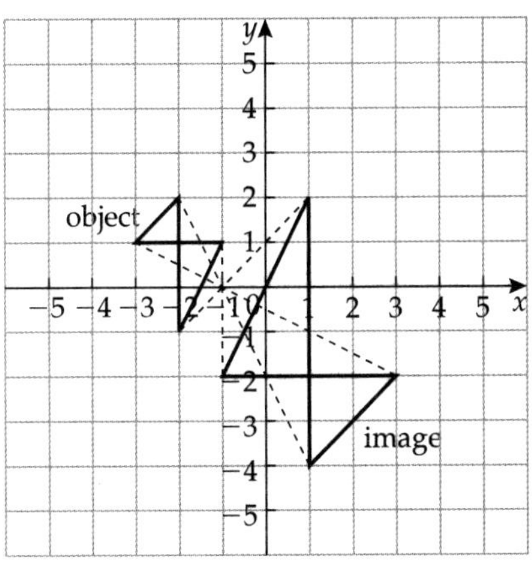

4

Scale factor	Side 1	Side 2	Side 3	Image rotated?
1	20 cm	10 cm	16 cm	No
-2	40 cm	**20 cm**	**32 cm**	Yes
$\frac{1}{4}$	**5 cm**	2.5 cm	**4 cm**	**No**
1.5	**30 cm**	15 cm	**24 cm**	No
-3.5	70 cm	**35 cm**	**56 cm**	**Yes**
$-\frac{1}{2}$	**10 cm**	**5 cm**	8 cm	Yes

5 14 cm

6 **a), b)**

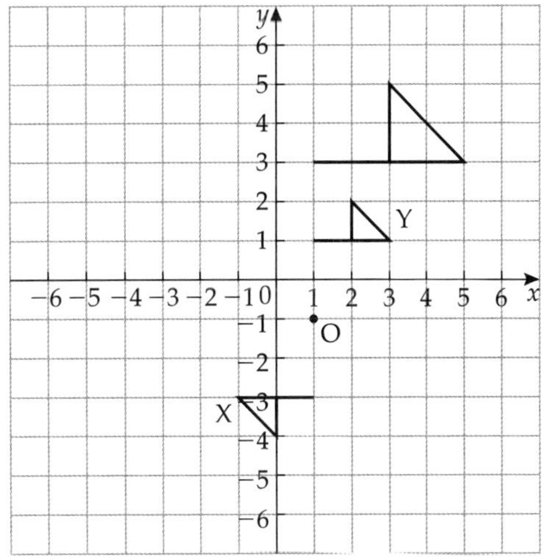

c) Rotation 180° with centre of rotation $(1, -1)$

7 A 180° rotation about a point O is the same as an enlargement with scale factor -1 and centre of enlargement O.

8 $\frac{1}{4}$

Page 144

1 **a)** 36 cm
b) Picture A = 64 cm, picture B = 192 cm, ratio = 1 : 3
c) The perimeter is based on the length of the sides and so the ratio of the perimeter changes with the ratio of the lengths.

2 **b)** $\angle A = 63°$, $\angle B = 90°$, $\angle C = 27°$
d) The angles are the same.
e) The sides are in proportion so the angles will be the same.

Looking back (page 145)

1 **a)** No: Corresponding sides not in same ratio
b) Yes: Two corresponding sides in same ratio and included angle equal
c) Yes: Angles of one triangle equal to angles of other

3 **a)–f)** yes

5

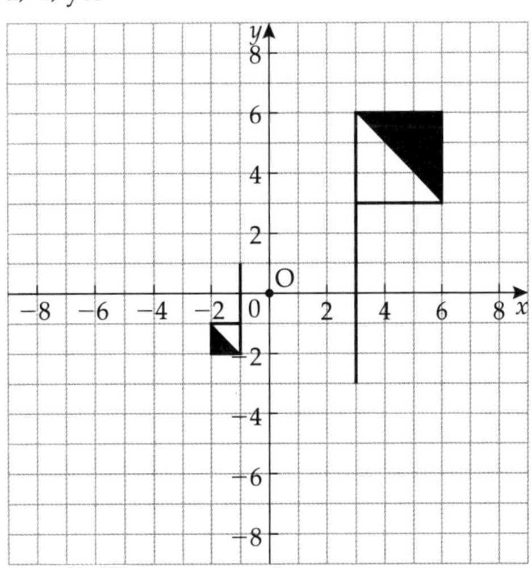

Revise and Consolidate 2

Page 149

1 a)

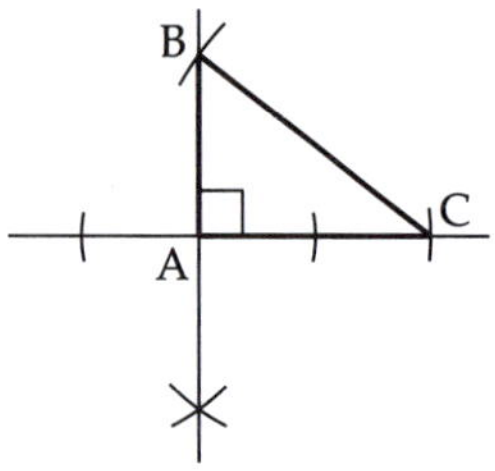

b) 3 cm

c) A = 90, B = 40°, C = 50°

2 a) \$42 **b)** \$15.67

3 \$20 700

4 a) 2 : 3 **b)** 1 : 1.5 **c)** 6.45 cm

5 a) $x = \pm 5$ **b)** $a = \pm 7$

c) $r = \pm 4$ **d)** $b = \pm 11$

6 a) $6x + 6 = 8x + 2$ **b)** $x = 2$

c) $\frac{7(3x + 2)}{2}$ **d)** 28 cm², 18 cm²

7 c) Rotation through 180°, centre origin

f) Reflection in x-axis

h) (0, 0) (4, 0) (0, 4)

8 a) $7x : 21x = 1 : 3$ **b)** 3

c) $3(x + 1)$ or $3x + 3$ **d)** $c = 35°$

e) A = 11.24, 71.68 B = 33.72, 215.04

9 \$365

10 a) 32.5 m **b)** 79 km² to 2 s.f.

11 a) $x = 4, y = 8$ **b)** $x = 3, y = 2$

c) $s = 4, t = 1$

12 a) i) 8.54 cm **ii)** 7.48 mm

b) No

13 a) 28% **b)** \$75

14 $\$1000 \times 1.04^n = \$1601.03$

$\$2^1 + \$2^2 \ldots = \$8192$

15 a) $9ab + 7b + a$ **b)** $15ab^2$

c) $a = \frac{5}{3}, b = \frac{5}{6}$

16 b) 2 **c)** $2 \times \text{sf}\,3 \times \text{sf}\,\frac{1}{2}$

d) 2 : 3 **e)** 2 : 3

f) 4 : 9 **g)** 15.3

17 6.9 m

18 69° – it is safe

19 a) 256 cm³ **b)** 266.4 cm²

20 d) 1 : 3 and 1 : 4

Chapter 11

Page 154

1 a) $A \cap B \cap C = \{\}$;
$A \cup B \cup C$ = {chair, table, stool, bed, dresser}

b) $A \cap B \cap C$ = {O, E, R};
$A \cup B \cup C$ = {B, O, T, H, E, R, L, D, Y, U, N, G}

c) $A \cap B \cap C$ = {E, L, I, A};
$A \cup B \cup C$ = {D, E, L, I, A, S, H, O, P}

d) $A \cap B \cap C = \{6\}$;
$A \cup B \cup C = \{1, 2, 3, 4, 5, 6, 7, 8, 9, 10, 12\}$

e) $A \cap B \cap C = \{\}$;
$A \cup B \cup C$ = {Donna, Sarah, Sharon, David, Jonathan, Sandra, Ricky}

2 a) i) {B, I, O, L, G, Y, H, S, T, R}

ii) {H, I, S, T, O, R, Y, C, E, M}

iii) {I, Y}

iv) {B, I, O, L, G, Y, H, S, T, R, C, E, M}

b) i) {G, R, A, P, E}

ii) {P, E, A, R, L}

iii) {A, P, E}

iv) {G, R, A, P, E, L}

c) i) {0, 2, 3, 4, 6, 8, 9, 10}

ii) {0, 1, 2, 3, 4, 6, 9}

iii) {6}

iv) {0, 1, 2, 3, 4, 6, 8, 9, 10}

d) i) {1, 2, 5, 10, 15, 20, 25}

ii) {1, 2, 4, 5, 9, 10, 16, 25}

iii) {25}

iv) {1, 2, 4, 5, 9, 10, 15, 16, 20, 25}

e) i) {1, 2, 3, 4, 5, 6, 7}

ii) {1, 2, 3, 5, 7}

iii) {3, 5}

iv) {1, 2, 3, 4, 5, 6, 7}

f) i) {1, 5, 7}

ii) {1, 2, 3, 6, 7, 14}

iii) {1, 7}

iv) {1, 2, 3, 5, 6, 7, 14}

g) i) {5, 10, 20, 25, 30, 40, 50}

ii) {5, 10, 20, 25, 30, 40, 50}

iii) {10, 50}

iv) {10, 20, 25, 30, 40, 50}

h) i) {1, 2, 3, 4, 6, 9, 10, 12, 15, 18}

ii) {1, 2, 3, 4, 6, 8, 9, 12, 18}

iii) {1, 3, 6}

iv) {1, 2, 3, 4, 6, 8, 9, 10, 12, 15, 18}

i) i) {2, 3, 5, 7, 11, 13}

ii) {1, 2, 3, 5}

iii) {3}

iv) {1, 2, 3, 5, 7, 11, 13}

j) i) {6, 8, 10, 12, 14}

ii) {6, 8, 12}

iii) {12}

iv) {6, 8, 10, 12, 14}

Page 156

1 a)

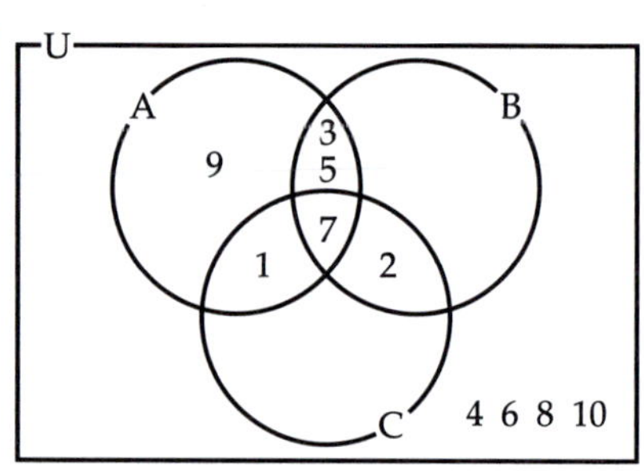

b)

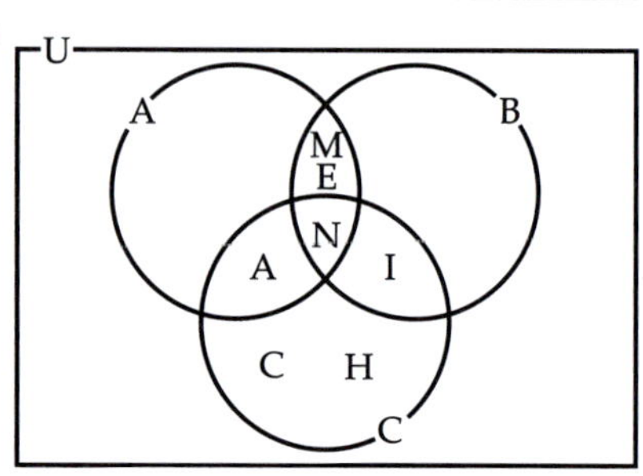

c)

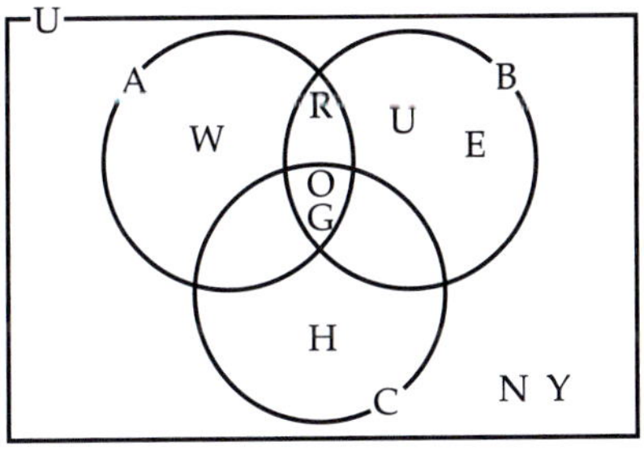

d)

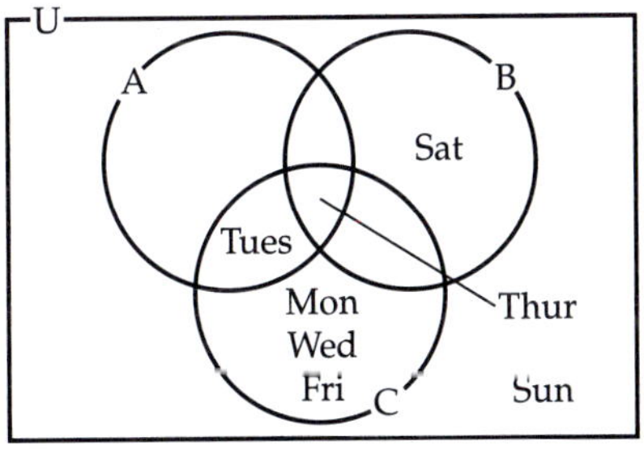

e)

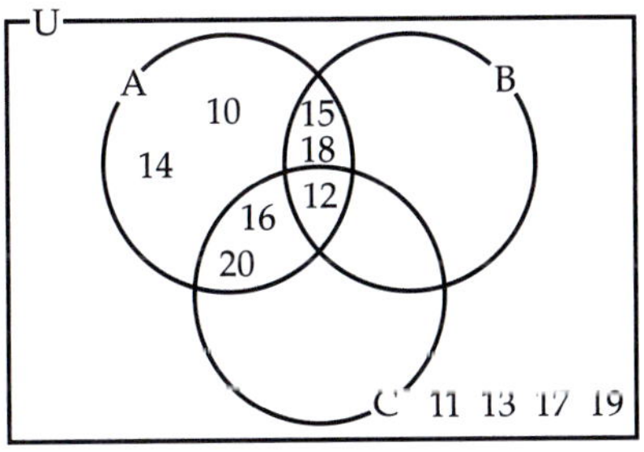

f)

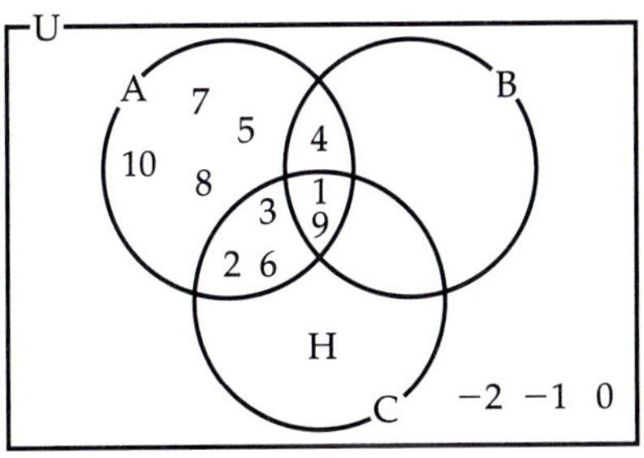

g)

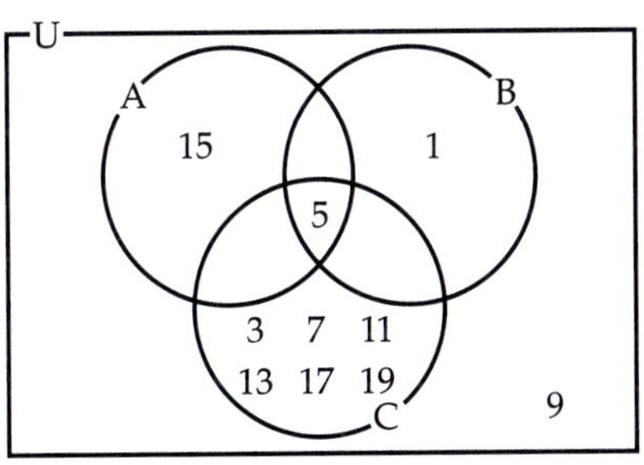

2 a) i) A ∪ B ∪ C = {1, 2, 3, 5, 7, 9}
ii) (A ∪ B ∪ C)′ = {4, 6, 8, 10}
b) i) A ∪ B ∪ C = {M, A, C, H, I, N, E}
ii) (A ∪ B ∪ C)′ = { }
c) i) A ∪ B ∪ C = {H, U, G, R, O, W, E}
ii) (A ∪ B ∪ C)′ = {N, Y}
d) i) A ∪ B ∪ C = {Monday, Tuesday, Wednesday, Thursday, Friday, Saturday}
ii) (A ∪ B ∪ C)′ = {Sunday}
e) i) A ∪ B ∪ C = {10, 12, 14, 15, 16, 18, 20}
ii) (A ∪ B ∪ C)′ = {11, 13, 17, 19}
f) i) A ∪ B ∪ C = {1, 2, 3, 4, 5, 6, 7, 8, 9, 10}
ii) (A ∪ B ∪ C)′ = {−2, −1, 0}
g) i) A ∪ B ∪ C = {1, 3, 5, 7, 11, 13, 15, 17, 19}
ii) (A ∪ B ∪ C)′ = {9}

3 a) June b) February

Looking back (page 157)

1 a) orange
b) apple, orange, grape, melon, banana, lemon
2 a) none; 1, 2, 3, 4, 5, 7, 9
b) 4; 1, 2, 3, 4, 6, 8, 9

Chapter 12

Page 159

1 a)–e) are functions.
2 a)+b) one-to-one c) many-to-one
d)+e) one-to-one
3 a) $f(x) = x^3$ b) $f(x) = 3 + x$
c) $f(x) = x^2 + x$ d) $f(x) = \frac{5}{x}$
e) $f(x) = 2 - x^2$ f) $f(x) = 2x - 11$
g) $f(x) = \frac{2}{x + 1}$
4

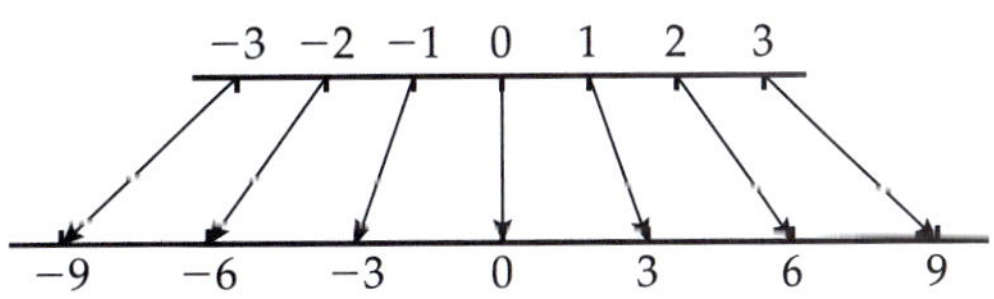

yes; one-to-one
5 a) 26 b) −1
c) 134 d) 10 034
6 a) 14 b) 21
c) 145 d) 1 000 020
7 a) −1, −125 b) 2, −2 c) 0, 20
d) −5, −1 e) 1, −23 f) −13, −21
g) ∞, $-\frac{1}{2}$

Page 162

1 a) $y = 3x + 1$ b) $y = -x + 10$
c) $y = x^3 + 6$ d) $y = x$
2 a) (1, 1) (2, 0.5) (3, $0.\dot{3}$) …
b) (−1, −10) (0, 0) (1, 8) …
c) (−1, −16) (0, −2) (1, 12) …
d) (−1, 15) (0, 11) (1, 7) …
4 a) (−1, −5) (0, −8) (1, −9)
b) Students' graphs
5 a) (−3, 8) (0, −1) (3, 26)
b) Students' graphs

Page 165

1 Students' graphs
2 a) $y = -4x + 2$ b) $y = x^2 - 9$
c) $y = x^2 - 3x - 4$
3 a) y-intercept = 8 b) roots are 0, 3
c) y-intercept = 15

Looking back (page 166)

1 **a)** 6 **b)** 24
c) 126 **d)** −6

2 **a)**

x	−4	−3	−2	−1	0	1	2	3	4
x^3	−64	−27	−8	−1	0	1	8	27	64

b) Students' graphs

3 Students' graphs

4 **a)** $y = x - 10$
b) $y = x^2 - 7x + 6$

Chapter 13

Page 168

1 **a)** $x \leqslant 11$
b) $y \geqslant -3$
c) $t \leqslant 27$
d) $m \geqslant 6$

2 **a)** −10, −9 … 1, 2, 3, 4
b) −4, −3 … 8, 9, 10
c) −10, −9 … 7, 8, 9
d) 0, 1, 2 … 8, 9, 10
e) −10, −9 … 0, 1, 2

3 **a)** $x < 4$
b) $x \leqslant 4$
c) $x \geqslant 4$

4 **a)** −2, −1, 0, 1, 2, 3, 4, 5; 3, 2, 1, 0, −1 …
b) $-2 < x < 5$ $x < 3$

5 $x < 4$, $x \leqslant 4$, $-2 < x \leqslant 4$

Page 171

1 A: $x \leqslant 0, y \geqslant 0$
B: $0 \leqslant x \leqslant 3, y \geqslant 0$
C: $x \geqslant 3, y \geqslant 0$
D: $x \leqslant 0, y \leqslant 0$
E: $0 \leqslant x \leqslant 3, y \leqslant 0$
F: $x \geqslant 3, y \leqslant 0$

2 Answers will vary

3 **a)**

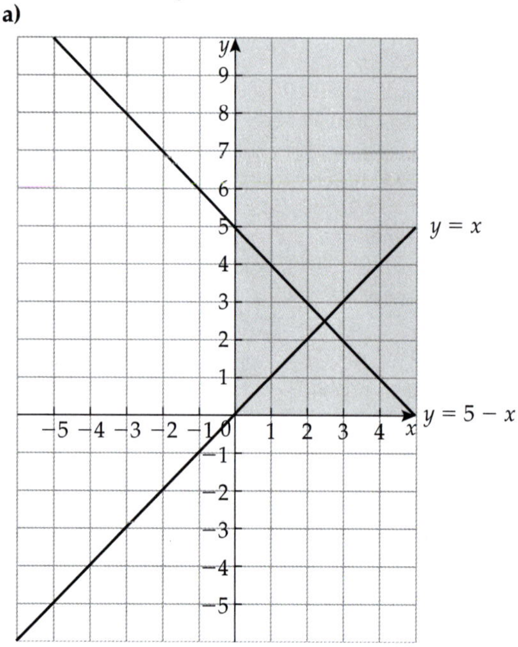

b)

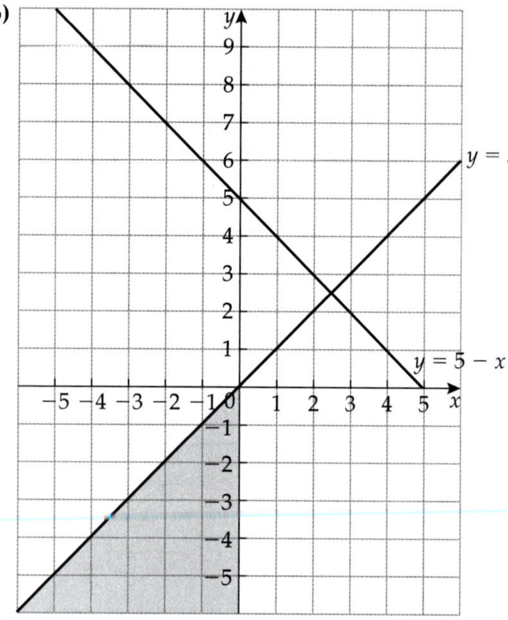

c)

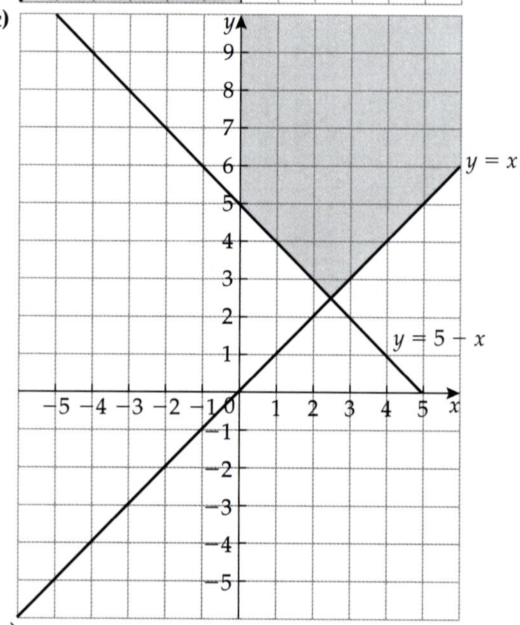

4 **a)**

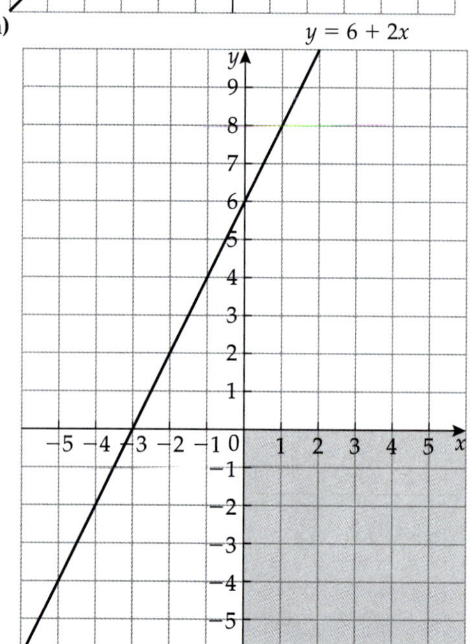

b)

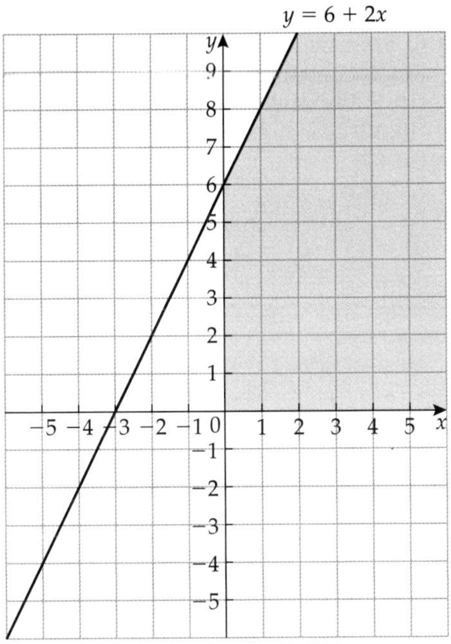

c)

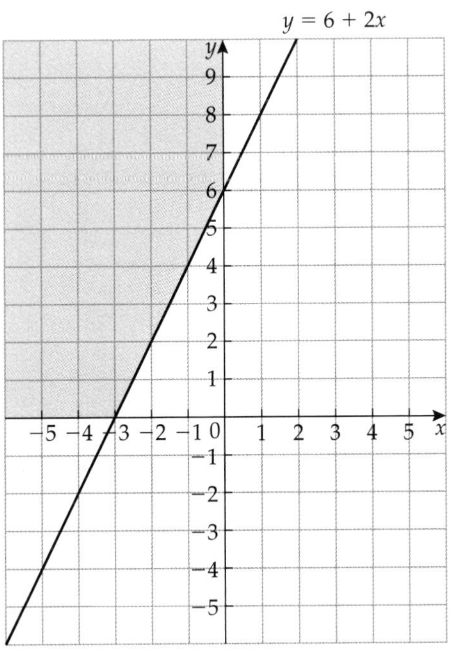

d)

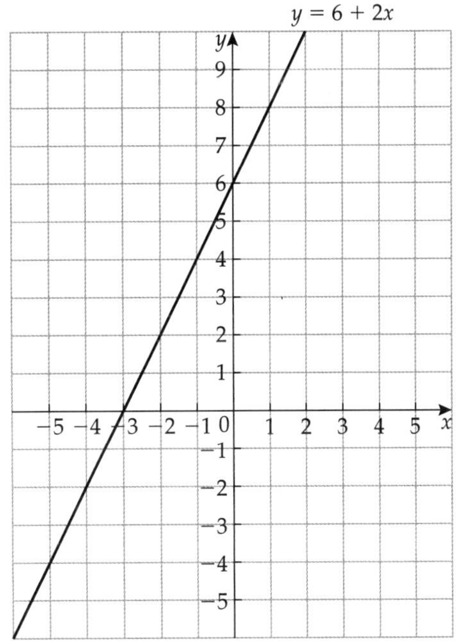

5 a)

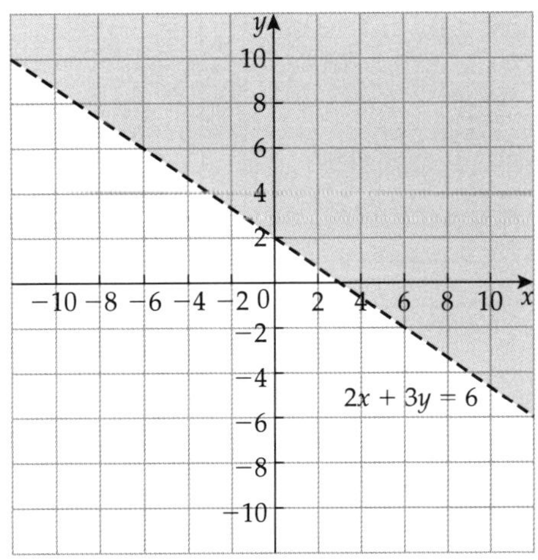

b)

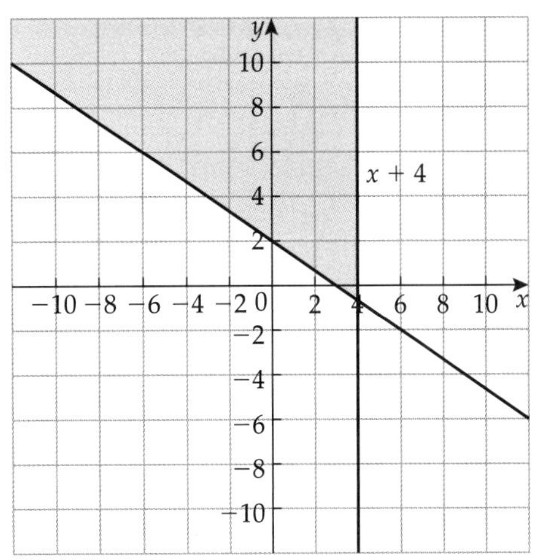

c)

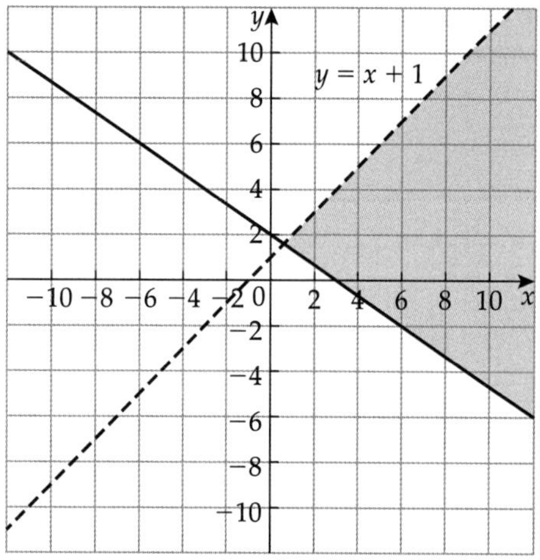

Page 173

1 a) x = number of T-shirts, y = number of shorts, $600 \geqslant 100x + 90y; x \geqslant 1; y \geqslant 1$

b)

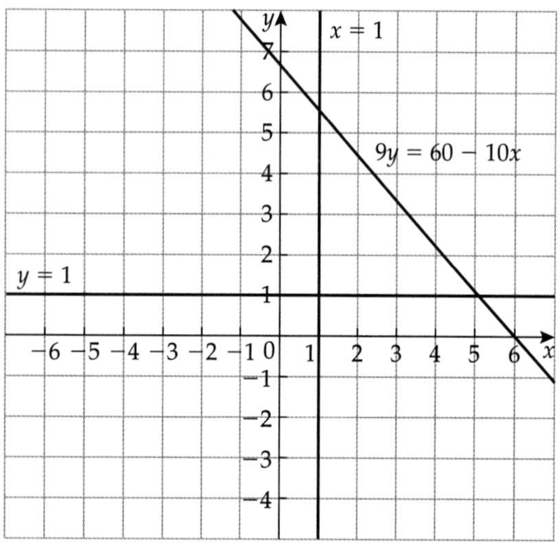

c) 3 d) \$30

2 a) x = cost of pens, y = cost of books
$5x + 2y = 50, 6x + y = 45$

b)

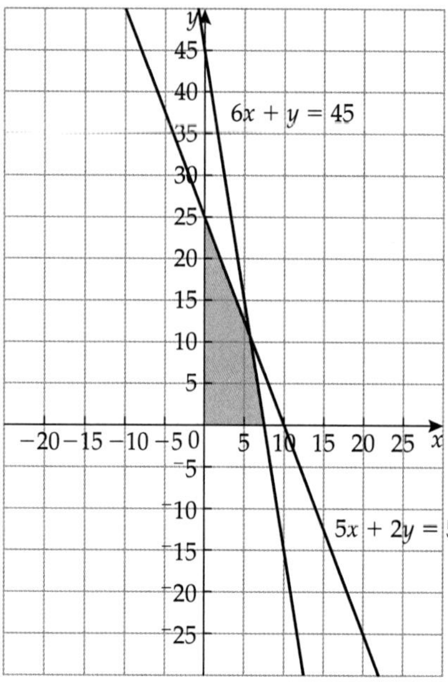

c) $x = \$5.71, y = \$10.71$

3 a) (0, 5) 5 man hours (all by Jack)
(0, 12) 12 man hours (all by Jack)
(10, 5) 15 man hours (10 by James, 5 by Jack)
(10, 12) 22 man hours (10 by James, 12 by Jack)

b) (10, 12)

c) 22 hours

d) James works 10 hours, Jack works 12 hours

4

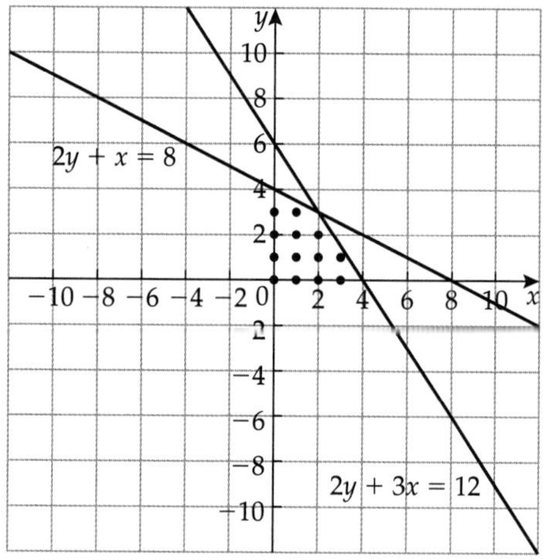

$1m, 3s; 2m, 2s; 3m, 1s; 4m, 4s$

Page 175

1 a) $x < -1$ or $x > 4$ b) $x < -3$
c) $x \geqslant 9$ d) $x > 4$
e) $-7 > x > 5$ f) $x \leqslant -1$ or $x \geqslant 5$
g) $x < -3$ or $x > 2$ h) $1 < x < 4$
i) $x > 3$ j) $x < 2$ or $x > 3$
k) $2 < x < 3$ l) $x < -2$ or $x > 1$
m) $-2 < x < 2$

4 a) $x < 1$ or $x > 3$

5 a) $-1 < x < 1$

6 a) $x \geqslant \frac{2}{3}$ or $3x \geqslant 2$ b) $x > \frac{5}{2}$ or $2x > 5$

7 $x < -4$ or $x > 2$

Page 178

1 a) $x < 3$ b) $x > 3$ c) $x < 2$
d) $x \geqslant 0$ e) $x \geqslant 3.5$ f) $x \geqslant 15$

3 a) $x < 4$ b) $x > 0.5$ c) $x > 3$
d) $x > -1$ e) $x > 1.25$

4 a) $x > 2$ b) $x < 3$ c) $x > -2$
d) $x < 4$ e) $x > 1$ f) $x < 1$
g) $x \geqslant 2$ h) $x \geqslant -\frac{1}{2}$

5 a) $n > 9$ b) 10

6 $4(x + 3) < 60, x < 12$

7 $\dfrac{11 - x}{2} < 5, x > 1$

8 $2(1 + x) < 4, x < 1$

9 $x > -\frac{14}{3}, 5$

10 a) $20x < 10x + 70$ b) $x < 7$

Looking back (page 179)

1 a) 1, 2, 3, 4 b) 0, 1, 2, 3, 4
c) 4, 5 d) −3, −2, −1, 0, 1, 2
e) −2, −1, 0, 1 f) −4, −3, −2
g) 4, 5 h) −2, −1, 0, 1, 2

2 **a)** 2, 3, 4 **b)** −2, −1, 0, 1
c) −3, −2, −1, 0 **d)** −3, −2, −1, 0, 1
e) 1, 2, 3 **f)** −2, −1, 0, 1, 2

3

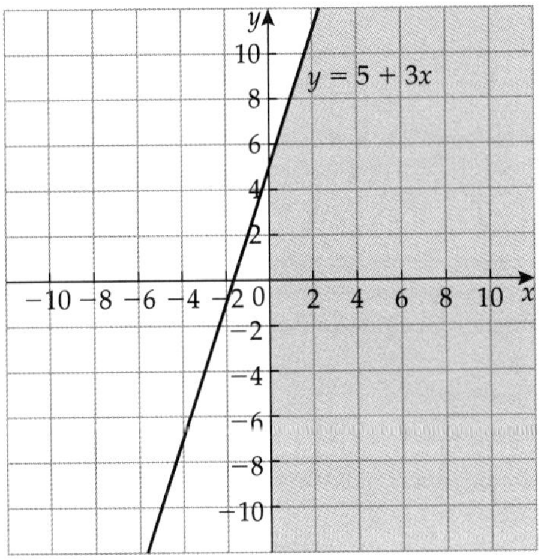

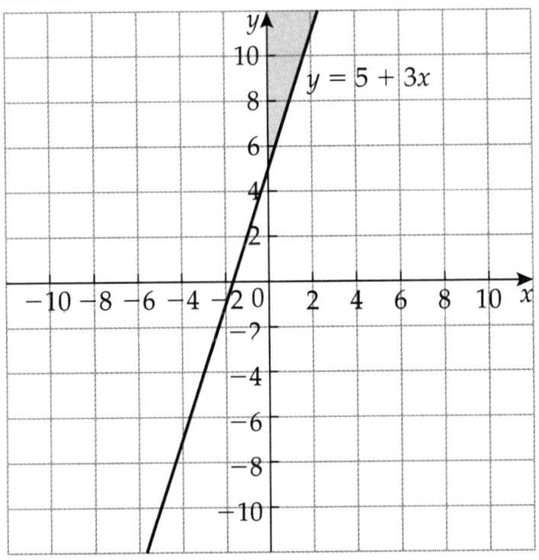

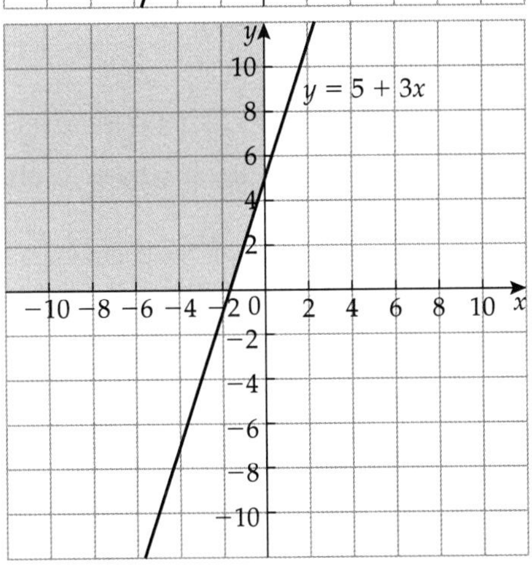

4 **a)** $8p + 10s = 70$
$6p + 9s = 57$

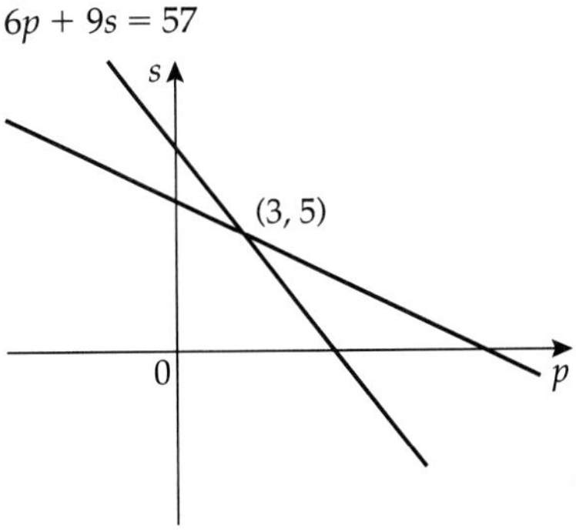

$s = 3, p = 5$

Chapter 14

Page 180

1 continuous 2 continuous 3 discrete
4 discrete 5 discrete 6 discrete
7 discrete 8 continuous

Page 182

1 **a)**

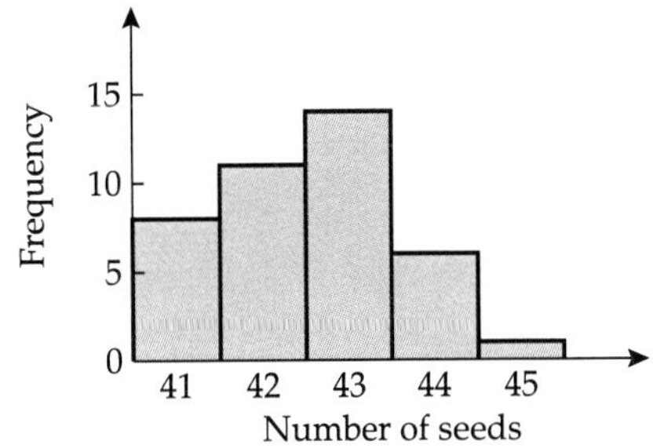

b) Mean = 42.5 to 1 dp
Mode = 43
Median = 43
Yes, the claim is reasonable

2 Students' bar charts or bar line graphs; values are not part of a whole, so pie chart is not suitable.

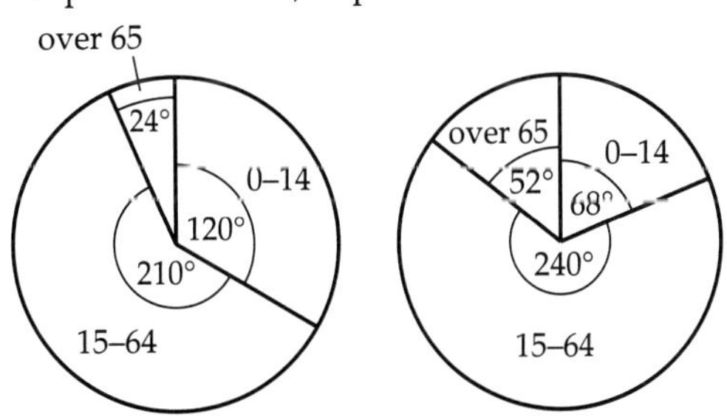

Jamaica Luxembourg

Jamaica has a much greater percentage of 0–14 year olds, and a much lower percentage of over 65 year olds.

Page 184

1

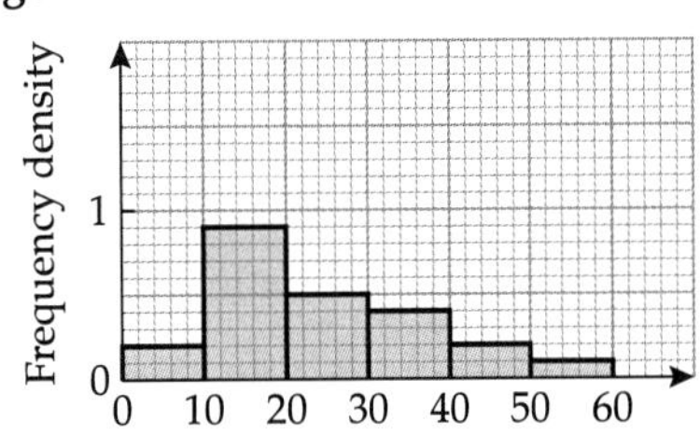

2

Lengths, cm	$50 \leqslant l < 100$	$100 \leqslant l < 150$	$150 \leqslant l < 200$	$200 \leqslant l < 250$
Frequency	3	5	3	3

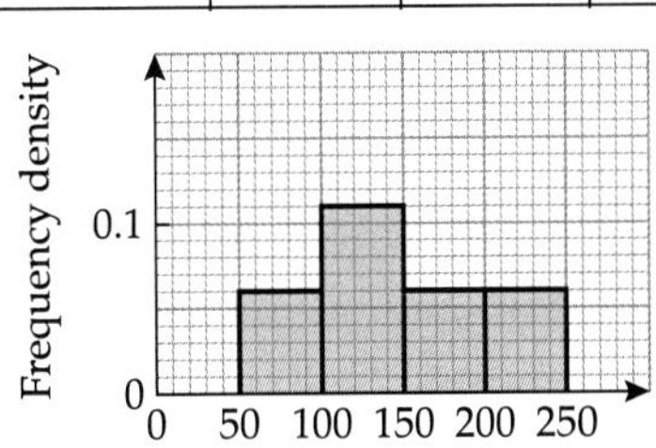

3

Masses, *g*	$0 \leqslant m < 30$	$30 \leqslant m < 60$	$60 \leqslant m < 90$	$90 \leqslant m < 120$
Frequency	1	4	12	3

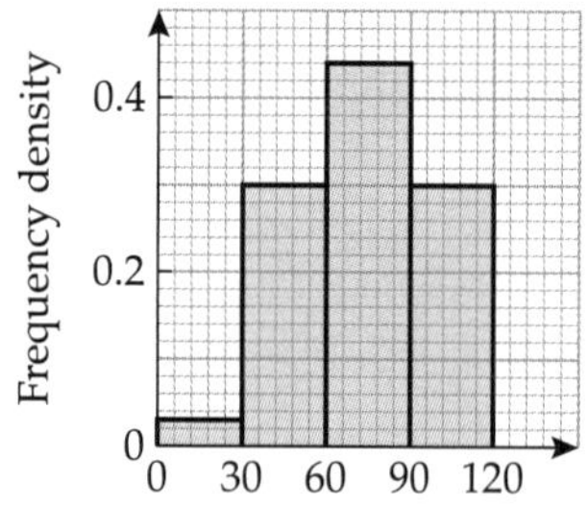

Page 186

1 a) Mid interval values: 5, 15, 25, 35, 45

b)

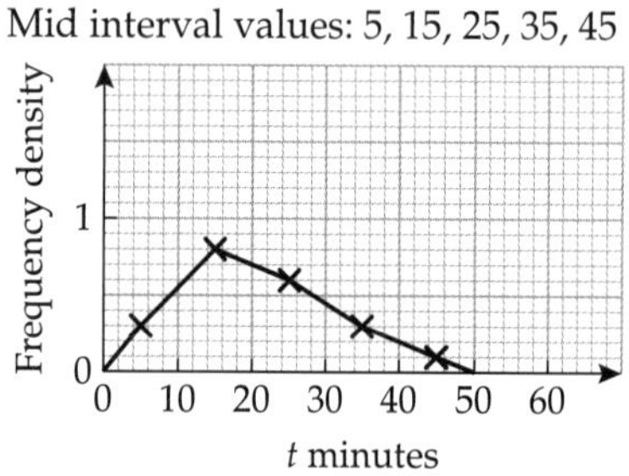

2

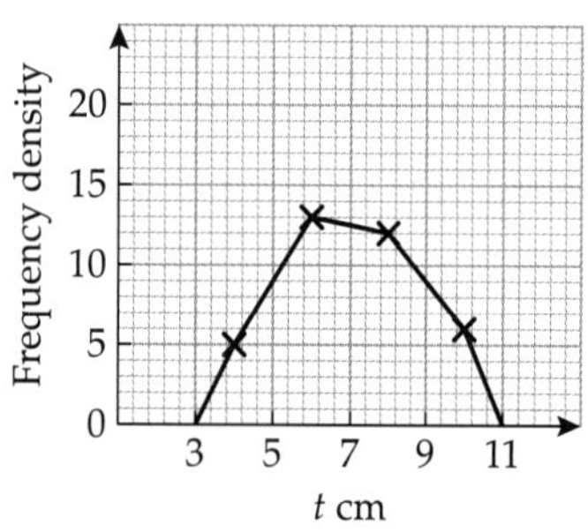

3

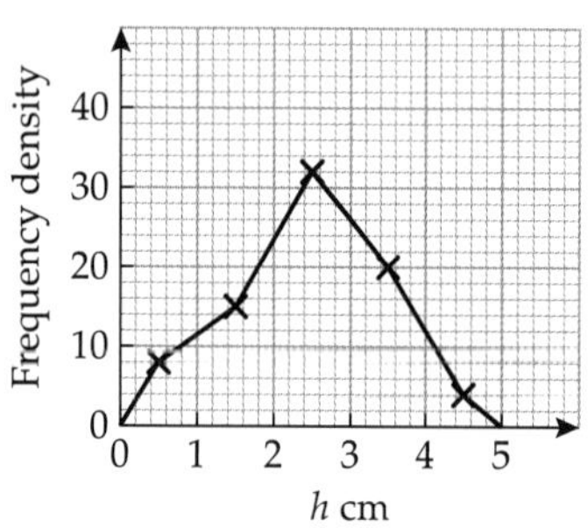

4 a) 6

b) 46

c) $12 is in the middle of a class

d)

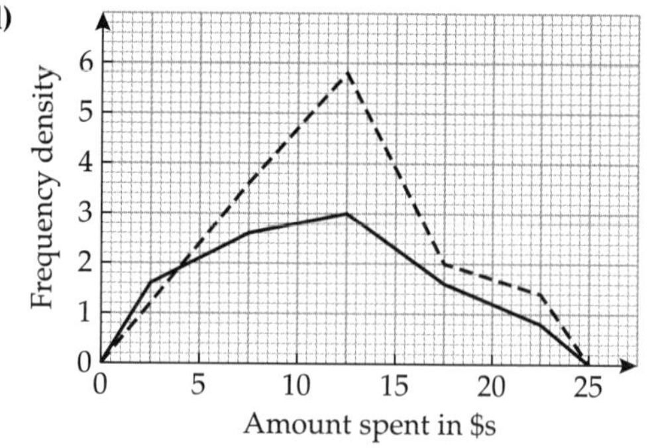

Page 189

1 $2.\dot{3}$

2 $48.4 to 1 dp

3 19 minutes to 2 sf

4 7.2 cm to 1 dp

5 7.5 minutes to 1 dp

6 a)

Mid-interval values, *x* (minutes)	Frequency, *f*	*fx*
2.5	5	12.5
7.5	9	67.5
12.5	50	625
17.5	21	367.5
22.5	9	202.5
	$\sum f$:94	$\sum fx$: 1275

b) 94

c) 13.6 minutes

7 10.8 cm

8 153.1 cm

Page 192

1 a) mean: 65.5, median class: 60–79, modal class 60–79

b) mean because it includes all the marks

2 a) mean: 16, median class $15 < t \leqslant 20$, no single modal class

b) median because it is not affected by the two low times

3 a) mean: 4.4 mm, median class $3 \leqslant l < 5$, modal class $5 \leqslant l < 7$

b) mean because it includes all the beans

4 a) mean: 0.79 km, median class $0.5 \leqslant d < 1$, modal class $0.5 \leqslant d < 1$

b) mode or the median because they are not affected by the few longer distances

5 The mean is not very useful as it could include very young and very old members.
The median is useful because it indicated where her daughter will be in the age range.
The mode is useful as it gives an indication of the ages of the largest group of members.

Looking back (page 193)

1 **a)** Continuous

b)

Batting average, b	$0 \leqslant b < 10$	$10 \leqslant b < 20$	$20 \leqslant b < 30$	$30 \leqslant b < 40$	$40 \leqslant b < 50$	$50 \leqslant b < 60$
Frequency	3	5	8	9	3	2

c), d)

Histogram and frequency polygon of batting average

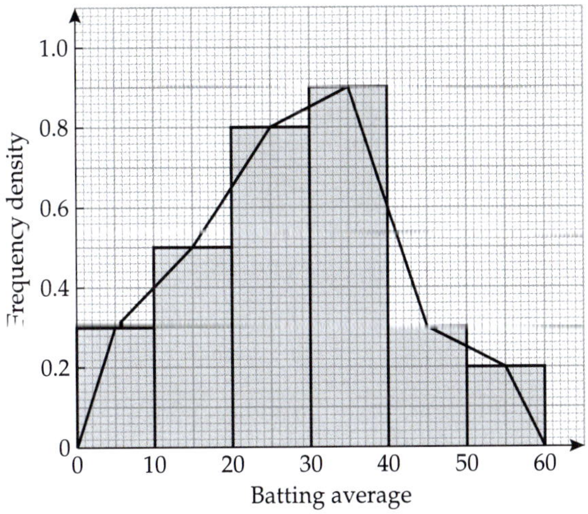

2

Top speed, v km/h	$100 \leqslant v < 120$	$120 \leqslant v < 140$	$140 \leqslant v < 160$	$160 \leqslant v < 180$	$180 \leqslant v < 200$
Frequency, f	4	5	21	8	2
Mid-class value, x	110	130	150	170	190
fx	440	650	3150	1360	380

a) Mean: 147 km/h
Median class: $140 \leqslant v < 160$
Modal class: $140 \leqslant v < 160$

b) As there are many values in one class, the modal class gives a good representation of the distribution.

Revise and Consolidate 3

Page 196

1 a) $-1, 0, 1, 2, 3, 4, 5$
b) $1, 0, -1, -2, -3, -4, -5$
c) $-3, -4, -5$
d) $-3, -2, -1, 0, 1$

2 b) Geography

2	589
3	0456788
4	01356

Science

2	1145699
3	122237
4	00

d) 25–27

3 a) 0.566 seconds
b) mode = 0.49, median = 0.56, range = 0.2
c) 15–20 years

4 a) Distance thrown is between 60 m and 63 m.
c) $69 \leqslant D \leqslant 72$
d) Minimum and maximum values lie in a range.

5 a) median = 46 **c)** 67

6 b) B

7 a) 252 cm
b) $240 < l < 270$
c) $\frac{40}{101}$

8 b) i) $x = -\frac{1}{2}, \ y = -\frac{1}{2}$
$x = 1 \quad y = 1$

ii) $x = -1, \ y = -1$
$x = 0, \quad y = 0$
$x = 1 \quad y = 1$

c) There are 2 solutions to the simultaneous equations that include a quadratic, and 3 solutions when there is a cubic.

9 Students' number lines

10 Students' answers will vary

11 d) (1, 4) double shaded
(3, -2) single shaded
$(-1, -3)$ double shaded
$(-2, -3)$ single shaded

12 a) $-8 < a$ **b)** $b \geqslant 12$
c) $x > 4$ **d)** $x \geqslant 3$

Test your knowledge

Test 1

1 c) **2 c)** **3 c)**
4 a) **5 c)**
6 120
7 a) $1\frac{7}{8}$ **b)** $\frac{3}{4}$ **c)** $1\frac{4}{15}$
d) $3\frac{29}{45}$ **e)** $5\frac{1}{4}$ **f)** 1
8 a) \$5340 **b)** \$20
9 39

10 a) 306 **b)** 506 **c)** 930
d) 272 **e)** 899

11 a) 2^5 **b)** 3^3 **c)** $2^{-1} \times 3^2$
d) 5^7 **e)** 7^6

Test 2

1 b) **2 a)** **3 b)**
4 c) **5 c)**

6 a) (7, 0) **b)** 21 units^2

7 6750 cm^3

8 a) Surface area = 184 cm^2
Volume = 120 cm^3
b) Surface area = 47480 mm^2
Volume = 223872 mm^3
c) Surface area = 44.72 m^2
Volume = 11.04 m^3
d) Surface area = 256.86 cm^2
Volume = 36.27 cm^3

9 a) 75 **b)** 78.5% (1 dp)

10 25 cm

Test 3

1 c) **2 a)** **3 b)**
4 b) **5 a)**

6 a) $\frac{pq + 30}{10q}$ *b)* $\frac{7t + 12s}{st}$
c) $\frac{40 + jk}{10j}$ *d)* $\frac{2xy - 5}{5y}$
e) $\frac{x^2 + y^2}{xy}$ *f)* $\frac{4pq - q}{2pq}$

7 a) $10f + 10g$ **b)** $7m + 42n$
c) $2x^2 - 10xy + 4y^2$ **d)** $-35j^2 + jk^2 + 3j^2k$

8

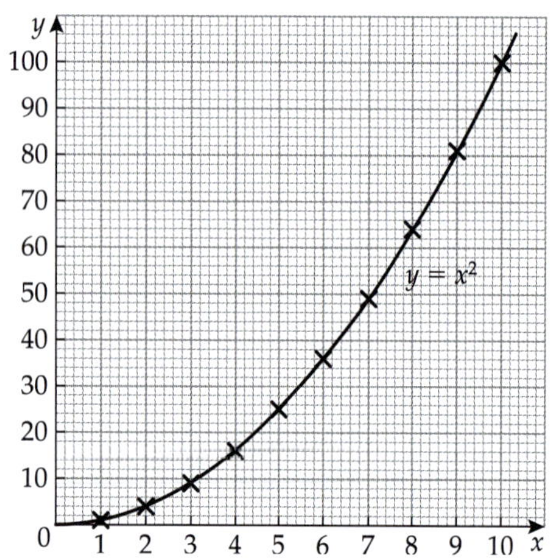

a) 13 ± 1 **b)** 61 ± 1
c) 8.4 ± 0.2 **d)** 9.7 ± 0.2

9 $a = 5$ and $b = 3$

10 a) $x = 1$ and $x = -3$ **b)** $y = x^2 + 2x - 3$

Test 4

1 a) **2 b)** **3 b)**
4 b) **5 d)**

6 a) 4.36×10^2 **b)** 3×10^6
c) 2.4×10^4 **d)** 5.3×10^{-3}
e) 7.08×10^{-1}

7 a) 4 **b)** 10 **c)** z
d) $3m$ **e)** k^2 **f)** $4s^2$

8 1663.1 cm^2 to 1 dp

9 $j = 4$ and $k = 38$

10

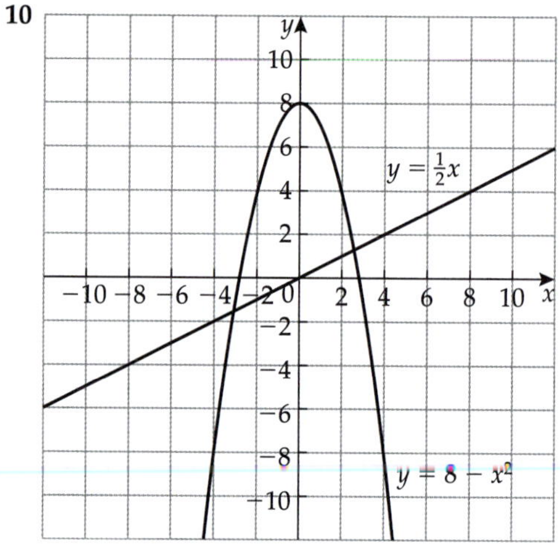

$x = 2.6$, $y = 1.8$ and $x = -3.1$, $y = -1.5$.
Allow ± 0.2 on all answers

11 a) \$132 **b)** \$564.30

12 a) \$20250 **b)** \$1350, 6.67%

Test 5

1 b) **2 c)**
3 a) **4 d)**

5 a) 13 m **b)** 18.4 cm
c) 4.6 m **d)** 16.4 mm

6 a) 22.9° **b)** 41.8°
c) 49.2° **d)** 68.8°

7 12.6 m

8 6.9 m

9 143.2 km

10 Accurate construction:

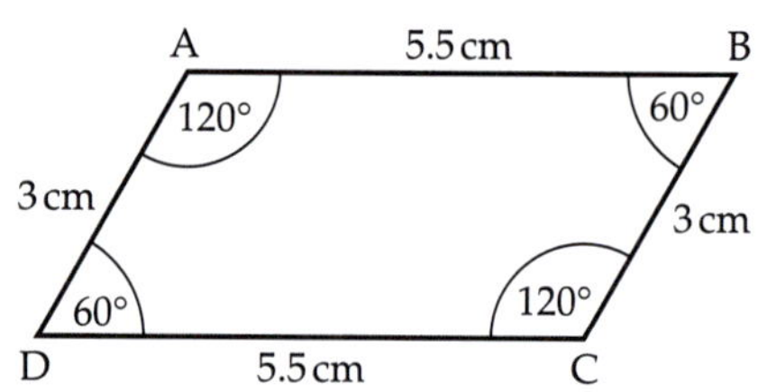

AC = 4.9 cm ± 0.2 cm

Test 6

1 b) **2 d)**
3 b) **4 b)**

5 a) P′(3, −1), Q′(9, −5), R′(3, −11)
b) P″(−1, 3), Q″(−5, 9), R″(−11, 3)

6 a) 4 **b)** 2 **c)** 0 **d)** 4

7 Rotation 90° (anticlockwise) centre (0, −2)

8 a) 9 m **b)** 25

Test 7

1 b) **2 d)**
3 c) **4 b)**

5

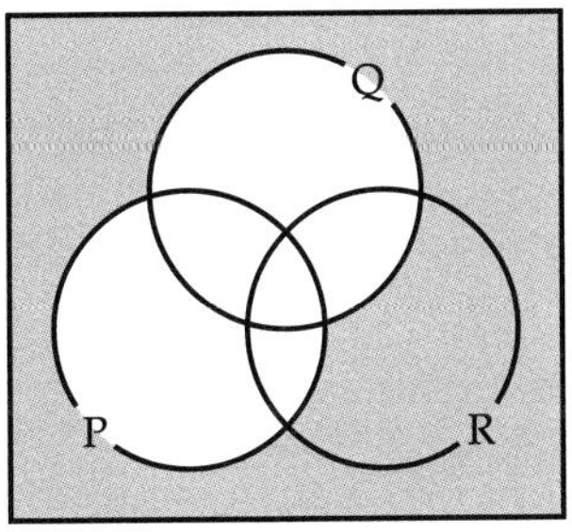

6 **a)** 7 **b)** 3 **c)** −1 **d)** −3

7

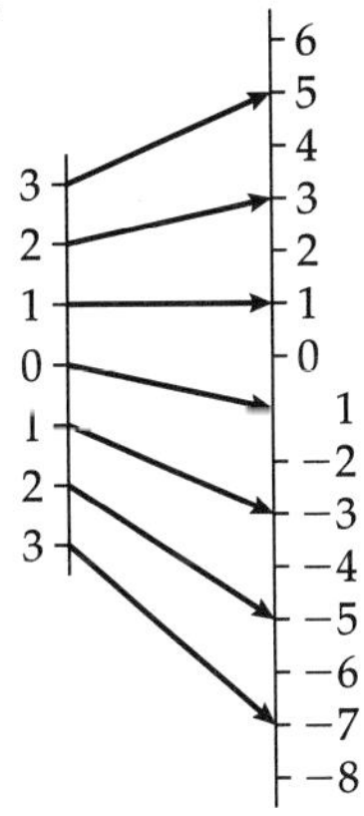

8 $x = 3, 4, 5, 6, 7, 8, 9$

9 **a)** $-4 \leqslant x \leqslant 3$ **b)** $x < 2, x > 9$

c) $x \leqslant -3, x > 0$ **d)** $-5 < x < 1$

10

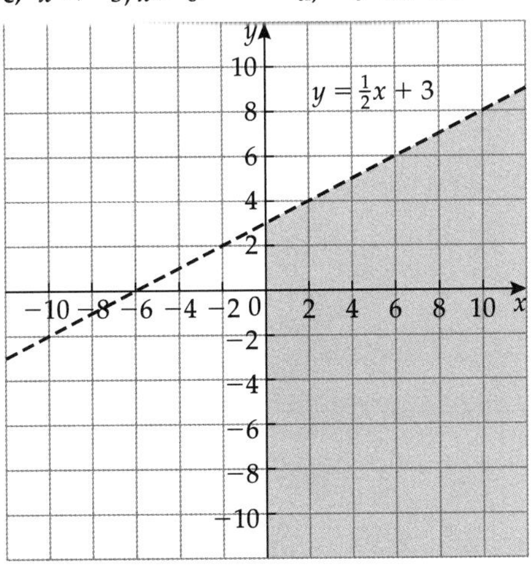

e.g. (0, 0), (2, 2)

Test 8

1 c) **2 c)** **3 a)**

4 a) $\frac{8}{27}$ **b)** $\frac{1}{16}$ **c)** $1\frac{24}{25}$ **d)** $\frac{5}{7}$

5 8

6 312 cm^3

7 $m = 15$ and $n = 10$

8 $17 496

9 55.2 m

10 $x < 5$,

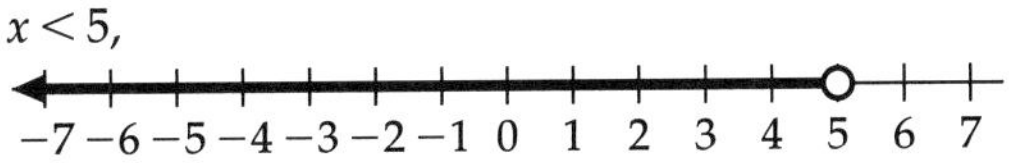

Test 9

1 a) **2 b)** **3 c)**

4 b) **5 b)**

6 $\frac{13}{33}$

7 **a)** 7.3 cm **b)** 4.9 m

c) 9.1 cm **d)** 55.0 m

8 9.9 cm

9 −1, 0, 1, 2, 3, 4, 5, 6

10 Estimate for the mean = 16.2 s

Median group = $16 \leqslant t < 18$

Modal group = $16 \leqslant t < 18$

Test 10

1 a) **2 b)**

3 d) **4 d)**

5 **a)** $\frac{a^2}{4}$ **b)** $\frac{8}{c^3}$ **c)** $\frac{1}{4a}{}_2$

6 **a)** $163 400 **b)** $120 850.64

7 **a)** $x = -9$ or 3 **b)** $m = -2$ or -5

c) $t = 15$ or -2 **d)** $n = 5$ or -5

8 Accurate construction:

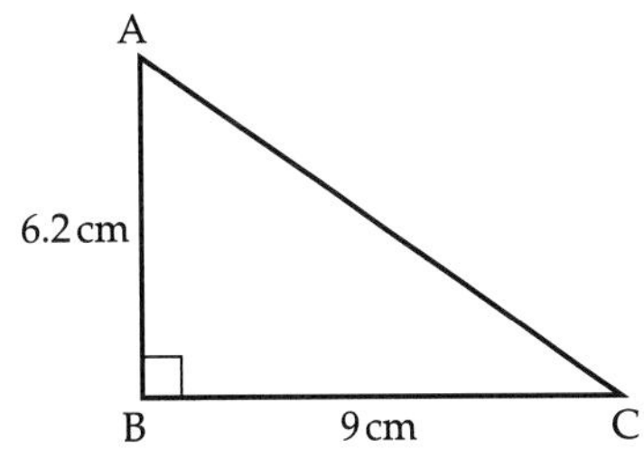

AC = 10.9 cm ± 0.2 cm, ∠A = 55.4° ± 2°,
∠B = 34.6° ± 2°

9 e.g.

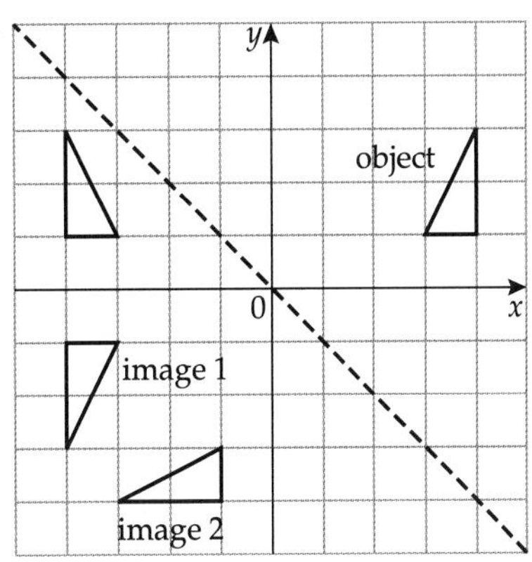

10 92 m

11 **a)** Continuous

b)

Height in cm, *x*	130 ⩽ *x* < 140	140 ⩽ *x* < 150	150 ⩽ *x* < 160	160 ⩽ *x* < 170
Frequency	3	2	4	1

c), d)

Histogram of heights

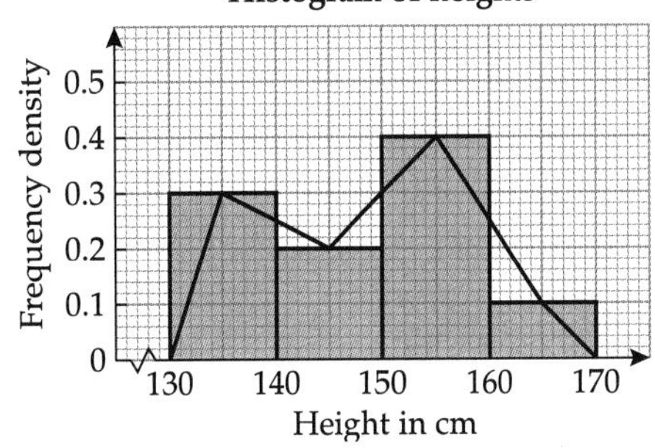

Index